This book questions the traditional "grand narratives" of science and religion in the seventeenth century. The binary oppositions underlying the story – between reason and faith, between knowledge and authority, between Scripture and the light of nature – have moulded it into a formative myth: the banner of modern rationalism, liberalism, and individualism. While deconstructing the oppositions behind the conflict, the book offers an analysis of the complex intellectual–institutional field in which the drama of Galileo and the church unfolded.

The well-known contradictions among the documents of Galileo's "trials" are reread as expressions of the contradictory nature of the Counter-Reformation church. A flashback into the formative years of Tridentine Catholicism demystifies its monolithic and brutally coercive tendencies. Rather, the church appears to have been torn between different cultural orientations and divided institutionally as well as theologically. The traditional intellectual elite of the Dominicans adopted an orthodox Thomist allegiance and refused innovation in the name of Thomist rationalism. Their reaction to the challenge raised by the Counter Reformation consisted in dogmatic Thomism. The Jesuits reacted to the same challenge by developing their vocation as educators of the entire Catholic society. In that role they reconstructed the Thomist synthesis by assimilating new scientific contents and reinterpreting its theology. Theirs was a pragmatic Thomism.

Galileo's Copernicanism emerged in the periphery of the cultural field newly organised by the Jesuits. The dispute on sunspots that took place between Galileo and the Jesuit astronomer Christopher Scheiner is the occasion signaling the emergence of a new discourse out of the Galilean–Jesuit dialogue.

The act of silencing exemplified in the trials of Galileo is in no need of demonstration. It has been so imprinted in our consciousness that to reassert it is to state the obvious. The author's story is not about the repression of truth by religious authority. It is the story of an encounter between different types of power–knowledge structures within the framework of a dialogical model.

Galileo and the Church

Galileo and the Church
Political Inquisition or Critical Dialogue?

RIVKA FELDHAY

CAMBRIDGE
UNIVERSITY PRESS

PUBLISHED BY THE PRESS SYNDICATE OF THE UNIVERSITY OF CAMBRIDGE
The Pitt Building, Trumpington Street, Cambridge, United Kingdom

CAMBRIDGE UNIVERSITY PRESS
The Edinburgh Building, Cambridge CB2 2RU, UK   www.cup.cam.ac.uk
40 West 20th Street, New York, NY 10011-4211, USA   www.cup.org
10 Stamford Road, Oakleigh, Melbourne 3166, Australia
Ruiz de Alarcón 13, 28014 Madrid, Spain

First published 1995
Reprinted 1999

Typeset in Sabon

Library of Congress Cataloging in Publication data
Feldhay, Rivka
Galileo and the church: political inquisition or critical
dialogue? / Rivka Feldhay
p.   cm.
ISBN 0 521 34468 9
1. Galilei, Galileo, 1564–1642 – Religion. 2. Religion and
science – Italy – History – 17th century. 3. Catholic Church – Italy –
Rome – History – 17th century. 4. Italy – Church history – 17th century.   I. Title.
QB36.G2F39   1995
261.5'5'094509032 –dc20   94-27117
CIP

A catalog record for this book is available from the British Library

ISBN 0 521 34468 9 hardback

Transferred to digital printing 2003

Contents

Contents

Part III Galileo and the Church

Preface and
Acknowledgements

There is something intensely alluring, and sinister too, in writing about the Galileo affair. The wish to solve the riddle plays against the consciousness that it may be insoluble. The monumental lifework of Antonio Favaro on Galileo led him to conclude, cryptically perhaps, that "not everything" was kept in historical memory. In the same vein Giorgio de Santillana became convinced that many aspects of the event were beyond the limits of historical knowledge. Our contemporary awareness of narrative techniques and narrative structures further points out the inherent constraints imposed on historians who strive to tell the true story. My long-term involvement with writing this book has only increased my anxieties in the face of those enormous difficulties. They have not, however, eroded my belief that though we may never know "how it *really* happened", we should abide by the duty to understand the *kind of thing that could have happened*. This book should be read with that intention in mind.

Many people and institutions have been involved in the process which led to the maturation of this project. Most significant was Yehuda Elkana, who instigated my first interests in the history of science, and who has accompanied me through the various stages of research and writing. I wish to thank Sabetai Unguru, Michael Heyd, and Amos Funkenstein, who read various versions of the manuscript and gave me their illuminating advice and comments. Father Creytens from the Pontificia Università di Santo Tomasso in Rome was of great help in showing me the way through the Dominican sources. So was the librarian at the old Dominican archive in Santa Sabina. I thank the staffs of the libraries at the Scuola Normale Superiore and the Sapienza in Pisa, as well as at the National Library in Florence. The Humanities Center at Stanford University and the Wissenschaftskolleg zu Berlin were very generous in granting me scholarships in the years 1987–1989, during which I was working on the manuscript. My friends Ora Limor and Daniel J. Spitzer have read the book and helped me with many problems of translation from the Latin and from other languages. Adi Offir, Azmi Bishara, Moshe Zuckermann, and Amnon Raz-Krakotzkin have been partners to many dialogues concerning the theoretical aspects of writing history. David Meisel, who edited the manuscript and translated some parts of it from an earlier Hebrew version, has also contributed significantly to the preparation of this book.

Many thanks to Daphna H. Shapiro and Neta Aloni, my research assistants from Tel Aviv University, who were helpful in many ways, including with the preparation of the index.

I am grateful to my parents for their moral and material support during the difficult years. But in my private story, the story of writing the Galilean story, my husband, Yoram, and my children, Eran and Michal, played a major role by taking upon themselves the burden of a writing wife and mother. Their "being there" enabled me to bring my task to completion.

Introduction

The story of Galileo Galilei (1564–1642), his scientific activity, and his struggle for legitimation has long ago passed beyond the boundaries of historical research to become one of the formative myths of modern science. Underlying this myth was the idea of a conflict between darkness and light, between reason and unreason. Within this conflict the theologians of the Catholic Church and the philosophers of the universities who supported them became symbols of cultural obscurantism seeking to defend the authority of the Scriptures to judge the conclusions of scientific thought. Galilean science, on the other hand, represented the new experimentalism based on observation of phenomena and application of mathematical methods to the investigation of nature. At the same time, Galilean science was also perceived as the product of a lone genius who defended the principle of independent thought against repression. Only too often has the Galilean episode been interpreted simply as a necessary step towards the liberation of Western thought from the yoke of an authoritarian tradition.

It is characteristic of myths that they force the imagination to use and abuse them in its attempt to fashion cultural identity by an interpretive process. The story of Galileo is no exception to this familiar pattern. The success of science in Western culture has guaranteed the story of Galileo its symbolic significance, engendering a rich literature[1] concerning its

1. The scope of the bibliography in a field now called "Galilean studies" amounts to thousands of items, see: E. McMullin, "Bibliographia Galileana" (1940–1964), Addenda to A. Carli-A. Favaro, *Bibliographia Galileana* (1564–1895, pub. 1896), and to G. Boffito "Bibliographia Galileana" (pub. 1943), in E. McMullin (ed.), *Galileo Man of Science*, New York and London 1967, appendixes A, B. For a good general survey of the literature since 1967, see W. A. Wallace's "Literature and Translations", in *Galileo and His Sources: The Heritage of the Collegio Romano in Galileo's Science*, Princeton 1984. See also the references in M. Biagioli's *Galileo, Courtier: The Practice of Science in the Culture of Absolutism*, Chicago 1993, pp. 363–391.

many different facets. Thus, the conceptual content of the church's out-look was defined not only in theological terms but also in terms of the prevailing Aristotelian philosophy. Likewise, the conceptual content of Galilean science was reassessed not only in the light of its experimental results but also in accordance with its underlying metaphysical assumptions. The political needs of a pope secretly involved in the Thirty Years War (1618–1648) were scrutinized, and the propagandist aspects of the Galilean campaign for Copernicanism were also examined. Certain biographies of prominent figures in the Counter-Reformation church result from persistent interest in the Galileo affair. Concurrent portrayals of the colourful and controversial Galileo reveal the same kind of historical "obsession". The Galilean episode also served as a basis for literary works representing modern man and his anxieties. Above all, the documents of Galileo's "trials" were published, making possible a reconstruction of the legal logic underlying the policies of the Inquisition, as well as of Galileo's "real" attitude towards religious authority.

The abundant academic literature concerned with the Galilean episode has contributed a great deal to a modification of the myth. The apparently "irrational" attachment of the church to biblical cosmology was seen, after all, to have been accompanied by a deep commitment to a body of "rational" philosophical knowledge. The Galilean struggle for scientific "truth" was revealed as full of scientific errors as well as political rhetoric and tactical blunders. The overemphasis on the obscurantist and reactionary nature of the church, no less than the unqualified notion of a wholly progressive and innovative science, gave way to a subtler understanding of the relationship of religion and science at the dawn of the Scientific Revolution.

And yet, whenever an attempt is made to retell the story of Galileo, some basic elements of the myth prove very much alive and well. Take, for example, a relatively recent article entitled "Galileo and the Church".[2] Here, a distinguished Galileo scholar begins by claiming that the condemnation of Galileo must, by now, assume a different meaning from the one long attributed to it, and that "it must be seen in historical perspective".[3] A long account follows, in which the political and religious problems of the Counter-Reformation church are discussed, the nature of Galilean science is said to be just as dogmatic as that of contemporary theology, and Urban VIII's personal and political likes and dislikes are considered. At the end of all this the author arrives at the conclusion that all these com-

2.   W. R. Shea, "Galileo and the Church", in D. C. Lindberg and R. L. Numbers (eds.), *God and Nature*, Berkeley and Los Angeles 1986.
3.   Ibid., p. 114.

plexities essentially boil down to a "conflict between the authoritarian ideal of the Counter Reformation and the nascent desire and need for freedom in the pursuit of scientific knowledge".[4] This appears to be a re-assertion of the symbolic meaning of the affair in the most traditional terms. In spite of a perfectly sincere intention to insist on the subtler aspects of the story, the old explanatory framework always seems to creep back in order to account for the undeniable "hard fact" of the condemnation.

What is so irresistible about this myth?

Throughout its long history the Galilean myth has been nurtured by implicit assumptions about the nature of science and the nature of religion which are deeply rooted in modern beliefs.

Modern culture, characterized as it is by a strong positivistic core, perceives scientific knowledge as a representation of "truth" whose growth consists in progressive accumulation. According to this view, science is characterized by its rationality, defined as the capacity to discover in a logical, inductive manner causal relationships between a phenomenon and its reasons, or to infer the phenomena deductively from self-evident general principles.

Religion, for its part, is distinguished from science by the absence of this rational basis. Religion is not based on proven principles, but on a total faith in principles which in part, at least – and certainly as regards the more important ones – are not susceptible to rational proof. The source of faith is not the intellect or the evidence of the senses, but revelation. Faith cannot be tested by its logical consistency but by its conformity to authority, which is above human understanding.

The traditional Galilean myth is entirely based on a mirror image of this kind of science and religion. Galilean science, according to this view, attempted to construct a picture of the universe based upon evidence of the senses and logical proofs. As against this, the church tried to preserve its authority through a reliance on an outmoded cosmology and on the evidence of the Scriptures. In this framework of thought, the notion of a conflict has proven irresistible to all historical explanations.

In the nineteenth century this conflict was often interpreted as a total clash of values and ideas, expressed in military terms ("the warfare of science and theology").[5] Although subsequent scholars discredited such language, the notion of conflict refused to die, and has persisted even in

4. Ibid., p. 132.
5. The most famous among these studies are: J. W. Draper, *History of the Conflict between Religion and Science*, New York 1874; A. D. White, *A History of the Warfare of Science and Theology in Christendom*, New York 1876.

the subtler representations of Galileo's dealings with the church. G. de Santillana's *The Crime of Galileo*[6] provides a good example. In the preface to his canonical study, Santillana states what he believes is necessary to change the mythical terms of the discussion. One needs to remove "the spell of a misunderstanding tacitly accepted by both sides": namely, "the idea of the scientist as a bold freethinker and 'progressive' facing the static resistance of conservatism".[7] Clearly, Santillana aims at getting rid of the dichotomy of Christianity and science as one of darkness and light, obscurantism and enlightenment, bad and good. However, what Santillana is unaware of is the need to discard the notion of "conflict" as a totalizing organizational principle. His study, on the contrary, is built entirely around the notion of a conflict and the need to explain it. In order to establish the relevance of his endeavour, he invokes a number of modern conflicts – genetics in Russia, the Oppenheimer case in America – which seem to manifest the same structure: "In such vastly different climates of time and thought, whenever the conflict appears, we find a similarity of symptoms and behavior which points to a fundamental relationship".[8] Against the classical studies of science and religion, which insisted on the essential nature of the conflict, Santillana invokes contingent elements of the story which had been unknown, or simply ignored, and manages to explain the conflict not as an "inevitable clash", but as an unhappy mistake, a kind of historical accident: "the tragedy was the result of a plot of which the hierarchies themselves turned out to be the victims no less than Galileo – an intrigue engineered by a group of obscure and diverse characters in strange collusion who planted false documents in the file, and who later misinformed the Pope and then presented to him a misleading account of the trial for decision".[9]

Santillana's explanation of the affair has been the only alternative to that of the inherent "warfare" of science and religion proposed by nineteenth-century scholars like Draper and White. As against them, Santillana successfully refuted the claim of a fundamental clash of values between science and religion. Nor could he accept any naive assumptions about the obscurantism, backwardness, or ignorance of religion versus the modernity, enlightenment, and liberalism of science. Yet, being unable to free himself of the notion of "conflict" as an organizing principle, he maintained the idea of the binary opposition between the political interests of the church on one hand and the science of Galileo on the other,

6. G. de Santillana, *The Crime of Galileo*, Chicago 1955.
7. Ibid., p. vii.
8. Ibid., p. viii.
9. Ibid., pp. xii–xiii.

while explaining it in terms of power versus knowledge: "the scientific mind . . . with its free-roaming curiosity, its unconventional interests, its detachment, its ancient and somewhat esoteric set of values . . . surprised by policy decisions dictated by "Reasons of state . . ."[10]

The concept of a "conflict" has lost nothing of its vitality in historical research in the last few years. Think of P. Redondi's *Galileo Heretic,*[11] for example, a most daring revision of the story, accomplished on the basis of a hitherto unknown document. From the pretext of the condemnation to the power-politics within the Curia, it revises all but one thing: the uncompromising antagonism between the new science and Jesuit "science", between the science of Galileo and the official science of the church.

The "clash", indeed, is an element of the myth which seems almost immune to change. Moreover, it seems capable of deviously imposing itself on an ever-growing store of previously unknown facts. It is precisely this quality which indicates that the "clash" does not simply refer to a series of historical events. Primarily, the "clash" is an organizing concept, the basis for a narrative structure that has become a much too forceful explanatory model[12] in the hands of traditional historiography. An exposition and questioning of this underlying structure is necessary in order to fully restore the Galileo affair to its proper place in history. An alternative narrative structure may then be suggested, which would make possible an equally true and no less interesting account of the dialogue of Galileo and the church in the seventeenth century.

Accounts of Galileo's dealings with the church have been marked by several structural elements. Most clearly, they have been based on an overriding binary opposition: that of the church versus science. A closer look obliges us to break down the main binary structure into smaller binary units: theology versus natural philosophy, biblical interpretation versus Copernican cosmology, authority versus free thought, power versus knowledge, etc. The plot proceeds linearly towards the climax of the conflict (i.e., the trials of Galileo), passing through a series of separate incidents which deal alternatively with one pole of the opposition or the other. This way of telling the story eliminates a priori even the possibility of significant communication between the poles. The meaning of the

10. Ibid., p. viii.
11. P. Redondi, *Galileo Heretic,* trans. by R. Rosenthal, Princeton 1987.
12. On narrative structures and their cognitive role, see: W. J. T. Mitchell (ed.), *On Narrative,* Chicago 1981.

story, then, lies in the ironic discrepancy between the apparent triumph
of one pole within the timespan of the plot, and the real triumph of the
opposite pole in the long-term perspective of the reader.

From the outset, let me make it clear that my questioning of this narrative did not originate with an attempted refutation of the accepted interpretations. Rather, it stems from a vague sense of dissatisfaction with a
historical account which still lacks coherence and relies on over-arbitrary
assumptions. One major difficulty should be mentioned, just for purposes
of illustration: the Inquisition's files contain two contradictory documents
referring to the interdictions placed upon Galileo by the church. The standard interpretation today[13] still depends upon the rather arbitrary hypothesis[14] that one of these documents is forged. Reliance upon this kind
of assumption seems to me symptomatic of the difficulty of giving a coherent and plausible account of Galileo's dealing with the church within
the structural constraints of binary oppositions, which still confine our
historical imagination.

No less disturbing is the failure of the traditional narrative to project a
more critical, and hence more complex image of the nature of science and
religion, in conformity with the theoretical insights of the philosophy of
science and the sociology of religion. The long philosophical debate on
"scientific rationality",[15] as well as a whole anthropological-sociological
tradition concerned with the religious origins of rationality,[16] should in-

13. Santillana, *The Crime* . . .
14. There is no external evidence to support the hypothesis of forgery. The only
    cogent reason to hold this view is the fact that the document is unsigned.
    But there are many other unsigned documents in the Inquisition's files. A
    hypothesis which mainly serves to save the coherence of the story is hardly
    acceptable.
15. Mainly I refer here to the critique of positivism launched by Quine, Feyer-
    abend, Kuhn, and their followers who maintain the indeterminacy of
    theory and point out the impossibility of detaching factual statements from
    theoretical networks enmeshed in a web of beliefs.
16. This tradition originates in a certain line of thought in E. Durkheim's *The
    Elementary Forms of Religious Life*, trans. by J. W. Swain, London 1968,
    6th impression. It is elaborated by M. Weber, *The Protestant Ethic and the
    Spirit of Capitalism*, trans. by T. Parsons, New York 1958; R. K. Merton,
    Science, Technology and Society in Seventeenth Century England, New
    York 1970; and R. Horton, "African Traditional Thought and Western Sci-
    ence" *Africa* 37 (1967), pp. 1–2, 50–71, 155–187. See also the two collec-
    tions: B. Wilson (ed.), *Rationality*, Oxford 1970, and M. Hollis and S. Lukes
    (eds.), *Rationality and Relativism*, Cambridge, Mass. 1982.

fluence our structures of historical narration, guarding them against the limitations of over-polarized typologies.

Actual science consists of a complex interaction between rationally proven elements and practical, linguistic, dogmatic, and authoritarian ones. The problematics of transcendental religions, on the other hand, is not only bound up with a need to defend the irrational core of revelation, but also with the creation of a bridge between a mundane and a transcendental sphere, between a rational and a revealed truth.[17] To understand the structure of Galilean science, therefore, means to perceive its encounter with authority not only in terms of defiance but also in terms of interaction. Likewise, to explicate the church's scientific policies means to recognize the autonomy of a reflective dimension in the sphere of religious thought, and the role of clerical intellectual elites in its conceptualization and institutionalization.[18] It is only when a binary oppositional structure finally gives way to a dialogic model that the complexities of science and religion in the story can be truly manifested, with their peculiar mixture of praxis and theory, freedom and authority, reason and faith, and with their social-institutional facets as well as intellectual ones.

The rejection of "conflict" as an organizing principle is paralleled by the questioning of such notions as "truth" and "authority" in other works in the history, philosophy, and sociology of science. Once it is admitted that scientific facts are not simply waiting to be discovered, but grow out of contingent social situations,[19] then the question of "authority" can no longer be considered in opposition to the question of "truth". Scientific truth as a cultural phenomenon has recently been described in the following manner: "Intellectual innovation in science is coming to be understood as a result of complex negotiations among groups of actors in particular local contexts with competing interests, constrained yet not determined by the collection of natural phenomena under investiga-

---

17. See S. N. Eisenstadt and I. Friedrich Silver (eds.), *Cultural Traditions and Worlds of Knowledge: Explorations in the Sociology of Knowledge*, Greenwich, Conn. 1988.

18. Eisenstadt, "The Axial Age, Rise of Transcendental Visions and the Emergence of Intellectuals and Clerics", *European Journal of Sociology* 23 (1982), pp. 294–314.

19. This assumption is shared by a growing number of historians, philosophers, and sociologists, who are seeking a theory of the construction of knowledge which would take into account both natural and cultural constraints on the creation of theories and experimental practice. Among them are: Ian Hacking, Nancy Cartwright, Bruno Latour, Peter Galison, Steven Shapin, Simon Schaffer, and many others.

tion".[20] My personal view about how Galileo's conception of scientific truth could possibly have come into being agrees with this description. However, the opposite is also true. Political and religious authority can seldom base itself on sheer brute force. Authority's acute need to anchor itself in the world of knowledge is forcefully argued in many of Michel Foucault's historical and theoretical works.[21] His suggestion concerning the interdependence or interpenetration of power and knowledge has served as a basic assumption in my understanding of the Galilean affair. The church, to be sure, represented a sophisticated power structure, rooted in specific kinds of knowledge. Galilean science, on the other hand, had specific claims to authority, connected to the types of knowledge it attempted to promote. If we are to account for the Galileo affair in terms of power–knowledge structures, however, a binary-oppositional structure will not do. A dialogic model is certainly more adequate for representing such problematics than a conflictual one.

A questioning of the binary oppositions of the accepted interpretations, therefore, is my point of departure. Here lies the challenge to the traditional reading of the Galileo affair. My story is not about the repression of truth by religious authority. The act of silencing exemplified in the trials of Galileo is in no need of demonstration. So much has it been imprinted in our consciousness that reasserting it is like stating the obvious. Silencing as a plausible result of the pursuit of rational truth, forever entangled in power play of groups with competing interests, is however something much less obvious which touches upon the realities of our life today and deserves to be fully examined. Scientific truth, for its part, does not appear in this story as the transparent reflection of logical processes, entirely rooted in the sphere of eternal ideas. Rather it emerges in the context of a process of interaction between logical demands and practical needs, closely related to power considerations. The story of Galileo, I believe, can tell us something much less obvious about the power–knowledge relationship than the repression of truth by brutal force. It can clearly indicate the mutual transformations of power and knowledge in a concrete historical situation. My aim is to reveal the institutional and intellectual conditions that structured those transformations at the time of Galileo. I have attempted to do so in two major steps: I have used the documents of the trials to expose the significant elements of my story on the basis of a rejection of "conflict" as a unifying concept. I have then told

20.   J. L. Sturchio, "Artifact and Experiment", *Isis*, 79 (1988), p. 298.
21.   Of particular interest are: M. Foucault, *L'ordre du discours*, Paris 1971; and *Surveiller et punir: Naissance de la prison*, Paris 1975.

the story of the encounter between different types of power–knowledge structures within the framework of a dialogic model.

The "trials" of Galileo have been registered in historical memory through the documents of the Roman Inquisition. It is to these documents, therefore, that attention should first be drawn.

As is well known, the bone of contention between Galileo and the church was the Copernican cosmology. According to the Galilean myth, the church forbade Copernicanism, but Galileo persisted in believing in it despite that prohibition. The more sophisticated historical accounts show, however, an awareness of the complex position of the church, which did not forbid the use of the Copernican theory as an astronomical hypothesis, despite the ban on believing in it as an absolute truth.

A careful scrutiny of the documents of the trials, recently republished by the Vatican,[22] reveals two different meanings assigned to the term "hypothesis". It appears that all the documents associated with the Jesuit Cardinal Robert Bellarmine (1542–1621) interpreted the expression "astronomical hypothesis" differently from those associated with Dominican theologians or members of the Inquisition. Moreover, throughout the legal proceedings which took place in 1633, Galileo consistently refused to admit that he regarded the Copernican doctrine as absolutely true, thereby distinguishing his position from that of Aristotelian philosophers who shared a notion of "absolute truth" pertaining to real scientific knowledge.

The different meanings assigned to one key term hold the clue for the alternative narrative structure suggested here. The principles constituted the intellectual basis for the dissemination of knowledge in Dominican and Jesuit schools. The principles have therefore been reconstructed from the Thomistic texts.

In spite of a rather broad Thomist consensus, Thomism itself became subject to different interpretations on the part of the two intellectual elites, the Dominicans and the Jesuits, who operated within different institutional contexts and between whom there was a rather strong rivalry. These divisions culminated in different conceptions of the organization of knowledge as well as in alternative theologies which became a focus of debate and power struggle within the church.

Thus, the background relevant to the Galilean affair cannot be limited

22. S. M. Pagano (ed.), *I documenti del processo di Galileo Galilei*, Vatican 1984.

to a description of the church's monolithic, coercive character with its presumed hostility to science. Neither should the Galilean affair be analysed solely in terms of the contents of scientific theories and the difficulty of adapting them to the common interpretation of the Scriptures. Instead, more emphasis should be placed on the church's inability to formulate a coherent attitude to Galileo's science, being torn between two opposing cultural orientations and divided institutionally as well as theologically.

The traditional elite of the Dominicans developed epistemological scepticism and tended to view Copernicanism not only as an *unproved* theory but also as an *unprovable* one. The Dominicans' rejection of Galileo's science is explained in terms of the epistemological stances they developed in the context of their theological and institutional rivalry with the Jesuits. The Jesuits, however, developed a dialogue with Galileo but attempted to control his science institutionally by suspending the philosophical implications of Copernicanism. The last section of the book reconstructs the cultural field common to Galileo and the Jesuit astronomers, and offers an analysis of their respective discourses, their similarities, and their differences.

The reconstruction of the dialogue between Galileo and the Jesuits accounts for the danger the Jesuits were exposed to with the publication of the *Dialogue Concerning the Two Chief World Systems* (1632). Not only did it disclose the affinity between Galileo's ideas and the Jesuits', but it also pushed these ideas beyond the limits the Jesuits resolved to sustain. Finding themselves under pressure from two extremes – the Dominicans who accused them of dissent from orthodoxy on the one hand and Galileanism which strove to push them toward the full consequences of their own intellectual tendencies on the other – the Jesuits were forced to retreat from backing Galileanism, and actually became its most dangerous opponents.

PART I

The "Trials" of Galileo

1

The Galileo Affair

Interpretation of a Historical Event

This book is an attempt to free the historical narration of the Galilean affair from the myth which has dominated it so far: the myth of the perennial clash between reason and unreason. There is no better point for beginning such a project than the place where most traditional interpretations end, with the trials of Galileo. For nowhere do the constraints on historical narrative and the obstacles to historical research, imposed by the myth, appear more clearly than in the case of Galileo's trials. Fortunately enough, we possess a rather large, although partial, body of documents[1] recording different stages of the legal process, from the first interrogation by the Roman Inquisition in 1611 to Galileo's confession and the sentence in 1633. From my own perspective on Galilean historiography, it is no less fortunate that the different readings of these documents run into difficulties and confusions. It is precisely these problems which justify a fresh look at the documents: once free from the domination of the myth and the presuppositions it imposes, one is able to explain away the difficulties and, at the same time, understand why they were unavoidable within the traditional framework of historical narration. A reexamination of the documents plays a double role in my argument. Indicating the difficulties run into by previous accounts, it at the same time identifies the source of their limitation, the myth of a clash between reason and unreason; ridding us of the presuppositions underlying the myth, it provides a clue to a new understanding of the interaction between power and knowledge, which the Galilean affair encompasses. But first, the readers must be reminded of the plot: the order of events inscribed in the documents.

---

1. A. Favaro (ed.), *Le Opere di Galileo Galilei,* 20 vols. in 21, Florence 1890 (repr. 1968). Vol. 19 contains all the documents related to the trial and organised in two different files; see: "The Documents and Their History", later in this chapter. Also see Pagano (ed.), *I Documenti . . .*

The Plot

The conflict between Galileo and the church took place in two stages. The first was in 1616, and ended with a warning to Galileo, although not with legal persecution. In popular literature and in certain historical works, this stage used to be called the "first trial". The second stage of the drama activated the elaborate judicial machine of the Inquisition, and has been known as the "second trial".

Act One

The first encounter begins some time after the publication of the *Sidereus nuncius* (1610), when Galileo's great telescopic discoveries have gained him and the Copernican cosmology a certain amount of fame, together with some hostility. In 1614–1615, complaints are raised in Florentine circles about the incompatibility of the new discoveries with certain verses from the Scriptures. The rumours reach Galileo through a letter from his disciple, the monk Benedetto Castelli. In reply, Galileo writes a letter to Castelli, expressing his ideas about the relationship of knowledge aiming at an understanding of the universe to knowledge connected with the attainment of salvation. The first kind of knowledge, he says, is based on evidence of the senses, and teaches us "how heaven goes", whereas the second, derived from the Scriptures, teaches us "how to go to heaven". Galileo questions Castelli about the validity and legitimacy of his explanations, and Castelli gives his assent with a recommendation to introduce some minor changes. Following this exchange, Galileo reformulates his ideas in the famous *Letter to the Grand Duchess Christina* (1615). The letter passes from hand to hand in circles sympathetic to the Galilean discoveries. Consequently, Galileo's name is twice denounced to the Inquisition. The Congregation of the Holy Office begins investigations, which culminate in official deliberations of the consultants of the Holy Office over the two main propositions of the Copernican theory: namely, that the sun is immobile in the centre of the universe, and that the earth is not the centre of the universe and moves around the sun at the same time performing daily revolutions. The deliberations end on 23 February 1616 with a decision to reject the Copernican theory and to deny it any philosophical or theological truth. The matter is now transferred to the Congregation of the Index, which issues a decree forbidding and suspending certain books whose subject-matter is connected with the Copernican theory. When the deliberations are at their height, Galileo goes to Rome to publicize his views and perhaps convince the church of the scientific truth of his opinions. However, on 26 February 1616, he is invited on the decision of the Congregation of the Holy Office to the residence of Robert Bellarmine, a prominent cardinal of the Inquisition, and is warned that he should abandon his position with regard to Copernicanism.

Act Two

The second act begins with Galileo's visit to Rome in 1630. He has recently finished the book which he had been working on for the prior twelve years, and in which he has presented all the arguments for and against the two great world systems – the Copernican and the Ptolemaic. Six years before (1624) he had a series of discussions touching on astronomical and cosmological matters with the new Pope Urban VIII (1568–1644), a liberal churchman well known for his love of the arts and sciences. In the course of these discussions Galileo was impressed by Urban's inclination to favour an astronomical dialogue. He has now come to Rome to ask for an imprimatur (permission of publication) for his book. The request is presented to Niccolo Riccardi, the Master of the Holy Apostolic Palace, who also functions as head censor of the Holy Office. Riccardi finds it necessary to consult with his friend and colleague Raffaello Visconti, who is better versed than himself in the mathematical sciences. Visconti recommends some corrections, all with the intention of emphasizing the hypothetical-mathematical character of the book, in the spirit of the wishes expressed by the pope. Galileo promises to follow Visconti's recommendations and to introduce the necessary changes. The imprimatur is given on condition that the new corrected version be submitted to Riccardi before printing.

Galileo is back in Tuscany, but finds it impossible to return with the new version to Rome owing to plague, quarantine, and poor health. From Tuscany he exerts enormous pressures in order to be allowed to have his book published in Florence. After a long period of hesitations and recurring procrastinations Riccardi gives in. The responsibility for a further reading of the manuscript is passed on to the Inquisitor of Florence, followed by precise instructions for the required introduction to the book. The manuscript is checked by the Florentine Inquisitor, is passed on to another censor named Giacinto Stefani for further examination, and after being approved by the two men finally goes to press and is out on the shelves by the spring of 1632.

Shortly afterwards, the first copies of the *Dialogue on the Two Chief World Systems* reach Rome. They are read by the pope, Riccardi, and some other church dignitaries, and are immediately confiscated. Further dissemination is forbidden. The pope appoints a special commission to decide upon the fate of both the book and its author, and Galileo is summoned to Rome. He arrives in the city in February 1633, after several delays. In March, he is incarcerated in the Inquisition building. It is there that the first two hearings of the "second trial" take place. Galileo is then transferred to the villa of Francesco Niccolini, the Tuscan ambassador to Rome. There he writes his self-defence, which is presented to the Commissary General of the Inquisition. This session is followed by one further hearing which finally leads to a sentence. Galileo is suspected of holding heretic opinions, which he is ordered to abjure; he is condemned to formal imprisonment at the pleasure of the Inquisition, his book is prohibited, and the sentence is sent to Inquisitors all over the Catholic world, to be read in full assembly and in the presence of the leading practitioners of the mathematical arts.

Rival Interpretive Strategies

Two interpretive positions, two reading strategies have suggested themselves to Galilean scholars: one being that of the champions of scientific rationality, the other belonging to the apologetic voices of the church and to some opponents of Galileo.[2] But at the heart of both readings there has always been one question: Was Galileo right? Was he right from a scientific point of view? Could he *really* prove what he claimed to discover? Was he right from a political point of view? In 1616, did he not, after all, give his word that he would abandon the Copernican doctrine?

2. The following discussion of the literature pertaining to the "trials" of Galileo does not aim at exhausting the subject, which is wide enough to require a separate research. In order to delineate the general directions of interest of this literature, a few representatives of each camp have been chosen. The researchers who have contributed most substantially to the pro-Galilean camp are: K. von Gebler, *Galileo Galilei and the Roman Curia*, trans by Mrs. G. Sturge, London 1879; E. Wohlwill, *Galilei und sein Kampf für die Kopernickanischen Lehre*, Hamburg and Leipzig 1909; G. de Santillana, *The Crime of Galileo*, Chicago 1955; G. de Santillana and S. Drake, "Arthur Koestler and His *Sleepwalkers*", *Isis*, 50 (1950), pp. 255–260. The researchers who have contributed most substantially to the apologetic campaign of the church are: J. Brodrick, SJ, *The Life and Work of Blessed Cardinal Bellarmine, S.J. 1542–1621*, 2 vols., London 1928; J. J. Langford, *Galileo, Science and the Church*, Michigan 1971, rev. ed.; A. Koestler, *The Sleepwalkers: A History of Men's Changing Vision of the Universe*, London 1959.

Was he right from an ecclesiastical point of view? How, after all, could the church be expected to give up the policies necessitated by the Protestant Reformation in order to stand up to his immense challenge?

Galileo was right, say the advocates, for he corroborated the Copernican theory and refuted Aristotelian cosmology and Ptolemaic astronomy. The church which opposed and censored him was blind to the effectiveness of his mathematical method and afraid of the truth of his physical hypotheses. But, whereas Galileo's correctness is usually described in terms drawn from his theories themselves, the church's position is explained in nondiscursive, political, and psychological terms. The church of the Counter Reformation, it is claimed, is known for its lack of tolerance towards unorthodox views, coupled with a persecution and oppression of intellectuals. Copernicanism was conceived as a direct threat to the traditional reading of the Scriptures and, as a consequence, to the whole authority of the church. The church could not afford to tolerate the sophisticated defence of a dangerous theory put forward by Galileo.

These institutional motives, however, were coupled with personal motivations. Galileo's bitter debates with some Jesuit astronomers made them into his enemies and pushed them into seeking vengeance by convincing the pope to bring Galileo to trial. Urban VIII, who had been criticized for his nepotistic policies and dubious involvement in the Thirty Years' War, vented his anger on poor Galileo, thereby trying to regain some lost political assets. So emotionally heated was the confrontation that some church officials deviated from common legal procedures, even going so far as to forge a document in order to justify their position.

Be the details as they may, the binary structure of the story is clear and simple: rationality is embodied in Galilean science, although not necessarily in the figure of the scientist; irrational modes of thought, institutions, and historical agents belong to the other side, the church, which is seen as an embodiment of political power.

This strategy of reading the events, however, has its counterpart. Galileo was wrong, say the adversaries, for he ascribed absolute truth to the Copernican world system. No scientific theory can be really, absolutely true, let alone the Copernican theory which had been faulty from the very beginning. The most one can say is that heliocentrism was to be preferred to geocentrism, and that the church was late recognizing this. Copernicanism, however, contained no more truth than its rival world system, and Galileo, in fact, did rather poorly in his attempt to prove the superiority of the modern theory. It was R. Bellarmine, cardinal of the Holy Office, who really understood the hypothetical nature of scientific theories, and whose point of view has become a cornerstone of modern scientific thinking.

On the other side, we find a stubborn man, very arrogant, whose claims to knowledge have been grossly exaggerated. It is hardly surprising if he deliberately disregarded Bellarmine's instructions and then acted dishonestly towards the authorities, asking for permission to bring his book to the printer. It was his inability to compromise, his lack of modesty which pushed the church to take drastic steps, to humiliate him and censor his book.

The story is thus inverted, and yet the structure remains the same. The church, with its prudent attitude towards ephemeral scientific theories, becomes the vicar of rationality, whereas the stubborn Galileo embodies an all-too-human irrationality. It matters little that the advocates speak about rationality in terms drawn from Galilean science, while the opponents do so in terms drawn from a Catholic "image of knowledge".[3] It matters little whether irrationality is ascribed to institutions or to individuals. What really matters is that both stories are but inverted interpretations of the same structure, the same clash between reason and unreason.

The Documents and Their History

We have a plot with diametrically opposed interpretations. And we have a body of documents with a story of their own.[4] The documents of the

3. In my effort to explicate the interaction between institutional and cognitive interests I have made some use of the concept of "images of knowledge" developed by Y. Elkana in his "A Programmatic Attempt at an Anthropology of Knowledge" in: E. Mendelsohn and Y. Elkana (eds.), *Sciences and Cultures. Anthropological and Historical Studies of the Sciences*, Dordrecht 1981, and elsewhere. In Elkana's terms every cultural system exhibits an interaction among three levels of thought and experience: the body of knowledge, "images of knowledge", and normative factors. The body of knowledge is the sum total of assumptions on the subject matter; the "images of knowledge" are a group of rational suppositions on human knowledge which do not necessarily stem from the body of knowledge but are rather embedded in the general culture, dependent on its social and institutional structures. Assumptions on possible sources of knowledge and the hierarchy among them (sense evidence, mathematics, divine illumination), the aims of knowledge (recognition of God, truth, utility), the legitimation of knowledge (logical proof, authority, etc.) are all "images of knowledge" through which normative factors such as ideologies, religious values, pedagogical interests infiltrate into the process of problem choice in the body of knowledge.
4. My main source for telling the story of the documents is A. Favaro, "I documenti del processo di Galileo", *Atti del Reale Instituto Veneto di Scienze, Lettere ed Arti 1901–1902*, vol. 61, II, Venice 1902, pp. 757–806.

trials were originally in the ownership of the Congregation of the Holy Office. The system of organization of the archives divided the documents into two groups, collected in two different kinds of files. One kind contained decisions of the General Congregation,[5] which met three times a week to supervise the work of the Italian Inquisition on behalf of the Roman Curia. Once a week, on Thursdays, the pope was also present at the meetings, and they then took place in one of his palaces in Rome. The decisions were recorded by the senior official of the Holy Office, the Assessor, and were then edited and copied by a notary and collected in chronological order in volumes marked *Decreta* (decisions). It was among the *Decreta* that after a long period were found the decisions relating to Copernicanism, Galileo's *Dialogue*, and other scientific subjects considered relevant to matters of religion and faith.[6] But the Inquisition also possessed files of another kind organised according to a name index. Those contained records of legal proceedings which took place in the law courts of the Inquisition: e.g., the courts' investigations of the accused and of other witnesses, correspondence concerning trials, legal decisions, abjurations and sentences, etc. This kind of file was called *Processus* (trials). Galileo's file in the archives of the Inquisition was marked volume 1181, and so it has remained.

Until 1810, the documents were preserved in the secret archives of the Inquisition in Rome, hidden from the eyes of strangers. Soon after the Napoleonic conquest, they were transferred, like the other Roman archives, to Paris.[7] The *Decreta* did not attract any particular attention, and were returned to the Holy See by the regime of Louis Philippe; but the historical value of volume 1181 was immediately recognized by the French, and projects were made for its translation and publication. E. Renan[8] mentioned a work based on the documents, of which only two passages survived in his own time: he considered it unscholarly and lacking in objectivity. The Napoleonic edition of the documents, however, was

5. On the organisation of the papal government see: H. O. Evennet, "The New Organs of Church Government", in *The Spirit of the Counter Reformation*, Cambridge 1968. For an excellent description of the structure and functions of the Congregation of the Index and the Holy Office, see: W. Brandmüller and E. J. Greipl (eds.), *Copernico Galilei e la Chiesa. Fine della controversia (1820): Gli atti del Sant' Uffizio*, Florence 1992, pp. 48–51.
6. Favaro, "I Documenti . . .", p. 759.
7. Ibid., p. 761.
8. E. Renan, "Nouvelles études d'histoire religieuse", Paris 1884, in Favaro, "I Documenti", p. 773.

never published.[9] In 1843, volume 1181 was finally returned to Rome, accompanied by rumours about a papal commitment to publish the documents and make them available to scholars. The revolution of 1848 brought professor Silvestro Gherardi, a liberal and a scholar of Italian science, to the palace of the Holy Office, only to discover that volume 1181 was still missing. In fact, after being returned to Rome, the volume was in the pope's private library until deposited by the fleeing Pius IX in the Vatican Archives,[10] where it remained until found and published, together with the *Decreta,* in the critical edition of Galileo's works by Antonio Favaro.

Even before Favaro, the widespread interest in the publication of the documents, coupled with a wish to exonerate the Italian Inquisition from unfounded suspicions of physical torture supposedly inflicted on Galileo, led the pope to grant a number of French, Italian, and German scholars permission to see some of the materials. Traces of their recorded impressions throw further light upon the nature of the documents.

Silvestro Gherardi, the first to copy the *Decreta* documents, was also the first to make a distinction between the official formulations of the documents and some raw material gathered during the meetings by the Assessor, of which nothing remained in the files.[11] In his letters to the German historian Emil Wohlwill, Gherardi mentioned he had seen – in addition to the official decisions – suggestions, rough drafts, sketches of protocols prepared before the meetings for members of the Congregation, papal instructions, and proposals for changes which were added to the manuscripts of the Assessor after the meetings.[12] Gherardi made lists based on this material,[13] which lists were later found by Favaro who scrutinized them, and decided not to include them in his critical edition.[14] Gherardi's recollections are significant, however, for they testify to the multiplicity of voices which exist beneath the unified surface, all doomed to silence by the legal language of the documents. Gherardi's memories of masses of raw materials which have become inaccessible to the historian should make one question not only the meaning of the written words, but also the identity, position, and interests of those responsible for them.

9. See Redondi, *Galileo . . . ,* p. 154.
10. Favaro, "I Documenti . . .", p. 774.
11. Ibid., p. 780.
12. Ibid., p. 794.
13. *Carte Gherardiane,* in Favaro, "I Documenti . . .", p. 792, n. 2.
14. Ibid., pp. 795–798.

Gherardi was also the first to raise suspicions concerning the authenticity of certain documents. In this he did not remain alone. Emil Wohlwill opened a Pandora's box when he suggested that one of the documents in file 1181 was forged.[15] This marked the beginning of a long controversy, in the course of which the authenticity of other documents was also questioned. The controversy reached its climax when the suggestion was made that the entire Galilean trial was a forgery of the Inquisition.[16]

Presumed forgery is a recurrent theme among historians of the trials of Galileo. This is a measure by which to assess the difficulty in constructing a coherent and plausible story on the basis of what remained accessible. And yet, there is no real reason to doubt the authenticity of the corpus as a whole. Rather, this corpus may be treated as a unit, signifying a wish to conceal no less than a need to disclose. The contradictions this corpus reveals may have been part of the historical reality, no less than the unanimous voice which intended to conceal them.

Favaro's experience with the documents would seem to support this way of reading the material. Time and again, Favaro pointed out that both files of the Inquisition were subjected to repeated examinations, manipulations, and natural disasters which corrupted their original form. Volume 1181 was repeatedly opened by the Inquisition itself, said Favaro, even before 1810, when the volume began its peregrinations. In 1632–1633, the first file opened in 1616 was examined, and at the end of the trial a sort of synthesis was made of the two files. After Galileo's death, the files were again examined when a proposal to erect a monument in his memory was discussed, and again when a reprint of the *Dialogue* was considered.[17] As for the *Decreta*, Favaro claimed that the method of recording the decisions altered in the course of time. In Galileo's time there was a tendency to shorten reports. Moreover, no record of the personal role of each cardinal has survived, contrary to the custom of earlier periods.[18] Favaro also related that the documents he found were not always arranged in chronological order. Some of them were missing in the original file, and then traced in different places. Significantly enough, the most famous among the missing documents were the sentence and the abjuration.[19] Both in the Napoleonic period and during the Second Roman Republic the archives went wandering from place to place, and some of the

15. E. Wohlwill, *Der Inquisitions process des Galileo Galilei*, Berlin 1870.
16. Favaro, "I Documenti . . .", p. 775.
17. Ibid., p. 760.
18. Ibid., p. 759.
19. Ibid., p. 791.

documents contained in them were probably lost and others possibly destroyed.[20] And above all, Favaro concluded, one must remember that, as a rule, "according to the immemorial practice of the Holy Office, not everything is preserved".[21]

The story of the unusual vicissitudes of the documents of Galileo's trials betrays the same kind of confusions and inconsistencies as characterize the attempts to interpret their meaning. There is an unresolved tension between the need to conceal and suppress, and the need to publish and disclose. What is disclosed is a formal, unitary facade of institutional authority and power, maintained by legal procedures. What is suppressed is a variety of voices representing contradictory interests, the sole evidence of which is the acts of manipulation revealed in the obvious incompleteness of the documents. The possible identities and interests of the representatives of these contradictory positions and the structured interrelationships between them have never been systematically investigated by historians. These were mainly interested in the contents of what survived, trying to fill in the gaps with hypotheses of forgery. My alternative strategy, however, is to treat the same gaps as clues to the reality expressed by the documents. But first the difficulties in the plot, the traditional interpretations, and the history of the documents should be examined, in order to expose the hypotheses common to all known readings of them, and to suggest new ones in their place.

Principles of Interpretation

Three main difficulties are common to most interpretations of the documents. They involve three inconsistencies, one for each layer of the story: (1) the conceptual, (2) the institutional, and (3) the psychological.

(1) Among some of the documents there are conceptual contradictions. The Inquisition files contain *two* versions of the warning of 1616 and its meaning.[22] There is a clear contradiction between the unsigned document containing the command of the Commissary of the Inquisition, in the presence of Bellarmine, prohibiting Galileo from *holding, teaching,* or *defending* the Copernican theory in any way whatsoever, and the admonition by Bellarmine himself which only forbade him to *hold* the Copernican theory as absolutely true. At least two conflicting notions of "the hypothetical" or "the probable" were implied by this inconsistency.

20. Ibid.
21. Ibid.
22. See ch. 2, pp. 45–50.

(2) The church was inconsistent in its attitude toward the publication of the documents. This may appear to be an external fact, but it is directly related to the way both the documents and Galilean science itself were understood by the church establishment. The tension between the interest in revealing and the interest in concealing suggests a far more significant tension between a negative and preventive role and the positive and constructive role which the church in fact played, or at least could have played, in the production of scientific knowledge.

(3) Galileo was inconsistent in his reaction to the demands posed by the church authorities. The Inquisition would not have treated Galileo so harshly had it not conceived of him as a realist (i.e., believing in the absolute truth of Copernicanism). Galileo could not possibly have abided by the church's demands and obtained the imprimatur had he not presented his claims in a nonrealist way.

Traditionally, the inconsistencies in the story have been resolved by means of the supposed opposition between reason and unreason, common to all interpretations. Inconsistencies pose no problem for the irrational side of the story. Thus, it is only Galileo's advocates who need worry about inconsistency (3). The opponents simply tend to say that Galileo acted as a realist when it pleased him to do so, and that the church could not afford such a game. Similarly, the inconsistency implied in (2) need worry the opponents only. Galileo's advocates are quick to point to the arbitrariness of the power politics of the church. It is inconsistency (1), however, which poses a serious problem for both sides. If both documents are authentic, then both Galileo and the church were giving at least two different meanings to key concepts like "the hypothetical" or "the probable". The strategies of advocates and opponents alike in dealing with this problem are equally unconvincing. If the church had conducted the trials with goodwill, say the advocates, there would not have been such an embarrassing discrepancy between two of its documents; if Galileo had truly abided by the first warning, say the opponents, there would have been no need for a second trial. In order to exonerate Galileo of either disciplinary or conceptual transgressions, most of the advocates tend to withdraw to an ad hoc hypothesis, namely, that the church forged the 26 February document: it was found unsigned, many years after the event, and totally surprised the accused, who had not been aware of such a strong interdiction against discussing Copernicanism.[23] The opponents, who insist upon Galileo's disciplinary error, resort to the hypothesis of duplicity on his side, which they try to corroborate by an analysis of his hidden intentions. This analysis is clearly at odds with everything Galileo

23. See Santillana, "The Problem of the Fake Injunction", in *The Crime* ...

explicitly and consistently expressed in writing and in his oral responses to the inquisitors.

Both the advocates and the opponents share three presuppositions, expressing the limitations of certain opposing concepts common to both parties: the concepts of reason vers us unreason, unity versus diversity, and knowledge versus power. These presuppositions are:

1. The church with which Galileo was dealing was monolithic in its reaction to his intellectual and institutional challenge.
2. The church was dealing with the same Galileo in 1616 and in 1632–1633.
3. The church, which actually represented political power, suppressed knowledge; it never created knowledge.

But what if we put aside the common categories – reason, faith, church, science – and tried to concretize them, to embody them in personalities, institutional positions and roles, specific patterns of action, particular policies? What would happen if we suspended judgment of the rational and the irrational, and disregarded the binary structure of the myth? We should then free ourselves of the above three presuppositions and start anew, with the following assumptions:

(1) The church was not monolithic. Although the Reformation gave birth to a political and cultural counter-movement, the Counter Reformation was merely a common ideal allowing rival forces within the Catholic Church to strive for hegemony. Therefore, always ask who is speaking and from where; try to identify personalities, institutional positions, specific intellectual and political interests, not only in the case of the seemingly heretical rebel, but also within the church itself.

(2) Both Galileo and the church may have changed their views on key issues at stake; there is no a priori reason to exclude such a possibility. Therefore, allow all parties to change views and positions. Look for possible changes and try to account for them in intellectual and institutional terms.

(3) The effects of power on the production of knowledge are not always negative; power is always constructive as well. Therefore, abandon the opposition between power and knowledge and look for interactions; look for the knowledge required in order to justify censorship and for the knowledge which censorship forces the censored to develop.

(4) Conflict and interaction are never restricted to a single domain: bodies of knowledge and images of knowledge, ideologies and positions within a structure of power relations are usually in question at the same time. Therefore, try to distinguish and then relate the different dimen-

sions, without, however, allowing one dimension to be reduced to any of the others.

My rereading of the documents will be both more abstract and more concrete than the traditional one. I shall sometimes violate the needs of the narrative by neglecting chronological sequence for the sake of thematic emphasis: not only shall I search for the meaning of words, but I shall also look for the identity of speakers, their possible intellectual traditions, their institutional affiliations. Some traditional features, however, will be preserved – the boundaries of the narrative dividing my story into two major sections, the events of 1616 and the events of 1633.

2

1616

The events of 1616 are recorded in six important documents which touch upon two aspects of the cultural policy of the church: its official attitude towards Copernicanism, and the application of this attitude to the Galilean case. Copernicanism, an innovative scientific theory bearing upon the interpretation of the Scriptures, required a reaction from the church following the denunciation of Galileo's name to the Inquisition. The first group of documents (1) from 24 February 1616; (2) from 5 March 1616; and (3) from 15 May 1620, allows us a glimpse at the concrete act of taking up a position vis-à-vis Copernicanism. This act marked the transition from an unstructured field of possible futures for the Copernican theory to a structured one. Although a scientific theory was not actually consigned to oblivion, Galileo's legitimate field of action was limited. The second group of documents, (4) from 25 February 1616; (5) from 26 February 1616; and (6) from 3 March 1616, defines the specific nature of these limitations. Two additional pieces of evidence cast more light on the above documents: the first is a letter of Cardinal Bellarmine concerning the problem of interpretation of the Scriptures, in which Galileo was mentioned; the second is a certificate Bellarmine left with Galileo after the warning of 1616, apparently without informing his colleagues in the Congregation of the Holy Office.

Official Catholic Position Concerning Copernicanism

My point of departure for discussing the church's official position vis-à-vis Copernicanism is the decision of a group of theologians – the consultants of the Congregation of the Holy Office – who formulated two theses deemed to encapsulate the essence of Copernicus's doctrine, on which they pronounced their theological verdict.

1. *Consultants' Report on Copernicanism (24 February 1616)*

The decision of the consultants of the Holy Office concerning Copernicanism was recorded as follows:

Assessment made at the Holy Office, Rome, Wednesday, 24 February
1616, in the presence of the Father Theologians signed below. Propositions to be assessed:

i. The sun is the center of the world and completely devoid of local
motion.

Assessment: all said that this proposition is foolish and absurd in
philosophy, and formally heretical since it explicitly contradicts in
many places the sense of Holy Scripture, according to the literal
meaning of the words and according to the common interpretation
and understanding of the Holy Fathers and the doctors of theology.

ii. The earth is not the center of the world, nor motionless, but it
moves as a whole and also with diurnal motion.

Assessment: All said that this proposition receives the same judgment
in philosophy and that in regard to theological truth it is at least
erroneous in faith.

Petrus Lombardus, Archbishop of Armagh.

Fra Hyacintus Petronius, Master of the Sacred Apostolic Palace.

Fra Raphael Riphoz, Master of Theology and Vicar-General of the
Dominican Order.

Fra Michelangelo Segizzi, Master of Sacred Theology and Commissary of the Holy Office.

Fra Hieronnimus de Casalimaiori, Consultant to the Holy Office.

Fra Thomas de Lemos.

Fra Gregorius Nunnius. Coronel.

Benedictus Justinianus, Society of Jesus.

Father Raphael Rastellius, Clerk Regular, Doctor of Theology.

Father Michael of Naples, of the Cassinese Congregation.

Fra Iacobus Tintus, assistant of the Most Reverend Father Commissary of the Holy Office.[1]

The consultants decided that philosophically – e.g., considered as physical truths – the Copernican hypotheses concerning the structure of the
universe were foolish and absurd. At the same time, they also related to

---

1.   *Opere*, XIX, pp. 320–321: "Propositiones censurandae. Censura facta in S.ᵗᵒ
Officio Urbis, die Mercurii 24 Februarii 1616, coram infrascriptis Patribus
Theologis. Prima: Sol est centrum mundi, et omnino immobilis motu locali.
Censura: Omnes dixerunt, dictam propositionem esse stultam et absurdam
in philosophia, et formaliter haereticam, quatenus contradicit expresse sen-
tentiis Sacrae Scripturae in multis locis secundum proprietatem verborum et

these propositions theologically, and found the first proposition heretical and the second erroneous. On one aspect of the Copernican theory the censors preferred to remain silent. They did not even mention the mathematical element predominant in the work of Copernicus. In fact, only the first book of Copernicus's De revolutionibus related to the physical aspects of the structure of the world. All the other books consisted of calculations and predictions of the motions of the heavenly bodies, much after the model of Ptolemaeus's Almagest.[2] In this silence lay the seeds of the church's multi-dimensional position.

Eleven consultants signed the decision of the Congregation. Most of them belonged to religious orders. Five, at least, were Dominicans. One member of the Society of Jesus was among the consultants.

2. Decree of Index (5 March 1616)

The censure of the consultants of the Holy Office was not even published. The task of implementing the decision was passed on to the secondary Congregation of the Index which, on 5 March 1616, issued a prohibition of certain works. The wording of the prohibition reflected the various levels of discussion of the Congregation:

> Therefore, in order that this opinion may not creep any further to the prejudice of Catholic truth, the Congregation has decided that the books by Nicolaus Copernicus [On the Revolution of Spheres] and Diego de Zuñiga [On Job] be suspended until corrected; but that the

---

secundum communem expositionem et sensum Sanctorum Patrum et theologorum doctorum. 2.a: Terra non est centrum mundi nec immobilis, sed secundum se totam movetur, etiam motu diurno. Censura: Omnes dixerunt, hanc propositionem recipere eandem censuram in philosophia; et spectando veritatem theologicam, ad minus esse in Fide erroneam. Petrus Lombardus, Archiepiscopus Armacanus. Fr. Hyacintus Petronius, Sacri Apostolici Palatii Magister. Fr. Raphael Riphoz, Theologiae Magister et Vicarius generalis ordinis Praedicatorum. Fr. Michael Angelus Seg.^{s}, Sacrae Theologiae Magister et Com.^{s} S.^{ti} Officii. Fr. Heronimus de Casalimaiori, Cosultor S.^{ti} Officii. Fr. Thomas de Lemos. Fr. Gregorius Nunnius Coronel. Benedictus Jus.^{nus}, Societatis Iesu. Dr. Raphael Rastellius, Clericus Regularis, Doctor theologus. D. Michael a Neapoli, ex Congregatione Cassinensi. Fr. Iacobus Tintus, socius R.^{mi} Patris Commissarii S. Officii." Translation quoted from M. A. Finocchiaro (ed.), The Galileo Affair: A Documentary History, Berkeley and Los Angeles 1989, pp. 146–147. Unless otherwise stated, all translations follow Finocchiaro and are quoted by permission of the University of California Press.

2. N. Copernicus, On the Revolutions, translation and commentary by E. Rosen, Cracow 1978.

book of the Carmelite Father Paolo Antonio Foscarini be completely prohibited and condemned; and that all other books which teach the same be likewise prohibited.[3]

Copernicus's book, which opened with some general physical assumptions, but was mainly mathematical, was suspended until corrected. The church did not entirely, or in principle, prohibit the use of this book. What was left unclear, however, were the precise limits of its freedom of use. In the same spirit, Diego de Zuñiga's commentary on the Book of Job, which had explained a scriptural verse by means of the Copernican theory, was suspended but not absolutely condemned. The book of the Carmelite monk Foscarini, however, which had attempted a systematic harmonization of the Scriptures and Christian doctrine with the idea of the immobility of the sun and the motions of the earth, was unequivocally condemned and all books which attempted to teach similar ideas were prohibited.

3. *Correction of Copernicus's* On the Revolutions *(15 May 1620)*

Thus, in 1616, the Congregation of the Index chose to leave the limits of the possible use of Copernicanism unclear. Only in 1620 did the Congregation issue another decree, in which an attempt was made to define those limits more precisely:

Although the Fathers of the Holy Congregation of the Index censured the writings *On the Revolutions of the World* of Nicolaus Copernicus, a respected astronomer-astrologer, to be wholly prohibited, for the reason that he does not doubt the principles of the location and motion of the terrestrial globe which are repugnant to the Sacred Scriptures and their true and Catholic interpretation (something which is not to be tolerated in any Christian), but construes them as entirely true and does not treat them hypothetically; nevertheless, because these writings contain many things which are very useful to the community, they were pleased in that decision to permit by unanimous consent the writings of Copernicus to be printed, as they had already permitted, but with certain corrections according to the

---

3. *Opere*, XIX, p. 323: "... ideo, ne ulterius huiusmodi opinio in perniciem Catholicae veritatis serpat, censuit, dictos Nicolaum Copernicum De revolutionibus orbium, et Didacum Astunica in Job, suspendendos esse, donec corrigantur; librum vero Patris Pauli Antonii Foscarini Carmelitae omnino prohibendum atque damnandum; aliosque omnes libros, pariter idem docentes, prohibendos ..." Finocchiaro, p. 149.

emendations below, in places where he discusses the location and motion of the earth not hypothetically but as an assertion.[4]

The decree of 1620 attested to the need to reconcile the theologians' decision of 1616 with the practical recognition of the usefulness of Copernicus's theory. It was asserted that the principles related to the position and motion of the earth could not be true since they contradicted the traditional understanding of the Scriptures. Therefore it was the duty of a Christian to doubt them a priori. The reason for the condemnation was presented as lack of sufficient doubt on the part of Copernicus. This seems an attempt to forbid Copernicus's claims not because they had not yet been proven scientifically, but because the very presupposition they could ever be proven was repugnant to the true faith. The problem with the Copernican theory was not only that it lacked proof, but mainly that it was an impertinent attempt to construe a sound theory upon suspicious principles. At the same time, the usefulness of the mathematical information contained in Copernicus's book was acknowledged as something beneficial for the community. For how could this be denied after the church had actually adopted it and made it the basis of the reform of the calendar?[5] Even the highly abstract language of the Congregation cannot conceal the practical significance of the decree, which touched upon the problem of the calendar, a primary instrument of social control over the lives of every individual in seventeenth-century Europe. The decree, then, permitted the use of Copernican knowledge, but insisted that the underlying principles of the theory should be regarded as only hypothetical. The decree of 1620 was signed by the Dominican friar Franciscus Magdalenus Capiferreus, a member of the Congregation of the Index.[6]

In 1616 the theologians of the Holy Office did not make a distinction between aspects of the Copernican theory which were useful and others which had to be rejected. The philosophical and theological terms in which the decision was cast, as well as the silence about the mathematical

4. Ibid., p. 400: "Quanquam scripta Nicolai Copernici, nobilis astrologi, De mundi revolutionibus prorsus prohibenda esse Patres Sacrae Congregationis Indicis censuerunt, ea ratione quia principia de situ et motu terreni globi, Sacrae Scripturae eiusque verae et catholicae interpretationi repugnantia (quod in homine Christiano minime tolerandum est), non per hypothesim [ea] tractare, sed ut verissima adstruere, non dubitat; nihilominus, quia in iis multa sunt reipublicae utilissima, unanimi consensu in eam iverunt sententiam, ut Copernici opera ad hanc usque diem impressa permittenda essent, prout permiserunt, iis tamen correctis, iuxta subiectam emendationem, locis, in quibus non ex hypothesi, sed asserendo, de situ et motu terrae disputat". (My translation, R.F.)
5. T. S. Kuhn, The Copernican Revolution, New York, 1957, p. 196.
6. Opere, XIX, p. 401.

aspects of Copernicus's work, testified to an unresolved problem which appeared on the surface later on. The decree of the Congregation of the Index of 1616, which translated the theologians' decisions into practical actions, already anticipated the distinction between the status of the Copernican theory as yet unproved, or probable position, and its assertion, as demonstrated truth. This distinction delineated the area of contention among different positions, interests, and groups within the church as well as among lay intellectuals. The second decree of the same Congregation issued in 1620, however, dealt with Copernicanism in terms of a distinction between forbidden, unproved or unprovable principles (the position and motions of the earth), which were obviously physical in nature, and useful implications, probably referring to the mathematical principles underlying the reform of the calendar.

This series of successive initiatives of the church establishment reveals an interplay between theoretical positions and practical interests in the process of taking up a position vis-à-vis Copernicanism. What remain hidden from the lay reader are the intellectual traditions by which the whole game of sanction and prohibition was governed. However, it is these intellectual traditions which permitted the game to proceed in relative tranquillity, and without our needing to assume a recourse to violent, extreme actions like forgery. The intellectual traditions transparent to the actors, but concealed from modern readers, actually constitute the conditions which made the game possible. This gap between the knowledge of contemporary actors and that of the modern reader calls upon the interpreter to explain the terms of the game that finally led to the condemnation of the *Dialogue*.

The epistemological basis for the distinction between the "hypothetical" and the "assertable" (e.g., the "true"), shared and manipulated by all parties to the game, is already to be found in the writings of Aristotle and his Greek, Hellenistic, and Arab interpreters.[7] But the distinction was

7. Many historians have discussed the emergence and development of an instrumentalist conception of astronomy, among whom P. Duhem, *To Save the Phenomena*, E. Doland and C. Maschler (trans.), Chicago 1969, is the best known; see also N. Jardine, "The Forging of Modern Realism: Clavius and Kepler Against the Sceptics", *Studies in the History and Philosophy of Science*, 10 (1979), pp. 141–173; and of the same author, *The Birth of History and Philosophy of Science: Kepler's "A Defence of Tycho against Ursus" with Essays on its Provenance and Significance*, Cambridge 1984; R. S. Westman, "The Astronomer's Role in the Sixteenth Century: A Preliminary Study", *History of Science* 18 (1980), pp. 105–147; idem, "The Copernicans and the Churches", D. C. Lindberg and R. L. Numbers (eds.), *God and Nature*, Berkeley 1986, pp. 73–117; idem, "The Melanchton Circle, Rheticus, and the Wittenberg Interpretation of the Copernican Theory",

further developed in the work of Thomas Aquinas, where it was given its most sophisticated logical formulation.[8] True knowledge of the natural world, according to Thomas, was knowledge of natural substances, obtained through abstraction from sense data and the application of deductive logic to establish the causal connections among them. For Thomas, Aristotelian physics represented a system of natural substances, coherently related by a network of logical relations. Hence, he considered it "scientia", i.e., true and assertable knowledge of nature. In contradistinction, astronomy was a science of natural substances which used mathematical methods to establish the connection between observed facts and the hypotheses which explain them, and to deduce additional facts from these hypotheses.[9] Epicycles and eccentrics, for instance, which accounted for the observed motions of the planets in Ptolemaic astronomy, were considered such hypotheses. On the basis of the distinction between physical concepts and methods of proof, and astronomical methods of connecting observation to hypothesis and predicting positions and motions of planets, Thomas defined the logical status of the hypotheses and distinguished it from the logical status of physical causes. Astronomical hypotheses, said Thomas, were possible, or probable, but could not be considered absolutely true.[10] This opinion was deeply rooted in the ontological and epistemological beliefs of the Aristotelian world of knowledge, which was Thomas's world as well. Particularly significant were two basic intuitions concerning the ontological nature of mathematical entities, and the logical criteria for accepting knowledge as "truth" (i.e., "scientia").

Isis 66 (1975), pp. 165–193; idem, "Proof, Poetics and Patronage: Copernicus's Preface to *De revolutionibus*", in D. C. Lindberg and R. S. Westman (eds.), *Reappraisals of the Scientific Revolution*, Cambridge 1990, pp. 167–205. For a more general discussion of the epistemological issues at play in the Galileo affair, see M. A. Finocchiaro, "The Methodological Background to Galileo's Trial" in W. A. Wallace (ed.), *Reinterpreting Galileo*, Washington, D.C. 1986; idem, "Toward a Philosophical Reinterpretation of the Galileo Affair", *Nuncius: Annali di Storia della Scienza*, 1 (1986), pp. 189–202. My focus is somewhat different from those historians' in that following G. Morpurgo-Tagliabue, *I processi di Galileo e l'epistemologia*, Milan 1963, I am trying to distinguish clearly between *two* currents of scientific scepticism within the Catholic Church which were at work in the interpretation of Copernican arguments. Also, I consider the epistemological stances as a point of a departure only. More important are the precise strategies by which these differing epistemologies were applied in practice by historical figures, and the tracing of their institutional affiliations.

8. See detailed discussion of Thomas's views, Chapter 4.
9. St. Thomas, *In II De Caelo*, lect. 17.
10. Ibid.

Continuing an Aristotelian line of thought, Thomas distinguished between mathematical entities, abstracted from time and place, and real substances, which made up the physical world. Epicycles and eccentrics were first and foremost mathematical entities. They could not therefore be considered necessary, physical causes of the observed motions of the planets.[11] An ontological assumption concerning the nature of mathematical entities created an "image of knowledge" which asserted that mathematics could not provide an absolute knowledge of nature. This doubt was coupled with a further logical consideration. Thomas claimed that the agreement which could be found between observed facts and astronomical hypotheses was established by means of an inference from effects to causes. Such an inference could not be considered a strong scientific proof,[12] for the logical concerns of Aristotelianism required that a scientific proof should be arrived at deductively from self-evident first principles. Astronomy, Thomas said, being a mathematical science, took its first principles from mathematics. The physical meaning of the suppositions which it used in order to explain the motions of natural substances – such as eccentrics and epicycles to explain the motions of the planets – was not established as fully proved physical truth. This assumption was again an "image" of astronomical knowledge, derived from Aristotelian epistemology and its logical canons.

On the basis of these images of knowledge, astronomy was assigned the status of a subalternated mathematical science,[13] deriving its first principles from mathematics but unable to prove its physical hypotheses. Astronomical claims were not absolute but merely "possible", or "probable", true as long as there was no better hypothesis to "save the phenomena". They could be used for calculating and predicting the positions of the planets and their future movements, they could be discussed as possible truths or be regarded as subject to proof, but qua hypotheses they were not to be treated as necessary physical truths.

At this point, the rationale of Thomas's distinction between hypothetical and true knowledge becomes transparent. It perfectly reflects the profound antimony that pervaded the Aristotelian-Thomistic world of knowledge, in which the accepted Ptolemaic astronomy used hypotheses which contradicted the principles of Aristotelian physics.[14] To illustrate this point, an essential law of Aristotelian physics demanded that the planets move in uniform velocity around a common center. Astronomical

11. *In Boeth. de Trin.*, q. 5, a. 3; *Sum. Theol.*, I, q. 85, a. 1–2.
12. *Sum. Theol.*, I, q. 32, a. 1–2; *In I Post Anal.*, lect. 25, n. 6.
13. *In Boeth. de Trin.*, q. 5, a. 3; *In II Phys.*, lect. 3, nn. 6–9; *Sum. Theol.*, I-II, q. 35, a. 8; II-II, q. 9, a. 2 ad 3.
14. Jardine, *The Birth of History and Philosophy of Science . . .*, ch. 7.

observations, however, never confirmed this principle of uniform velocity around a common center. Thus, Ptolemaic eccentrics and epicycles were invented to "save the phenomena" – namely, observed facts. However, motion in eccentrics and epicycles was neither uniform nor related to a common center for all planets. Thus, Thomas considered eccentrics and epicycles hypotheses, and the knowledge based upon their suppositions hypothetical. Thomas, however, never made a distinction between the truths of the Christian faith based on the Scriptures, and therefore absolute, and philosophical truths based on experience and the human intellect, and thus only "possible". Therefore, the "possible" status assigned to astronomical hypotheses did not imply the imposition of significant theological limitations on their discussion or elaboration.

The Thomistic framework of thought provided an interpretive option for reconciling the judgement of the theologians of the Holy Office [document 1], with the practical considerations of the real world of the Counter Reformation – a world of educational and technical requirements, of political pressures and interest groups. Indeed, the decree of 1616 [document 2], which ordered the suspension of Copernicus's book rather than its condemnation, was a clear sign that a way had been cleared for such a move. There is also a piece of historical evidence which demonstrates the inclination of one very prominent cardinal of the Inquisition – Robert Bellarmine – to favour a solution in the Thomist spirit. A few months before the theologians of the Holy Office came to their decision, the cardinal had written a letter to Foscarini, whose book was later condemned by the Congregation of the Index, in which he expressed the need to reject some interpretive strategies and to suggest more adequate ones instead. Galileo himself was mentioned in the letter.

> First, I say that it seems to me that Your Paternity and Mr. Galileo are proceeding prudently by limiting yourselves to speaking suppositionally and not absolutely, as I have always believed that Copernicus spoke. For there is no danger in saying that, by assuming the earth moves and the sun stands still, one saves all the appearances better than by postulating eccentrics and epicycles; and that is sufficient for the mathematician. However, it is different to want to affirm that in reality the sun is at the center of the world and only turns on itself without moving from east to west, and the earth is in the third heaven and revolves with great speed around the sun; this is a very dangerous thing, likely not only to irritate all scholastic philosophers and theologians, but also to harm the Holy Faith by rendering Holy Scripture false.[15]

15. R. Bellarmine to A. Foscarini, 12 April 1615, Opere, XII, pp. 171–172; Finocchiaro, pp. 67–69.

Bellarmine did not strive to present a rigorous philosophical analysis of the logical status of the Copernican theory, on behalf of which Foscarini and Galileo had felt the need for its accommodation to the Scriptures. After all, it was in the form of a letter that he chose to express himself. And yet, the language of the letter contains enough Thomist terminology to justify the claim that he favoured the Thomist position and to lead us to attempt its fuller reconstruction in such terms. Bellarmine advised both Foscarini and Galileo to remain within the boundaries of a hypothetical claim. Such a claim could be fully justified by the "ex suppositione" method typically used by astronomers. Obviously what Bellarmine was referring to, here, was the procedure by which astronomers used to predicate certain entities, such as eccentrics and epicycles, by means of which they could account for observed facts, and then predict further positions and motions on the basis of such suppositions. A hypothetical claim, moreover, was all the mathematician (i.e., the astronomer) needed in order to go on with his work. To assert the Copernican theory in absolute terms, however, meant to pass beyond the boundaries of the mathematicians and to move into the field of the philosophers and the theologians. Such a move had implications: it required a full, physical proof of the Copernican hypothesis, which could then serve as a basis for the reinterpretation of the Scriptures. But the need to reinterpret, in fact, seemed irrelevant to Bellarmine at the time of the composition of the letter. For he believed the Copernican theory had not yet acquired the status of a fully proved physical truth, in terms of the canons of proof stated by Thomas:

> Third, I say that if there were a true demonstration that the sun is at the center of the world and the earth in the third heaven, and that the sun does not circle the earth but the earth circles the sun, then one would have to proceed with great care in explaining the Scriptures that appear contrary, and say rather that we do not understand them than that what is demonstrated is false. But I will not believe that there is such a demonstration, until it is shown me.[16]

Clearly, there was no attempt on the part of Bellarmine to limit further scientific discoveries by denying the possibility that the Copernican hypothesis may be true in nature, and not only in the astronomical discourse. No limitation was put upon further attempts to develop astronomical hypotheses and to treat them as candidates for scientific truth. A clear interest in preserving the boundaries of mathematics, philosophy, and theology, however, is apparent.

In the pages that follow it will become obvious that the interest in the

16. Ibid.

disciplinarian boundaries expressed by Bellarmine had deep scientific, theological, and political implications. At this point, however, it is crucial to specify the interpretive position Bellarmine opted for, and the specific sensibilities he exhibited. Bellarmine's position did not imply theological suspicion of scientific progress, nor did it deny the legitimacy of treating the Copernican hypothesis as a possible opinion, not yet fully proved, but not entirely fictitious either. At the same time, it betrayed particular sensitivity to the institutional complications introduced by Copernicanism. Thus, an attempt to prove the motion of the earth might result in an encroachment on the domain of scholastic philosophers and theologians, who, in fact, had been unchallenged by the traditional form of astronomy. It could also be perceived as a threat to the monopoly of priests in the interpretation of the Scriptures which the decrees of the Council of Trent for the first time had anchored in canon law:

> Second, I say that, as you know, the Council prohibits interpreting Scripture against the common consensus of the Holy Fathers; and if your Paternity wants to read not only the Holy Fathers, but also the modern commentaries on Genesis, the Psalms, Ecclesiastes, and Joshua, you will find all agreeing in the literal interpretation that the sun is in heaven and turns around the earth with great speed, and that the earth is very far from heaven and sits motionless at the center of the world. Consider now, whether the Church can tolerate giving Scripture a meaning contrary to the Holy Fathers and to all the Greek and Latin commentators.[17]

In such times as these, Bellarmine seems to be saying, when the exclusive right of interpretation had already been challenged by the reformed church, the need to reinterpret presents difficulties and should be handled with maximum caution. But the need to reinterpret was not entirely excluded by Bellarmine either, since he said that we should rather say "that we do not understand them than that what is demonstrated is false".

There is no doubt that Bellarmine's interpretive position was rooted in Thomism. However, Thomism was not necessarily the sole choice open to churchmen in his position. The rich and variegated cultural environment of the Renaissance also offered other options for an interpretation of the distinction between a hypothetical treatment of Copernicanism and an assertion of the absolute reality of the motions of the earth. Morpurgo-Tagliabue already drew attention to this fact when he pointed out the ambiguity of the term "hypothetical" in the period of Galileo.[18]

An alternative tradition to the Thomist understanding of the "hypothetical" had already existed in the Hellenistic period, a period of pro-

17. Ibid.
18. Morpurgo-Tagliabue, I processi . . .

found scepticism, disbelief in the capacity of the human intellect to attain truth, and strong inclinations towards mysticism. The distrust of the human intellect brought about a tendency to regard any kind of human knowledge as provisional, or "hypothetical". This emphasis on the ephemeral, uncertain character of knowledge of the world was further reinforced by the voluntaristic theology and nominalistic philosophy of the fourteenth century. Voluntaristic theology, seeking to defend the omnipotence of the Christian God, stressed the contingent quality of the natural world and the incapacity of the human intellect to attain absolute knowledge with regard to it. This approach well suited the mood of the Protestant Reformation, as was demonstrated by the Protestant pastor Osiander's introduction to Copernicus's book. Astronomical hypotheses, for Osiander, were not "possible" in the sense of "true as long as there was no other hypothesis which corresponded better to the facts". Rather, these hypotheses were understood as abstract constructions, bearing no relation whatsoever to the real physical world, and invented solely for purposes of calculation.[19]

The decree of 1620 [document 3] opted for the more sceptical interpretation of the "hypothetical", and thus for a more restricted use of Copernicanism. The decree stated that it was forbidden to construe the principles of the location and motion of the earth as entirely true ("ut verissima adstruere"). The use of the term "adstruere" is peculiar to this text; it means to build an additional structure onto an existing building. Here a limitation was placed upon the freedom to further develop the Copernican hypothesis, and hence to treat it as a possible, provisional truth, subject to further verification. The Copernican doctrine, in this interpretation, did not merely "present" the Scriptures as mendacious (as Bellarmine had claimed in his letter from 1615); rather, Copernicanism was dangerous because it *was* contrary to the Scriptures. Bellarmine's anxiety had been expressed in institutional terms, showing concern with the possible erosion of the claim of the Catholic Church to an exclusive right of scriptural interpretation; the decree expresses fear of a genuine opposition between scriptural and philosophical truths. Hence, the decree tended to forbid discussion of the Copernican hypothesis as a possible natural truth. Yet, it still left open the option of using it as a convenient mathematical model, devoid of physical significance: an abstract construction, unrelated to the real world.

Once documents 1, 2, and 3 are allowed to speak, a nonmonolithic, nonunitary position of the church with regard to the Copernican question reappears on the surface. The discrepancies between the documents

19. Ibid., pp. 38–48.

clearly demonstrate that there was no agreed position of the church on
the Copernican question: rather, there was a variety of voices speaking
from different positions on behalf of different people, groups, and institutions. In view of such a reading it is plausible to suppose that a plurality of
subcultures was forced into suppression by an overriding need for cultural
unity vis-à-vis a common foe – the Protestant Reformation. Such a unity
was enforced by means of a rhetoric of uniformity and unanimity which
could barely conceal the vital dialogue of voices taking place simultaneously. The field of contention was more or less structured by the following
moves: The theologians of the Holy Office claimed the Copernican doctrine philosophically absurd and theologically heretical, but all the documents meant to translate this claim into practical terms, however, refrained from rejecting the astronomical teaching of Copernicus outright.
Everyone agreed that it was permissible to use this teaching as a hypothesis. The epistemological status of a hypothesis, however, was related to
the major scientific, philosophical, and theological debates of the period.
The interpretation of the concept "hypothetical knowledge" thus became
dependent on the theological and philosophical orientation of the interpreter.

   Some churchmen chose to interpret "hypothetical knowledge" in a
Thomist spirit, as meaning knowledge unsupported by full logical proof.
This was obviously Bellarmine's way. Such an interpretation did not pre-
clude relating to a hypothesis as a possible idea or attempting to confirm
or to prove it in Aristotelian terms, for, according to Thomism, the human
intellect was capable of apprehending absolute truths about the universe,
and such activities built up faith in the truths contained in the Scriptures.
At the same time, such an interpretation was rooted in philosophical
assumptions, images of knowledge, about the ontological character of
mathematical entities, the degree of certainty of the knowledge of nature
obtained through mathematical methods, and the possibility of demonstrating a logical causal connection between phenomena and the hypotheses which explain them. The theological boundaries permitted in astronomical discourse thus came to depend on these "images of knowledge"
about which there was no consensus in the church or among the philosophers, and each position with regard to them produced a different interpretation of these boundaries. Another possible interpretation of the formula "hypothetical knowledge" was closer to the voluntaristic spirit,
which considered the Scriptures to be the only source of absolute truths,
and regarded all knowledge of the universe obtained through experience
and the human intellect as "hypothetical" knowledge. The motivation for
this attitude was essentially theological and derived from the need to

avoid any contradiction between scriptural verses and rational truths. This theological and philosophical orientation pervading the decree of 1620 [document 3] tended to treat astronomical hypotheses as purely abstract constructions unrelated to reality, whose only conceivable use was for purposes of calculation.

If one assumes that there was no agreed position of the church on the Copernican question, but merely alternative orientations connected to a wider philosophical-theological world view and to institutional interests, the question of the identity of the interpreters becomes crucial for an exposition of the interplay of knowledge and power involved in the Galileo affair. One is struck by the massive presence of Dominican theologians among the signatories of the censorship decision. Their names signify nothing to the modern reader. Even an expert in Galilean studies is unlikely to recollect anything in connection with these names, since they have not been considered a clue to understanding, a way of unmasking the faceless surface of the church. From the point of view of Galilean scholars, a bunch of forgotten theologians had not been considered a suitable subject for further research, as they could throw no light on Galilean science. But, in fact, the list of theologians does contain at least one name which may be a clue, that of Thomas de Lemos.[20]

In the cultural milieu of Rome in the 1600s, Lemos was far from the nonentity he has since become. In 1600 he had come to the Catholic capital where he was to spend twenty-nine more years, the remainder of his entire life. The circumstance which brought over the Dominican friar of noble origins from the university of Valladolid in Spain, where he had taught theology since 1590, is well known. It was the greatest and most dangerous scandal witnessed by the Counter-Reformation church, threatening to bring about another schism among believers: the controversy *De auxiliis*. For over twenty years two major Catholic orders, the Dominicans and the Jesuits, had been fiercely debating the gravest questions of dogma concerning grace, free will, and predestination, accusing each other of new and heretical opinions, and seeking to have the Holy See put an end to the scandal by declaring one side condemned. Thomas de Lemos could not have remained anonymous in those years, for, in the absence of the penetrating eye of a public press, the drawn-out affair attracted the attention of the educated public by other means, among them

20. Biographical information on Lemos is very scarce. My main source for details on Lemos's life and writings is: J. Quetif and J. Echard, *Scriptores ordinis praedicatorum recenniti*, Tomus Secundus. Pars I, A.D. 1499–1639, New York 1959, pp. 461–464.

street posters which ridiculed the contentious parties, and added a hu-
morous flavour to a serious cultural division within the church elite.[21]
Having been selected by his order as the Dominican disputant, Lemos
became known as a tough and unyielding opponent, never giving in an
inch to his interlocutors nor ever losing his temper in disagreement. For
six consecutive years (1600–1607) he had argued with a series of Jesuit
disputants, who were all astonished at his incessant zeal. A long passage
about him in the *Scriptores ordinis praedicatorum* praises his extraordi-
nary erudition, rhetorical skills, memory, and argumentation.[22]

By 1607, at the end of the official stage of the *Congregatio De auxiliis*,
Lemos was invited to join the Minerva, the great Dominican convent in
Rome, and became a consultant of the Holy Office. His deep commitment
to the Dominican cause led him to record the proceedings of the congregations[23] and to write many other texts concerned with the problems of
grace and free will debated with the Jesuits. Among these various works,
one was a direct response to a text composed by a Jesuit theologian,
Cardinal Bellarmine,[24] who also played a prominent role both in the debate between the Dominicans and Jesuits and in the Galileo affair. As a
consultant of the Inquisition, Lemos took part in the deliberations on the
censure of the Copernican theses. His name among the bulk of Dominican theologians betrays a possible circumstantial connection between the
two cultural scandals of the Counter Reformation: the Galileo affair and
the *De auxiliis* controversy.[25] This controversy could not do otherwise
than give expression to institutional interests and power relations between the two elite groups, which may throw light on the mechanisms of
decision making within the church. Moreover, the controversy which
dealt with theological problems in terms of philosophical concepts such
as causality, knowledge of contingents, etc., must have created different
sensibilities towards what was considered dangerous to theology, philosophy, and the priestly monopoly of reading the Scriptures. It may therefore

21. J. Brodrick, *Bellarmine* . . . , II, p. 55.
22. *Scriptores* . . . , p. 462.
23. *Acta omnia congregationum ac disputationum, qua coram SS. Clemente
     VIII & Paulo V summis pontificibus sunt celebratae in causa & controversia
     illa magna de auxilliis divinae gratiae, quas disputationes ego F. Thomas
     de Lemos eadem gratia adjutus sustinui contra plures ex Societate*, Typis
     emenditis Lovanii, Aegidii Denique 1702 in fol. coll. 1364, in Quetif and
     Echard, *Scriptores* . . . , p. 462.
24. *Ad scriptum cardinalis Bellarmini Paulo V recens electo exhibitum Annotat-
     iones quarta junii MDCV scriptae*, in Quetif and Echard, *Scriptores* . . . ,
     p. 463.
25. See detailed discussion of the debate in Chapter 9.

be the case that the division between two cultural orientations that culminated in the *De auxiliis* controversy prevented the crystallization of a coherent and uniform position on those problems within the church. Both the Galileo affair and the *De auxiliis* controversy were highly embarrassing to the image of uniformity and sobriety envisaged by the official policy of the Counter Reformation; both invoked contradictory impulses related to the need both to publicize and to conceal their details. Lemos's name in the list of theologians who unanimously condemned the Copernican theses signals a hitherto undiscussed context of interpretation of the censure, as well as of the Index decrees. The fact that the decree of 1620, which gave the most stringent interpretation to the Consultants' censure, was also signed by a Dominican theologian is a further hint in the same direction.

It is not only Lemos's name, however, which suggests that the rivalry between the Jesuits and Dominicans may be a relevant context for understanding the institutional and intellectual sensibilities operating within the church. Reflection upon the origins, formation, career, and interests of another man, who seemed to express a much more liberal interpretation of the censure of the Copernican theses, strengthens such a hypothesis. Robert Bellarmine (1542–1621), a Jesuit cardinal, is well known to Galilean scholars, and there is no need to repeat the details of his biography here. A quotation from the introduction to the instructive biographical study of Bellarmine by J. Brodrick points to his prominence in the Catholic establishment of his time:

> He was the orator of universities and of the Papal Court, professor of almost every branch of theology, consultor of the majority of the Roman Congregations and of a Papal Legate, rector of the most important college of the Society of Jesus and superior of one of its largest provinces, Archbishop of Capua, and for twenty-two years a Cardinal. There was scarcely a single important ecclesiastical affair of his age in which he did not take a leading part, the struggle with heresy, the reform of the Calendar and Breviary, the revision of the Vulgate under Sixtus V and Clement VIII, the great controversy between the Dominicans and Jesuits about efficacious grace, the interdict pronounced against the Republic of Venice, the assault of King James of England and his theologians on the temporal prerogatives of the Holy See, the events leading up to the first trial of Galileo – these were but some of the more prominent.[26]

While avoiding too much repetition of well-known facts, the reader should nevertheless be reminded of some aspects of Bellarmine's life

26. Francis Cardinal Erhle's introduction to Brodrick's *Bellarmine* . . . , pp. IX–X.

which, although known, have not hitherto seemed particularly relevant to traditional readings of the Inquisitorial documents. Bellarmine's intellectual breadth, his versatility in humanistic studies, his interest in the artistic culture of his time, and his desire to keep abreast of the latest "scientific" developments have all been stressed by previous studies, which sought to refute the obscurantist image of the church by invoking Bellarmine's enlightened personality. In the literature on religion and science in the Catholic world Bellarmine has often been perceived as a "Catholic latitudinarian". However, this characterization only served to set Bellarmine apart from the other Catholic officials involved in the affair, and to emphasize the rarity of his type in the Counter-Reformation era. Thus, Bellarmine's spiritual stature and sweetness of temper were used to explain the church's essential moderation in the Galilean affair; an explanation which was bound to remain personal rather than structural, cultural, and contextual. Certain aspects of Bellarmine's "latitudinarian" approach, implied in his letter to Foscarini, may, however, be attributed to his institutional affiliation, no less than to his personal preferences. It is these aspects which should now be emphasized as possible foci for further inquiry and research.

In Bellarmine's career three lines intersected: the intellectual, the administrative, and the dogmatic. All three had the same roots, and went back to the same source: a deep commitment to a life shaped by and permeated with the Catholic faith. It is hard, if not impossible, to separate Bellarmine's intellectual concerns from the environment of the Jesuit world, where he was first sent to school by his humble parents to be acquainted with the rudiments of Latin grammar, rhetoric, and arithmetic. It was in the same institutional milieu that he completed his education in philosophy and theology, became a lecturer, a spiritual guide, and finally the rector of the Roman College (the Collegio Romano), the most important Jesuit university of his time. Bellarmine's career as an intellectual exemplifies the educational concerns of the Jesuits, who sought to reconcile the most modern intellectual innovations with a strictly religious way of life. The two most significant legacies which he, as an intellectual, bequeathed to the following generations were those items in the *Ratio Studiorum* for which, according to his biographer, he was probably responsible: one concerning Thomism and another concerning the mathematical sciences.[27] Most important of all, as an intellectual, Bellarmine's concerns were always connected with Jesuit institutions. It is hard to imagine his search, even in the sixteenth century, to define the relation of philosophy

27. R. P. Xavier-Marie Le Bachelet, *Bellarmine avant son Cardinalat 1542–1598. Correspondance et Documents*, Paris 1911, p. 463 ff.

to theology, or his special interpretation of the status of astronomical hypotheses, outside the context of a Jesuit university, with its peculiar synthesis of intellect and religious fervour. It is mainly as an intellectual, therefore, that Bellarmine expressed his Jesuit affiliation, living within the confines of the order and experiencing its possibilities together with its limitations.

As an administrator of faith, however, Bellarmine operated far more within the general context of his time than within the boundaries of the Jesuit order. It was in the "saeculum" that he confronted heretics, consulted the papal legate in France, and argued with King James's theologians; it was also outside the boundaries of the Jesuit order that he performed many of his official functions, practicing ecclesiastical politics at the papal court as cardinal, and caring for souls as archbishop. Here, his special understanding of human nature, coupled with an enormous practical talent, combined with intellectual insight, permitted an articulation of new ways of thinking about authority, rules of reasoning with opponents, contradictions, and methods of persuasion. No wonder that this aspect of his career was, in a way, summed up in a monumental work named *Controversies.*[28]

Among the numerous dogmatic questions which preoccupied Bellarmine the theologian, there was none as urgent as the problem of grace and free will. From the time of his first initiation to theological studies at Padua, where he could not agree with his teachers' opinions, to his deathbed, when he confirmed, for the last time, the views he had consistently held throughout his life, grace and free will had been haunting problems, never leaving his soul in peace and always calling for a resolution, despite the sense of mystery they necessarily inspired.

Undoubtedly, the problem of grace and free will was as central to Bellarmine's theological thought as it was to the Catholic theology of the Counter Reformation as a whole. He stated his views in many writings from the beginning of his career as a theologian in Louvain and in his lectures in Rome, his autobiography, and the third volume of the *Controversies,* as well as in numerous letters and memoranda referring to the debate in the Catholic universities of Louvain, Douai, Salamanca, and finally in front of two Popes in Rome. Not only was Bellarmine deeply involved in the controversy, but his writings contain enough information to enable us to perceive the continuity of the themes related to the debate from the Council of Trent, through the *Ratio Studiorum,* the *Congregatio De auxiliis,* and up to his statement on his deathbed in September 1621,

---

28. R. Bellarmine, *Disputantes de controversiis christiane fidei adversus hujus temporis haereticos,* Ingolstadt, 1586–1593.

which he asked his fellow Jesuit Father Eudaemon Joannes to record for posterity:

> As to the affair *De auxiliis Divinae Gratiae,* which is now a subject of dispute between the Society of Jesus and other Catholics, I was to say that he ratified and maintained as true all that he had written in his *Controversies,* and that he never changed his opinion.[29]

When the figures of Thomas de Lemos and Robert Bellarmine are set against each other, and their social origins, formation, and range of intellectual and institutional interests compared, one begins to see that the divisions within the Catholic Church could have given birth to two genuinely different interpretations of the censure of the Copernican theses which were rooted in structural conditions, rather than in mere personalities. Thomas de Lemos, of noble origins, had early in life committed himself to the most intellectual among the mendicant orders, the Dominicans. In that environment, he spent the remainder of his days, dedicated to the traditional Dominican way of life: teaching, debating, and defending the purity of the faith as a consultor of the Inquisition. Robert Bellarmine, of modest social origins, had joined a newly founded institution, the Jesuit order. His personal achievements as educator, politician, and theologian were closely connected to the reputation of his Order. However, the circles in which he moved were far wider than those of a traditional friar, and his interests extended from the dogmatic to the educational and political spheres. Both men were involved in both the *De auxiliis* and Galileo affairs, and could not do otherwise than express their institutional affiliations in decisions which had a bearing upon the intellectual, religious and political interests of both Orders.

Galileo and Copernicanism

Let us now turn to the Inquisitorial documents pertaining directly to Galileo, with the intention of ascertaining whether they lend themselves to a reading which corroborates the coexistence of two different lines of interpretation of the censure of the Copernican theses within the church establishment. One was expressed in the documents signed by Bellarmine and may have reflected the interests of the Jesuits; the other was expressed by officials of the Inquisition reflecting, perhaps, the interests of the Dominicans. The Galilean documents seem to reveal the practical implications of a difference between two "factions" in terms of different kinds of interdictions imposed on Galileo after 1616. Each faction gave its own inter-

29. Quoted in Brodrick, *Bellarmine . . .* , p. 445.

pretation to the Pope's decision following the censure, quoted in document 4, and accordingly imposed the kind of interdiction considered suitable [documents 5 and 6].

4. Inquisition Minutes (25 February 1616)

Pope Paul V's decision is mentioned in a document of 25 February 1616, in which Cardinal Mellini informed the Assessor and the Commissary of the Holy Office how the pope wished Galileo to be treated in the light of the censure:

> The Lord Cardinal Mellini notified the Reverend Fathers Lord Assessor and Lord Commissary of the Holy Office that, after the reporting of the judgment by the Father Theologians against the propositions of the mathematician Galileo (to the effect that the sun stands still at the center of the world and the earth moves even with the diurnal motion), His Holiness ordered the Most Illustrious Lord Cardinal Bellarmine to call Galileo before himself and warn him ["moneat"] to abandon these opinions ["ad deserendas dictam opinionem"]; and if he should refuse to obey, the Father Commissary, in the presence of a notary and witnesses, is to issue him an injunction ["praeceptum"] to abstain completely from teaching ["docere"] or defending ["defendere"] this doctrine and opinion or from discussing ["tractare"] it; and further, if he should not acquiesce, he is to be imprisoned ["carceretur"].[30]

The document testifies to three levels of Inquisitorial control on knowledge, as commonly practiced at the time of Galileo and anchored in the canon law.[31] These were admonition (*monitum*), injunction (*praeceptum*), and imprisonment (*carcere*). The three were correlated, in the pope's decision, to three possible degrees of violation of the proper status assigned by ecclesiastical judgement to the Copernican theses. The Cop-

30. *Opere*, XIX, p. 321: "Ill.^{mus} D.Cardinalis Millinus notificavit RR.PP.DD. Assessori et Commissario S.^{cti} Officii, quod relata censura PP. Theologorum ad propositiones Gallilei [*sic*] Mathematici, quod sol sit centrum mundi et inmobilis [*sic*] motu locali, et terra moveatur etiam motu diurno, S.^{mus} ordinavit Illmo. D. Cardinali Bellarmino, ut vocet coram se dictum Galileum, eumque moneat ad deserendas dictam opinionem; et si recusaverit parere, P. Commissarius, coram notario et testibus, faciat illi praeceptum ut omnino abstineat huiusmodi doctrinam et opinionem docere aut defendere, seu de ea tractare; si vero non acquieverit, carceretur". Finocchiaro, p. 147.

31. Details on the juridical aspects of Inquisitional procedures can be found in: I. Mereu, *Storia del'intollerenza in Europa: Sospetare e punire*, Milan 1979.

ernican opinion had to be *abandoned;* in other words one could not *hold* (*tenere*) it. The term was unambiguous, denoting an opinion which had been proved, according to the criteria of proof accepted in Aristotelian-Thomistic discourse. Since officially everybody in the church agreed that the Copernican opinion had not been proved, it was clear that one could not *hold* (*tenere*) it, e.g., it had to be abandoned. This judgement of the church, the pope decided, had to be transmitted to Galileo through an official act of *admonition* ("monitum") to be performed by Bellarmine. "Ex silentio", this decision meant that, without *holding* (*tenere*) the Copernican opinion, one could nevertheless *defend* (*defendere*) and *teach* (*docere*) it. For it was only in case Galileo refused to *abandon* the Copernican opinion, that a second and more severe degree of control had to be exercised. In such a case, an injunction (*praeceptum*) had to be issued, not by Bellarmine but by the Commissary of the Inquisition, in presence of a notary and witnesses, which would forbid not simply *holding* (*tenere*) the Copernican opinion, but also *defending* (*defendere*) it, *teaching* (*docere*) it, or *discussing* (*tractare*) it at all.

This second option for the controlling of knowledge would, in fact, lead to the complete silencing of Galileo, with much wider implications than an interdiction from holding the Copernican opinion but leaving him free, "ex silentio" to defend, teach, and discuss it. A few words should be inserted here about the meaning of *defending, teaching, and discussing* an opinion one *did not hold* in the Aristotelian-Thomistic discourse, a meaning shared by the pope, Bellarmine, and the consultors of the Holy Office. Epistemologically speaking, in the Aristotelian-Thomistic world of knowledge a clear distinction existed between *true*, scientifically legitimated, opinions and *possible* or *probable* ones; this distinction has already been mentioned in the context of the Thomistic interpretation of astronomical hypotheses, and will be further examined later in order to demonstrate its centrality to the Thomistic organization of the world of knowledge within a wider religious world view. Here, it is enough to point out that an *admonition* to abandon the Copernican doctrine, without an explicit warning not to *teach, defend,* or *discuss* it at all, created the conditions for the application of a limiting epistemological distinction, without, however, exercising any practical act of silencing. Moreover, such an epistemological distinction gained practical meaning in the educational context of the period, where teachers and students were currently engaged in developing disputative skills. One of the means to intellectual development, it was believed, was the acquisition of competence in the *modus disputandi* by *teaching* and *defending* not only opinions one did not share and believe, but also ideas considered absurd, on behalf of which an educated person was nevertheless expected to be able to argue.[32]

Within the context of Aristotelian-Thomistic discourse, therefore, the pope's distinction was clear, sensible, and unambiguous. Galileo had to be admonished to abandon the Copernican opinion, an admonition which would not necessarily put any constraint upon his freedom to *defend, teach*, or *discuss* it in the future. Only if he refused to pay heed to the epistemological and practical distinction between the *truth* and the *probability* of the Copernican hypothesis would his freedom of *teaching, defending*, and *discussing* be limited also, in a much more severe act of official injunction (*praeceptum*). However, the pope's decision included more than admonition and command. The pope also foresaw a possibility of refusal, on the part of the scientist, to be silenced. Then, the pope decided, a third and most severe degree of control had to be practiced; the stubborn scientist would, in such a case, be physically silenced by *imprisonment* (*carcere*).

Two documents from the files of 1616 still remain to be examined: they record the acts of Bellarmine and those of the officials of the Inquisition following the pope's decision. Not only were those acts less faithful to the decision itself than one might expect, but they also testify to a subtle reinterpretation of the decision by the executors, in two quite different directions. An additional document written by Bellarmine, which apparently had been kept secret from the Holy Office in 1616, corroborates the hypothesis, suggested here, of different factions within the church.

5. Special Injunction (26 February 1616)

One day after Cardinal Mellini's notification of the pope's decision, action was taken. This is recorded in a document of 26 February:

> At the palace of the usual residence of the said Most Illustrious Lord Cardinal Bellarmine and in the chambers of His Most Illustrious Lordship, and fully in the presence of the Reverend Father Michelangelo Segizzi of Lodi, O.P. and Commissary General of the Holy Office, having summoned the above-mentioned Galileo before himself,

---

32. See A. Kenny, "Medieval philosophic literature," in N. Krezmann, A. Kenny, and I. Pinborg (eds.), *The Cambridge History of Later Medieval Philosophy*, Cambridge 1982, pp. 19–29; P. Glorieux, "L'enseignement au moyen âge. Techniques et methodes en usage à la Faculté de Théologie de Paris, au XIIIe siècle", *Archives d'histoire doctrinale et littéraire du moyen âge*, 35 (1966), pp. 65–186.

the same Most Illustrious Lord Cardinal warned Galileo that the
above-mentioned opinion was erroneous and that he should abandon
it; and thereafter, indeed immediately, before me and witnesses, the
Most Illustrious Lord Cardinal himself being also present still, the
aforesaid Father Commissary, in the name of His Holiness the Pope
and the whole Congregation of the Holy Office, ordered and en-
joined the said Galileo, who was himself still present, to abandon
completely the above-mentioned opinion that the sun stands still at
the center of the world and the earth moves, and henceforth not to
hold, teach, or defend it in any way whatever, either orally or in writing; otherwise the Holy Office would start proceedings against him.
The same Galileo acquiesced in this injunction and promised to obey.

 Done in Rome at the place mentioned above, in the presence, as
witnesses, of the Reverend Badino Nores of Nicosia in the kingdom
of Cyprus and Agostino Mongardo from the Abbey of Rose in the
diocese of Montepulciano, both belonging to the household of the
said Most Illustrious Lord Cardinal.[33]

The decision of the pope was violated in this document in that Galileo
was committed both to an *admonition* (*monitum*) and to an injunction
(*praeceptum*) at the same time. This violation could not have occurred
without a mutual understanding between the executors, who, according
to the document, were both present in the quasi-official event which took
place at Bellarmine's residential palace. However, whereas Bellarmine delivered the admonition verbatim, formally forbidding Galileo to *hold* the

33. *Opere*, XIX, pp. 321–322: In palatio solitae habitationis dicti Ill.^{mi} D.
Card.^{lis} Bellarminii et in mansionibus Dominationis Suae Ill.^{mae}, idem Ill.^{mus}
D. Card.^{lis}, vocato supradicto Galileo, ipsoque coram D. sua Illma existente,
in praesentia admodum R.P. Fratris Michaelis Angeli Seghitii de Lauda, ordinis Praedicatorum, Commissarii generalis S.^{ti} Officii, praedictum Galileum
monuit de errore supradictae opinionis et ut illam deserat; et successive ac
incontinenti, in mei etc. et testium etc., paesente etiam adhuc eodem Illmo
D.Card.^{li}, supradictus P. Commissarius praedicto Galileo adhuc ibidem
praesenti et constituto praecepit et ordinavit [proprio nomine] S.^{mi} D.N
Papae et totius Congregationis S.^{ti} Officii, ut supradictam opinionem,
quod sol sit centrum mundi et immobilis et terra moveatur, omnino relin-
quat, nec eam de caetero, quovis modo, teneat, doceat aut defendat, verbo
aut scriptis; alias, contra ipsum procedetur in S.^{to} Officio. Cui prae-
cepto idem Galileus aquievit et parere promisit. Super quibus etc. Actum
Romae ubi supra, praesentibus ibidem R.^{do} Badino Nores de Nicosia in
regno Cypri, et Augustino Mongardo de loco Abbatiae Rosae, dioc. Politia-
nensis, familiaribus dicti Ill.^{mi} D. Cardinalis, testibus etc.". Finocchiaro, pp.
147–148.

Copernican opinion as a scientific truth, but still leaving the option of *defending, teaching,* and *discussing* it as a possible or probable one, the Dominican official of the Inquisition proceeded to exercise the second, and more severe means of control, delivering the injunction (*praeceptum*) immediately, without Galileo having the chance to commit any act of refusal, for which the sanction had originally been intended. Nevertheless, the Dominican official did not in fact deliver the injunction verbatim. For, according to the pope's decision, the second and more severe option for the control of knowledge meant complete silencing, prohibiting defending, teaching, or even discussing an opinion altogether. The Dominican executors of the injunction did not insist on complete silencing. Indeed, they explicitly forbade *defending* or *teaching* Copernicanism, the possibility for which was still left open by Bellarmine. They did not, however, explicitly forbid discussing it, for the term *tractare,* in spite of being mentioned in the pope's decision, is absent from the description of the injunction recorded in the document of 26 February. The clear distinction made by the pope between *holding* an opinion on the one hand and *defending, teaching,* and *discussing* it on the other, which had a specific epistemological and practical significance both for the organization of knowledge within a hierarchical system and for educational practice, had been distorted. Instead, a less clear, but still consistent and meaningful distinction had been introduced, between *defending* and *teaching* an opinion, which could lead to its recognition as scientific truth and was therefore forbidden, and *discussing* an opinion without considering it a candidate for scientific truth, which was still allowed.

My suggestion is that, although the Dominican executors of the injunction had an interest in putting further constraints on Galileo's possibilities, they did not go as far as completely silencing him at this stage. Intellectually, however, there could be only one meaning to the interdiction to *defend* and *teach,* combined with a permission, "ex silentio", to *discuss* Copernicanism: Copernicanism could be treated as an abstract construction, convenient for purposes of calculation; it could not, however, be considered a candidate for scientific truth. This interpretation could have been part of a sceptical climate of opinion, which allowed, and in fact encouraged, the imposition of theological constraints upon possible scientific interpretations of the physical world, and which was not uncommon in sixteenth-century Europe, especially among reformers. It is in this spirit that the decree of the Index committee of 1620, also signed by a Dominican, was in fact written. It was similarly the Dominicans, whose presence was prominent among the signatories of the censure, who executed the injunction, understood by them in much the same terms.

6. Inquisition Minutes (3 March 1616)

The same kind of consistency as has been shown to exist among the different documents associated with the Dominicans and intended to limit Copernicanism to an abstract construction, not even probable as a physical hypothesis, also exists among the documents associated with Bellarmine. This consistency suggests an inclination on Bellarmine's part to interpret the theologians' censure in a more liberal way, leaving the door open for the acquisition of a scientific status for Copernicanism, even though such status could not be assigned to it at the moment. An additional document supporting this interpretation dates from March 3rd, and records Bellarmine's report to the Congregation of the Holy Office, concerning his admonition to Galileo:

> The Most Illustrious Lord Cardinal Bellarmine having given the report that the mathematician Galileo Galilei had acquiesced when warned ["monitus"] of the order by the Holy Congregation to abandon the opinion which he held till then, to the effect that the sun stands still at the center of the spheres but the earth is in motion, and the Decree of the Congregation of the Index having been represented, in which were prohibited and suspended, respectively, the writing of Nicolaus Copernicus *On the Revolutions of the Heavenly Spheres,* of Diego de Zuñiga *On Job,* and of the Carmelite Father Paolo Antonio Foscarini, His Holiness ordered that the edict of this suspension and prohibition, respectively, be published by the Master of the Sacred Palace.[34]

In his report, Bellarmine refrained from mentioning the injunction imposed by the Commissary, restricting his account to the admonition alone. As far as he was concerned, then, Galileo was admonished to abandon the Copernican opinion. The interdiction against *defending* or *teaching* the same opinion was simply ignored.

However, it was not only by ignoring but also by more actively distorting the epistemological and practical distinctions that served as a basis for the pope's decisions, that Bellarmine's interpretive policy operated. In

---

34. Ibid., p. 278: "Facta relatione per Ill.[mum] D. Cardinalem Bellarminum quod Galileus Galilei Mathematicus, monitus de ordine Sacrae Congregationis ad deserendam opinionem quam hactenus tenuit, quod sol sit centrum spherarum et immobilis, terra autem mobilis, acquievit; ac relato Decreto Congregationis Iudicis, quo fuerunt prohibita et suspensa, respective, scripta Nicolai Copernici De revolutionibus orbium caelestium, Didaci Astunica in Iob, et fratris Pauli Antonii Foscarini Carmelitae; S.[mus] ordinavit publicari aedictum a Magistro Sacri Palatii huiusmodi suspensionis et prohibitionis respective". Finocchiaro, p. 148.

this sense, there was no difference between the Jesuit Bellarmine and the Dominican officials of the Holy Office: both made a tactical use of interpretation, with a clear cultural policy in mind. This becomes evident when, to the documents hitherto discussed, one last piece of evidence is added: a certificate written by Bellarmine to Galileo, to reassure him in the face of rumours about an abjuration supposedly imposed on him by the church, which Bellarmine had denied:

> We, Robert Cardinal Bellarmine have heard that Mr. Galileo Galilei is being slandered or alleged to have abjured in our hands and also to have been given salutary penances for this. Having been sought about the truth of the matter, we say that the above-mentioned Galileo has not abjured in our hands, or in the hands of others here in Rome, or anywhere else that we know, any opinion or doctrine of his; nor has he received any penances, salutary or otherwise. On the contrary, he has only been notified of the declaration made by the Holy Father and published by the Sacred Congregation of the Index, whose contents is that the doctrine attributed to Copernicus (that the earth moves around the sun and the sun stands at the center of the world without moving from east to west) is contrary to Holy Scripture and therefore cannot be defended ["difendere"] or held ["tenere"]. In witness whereof we have written and signed this with our own hands, on this 26th day of May 1616.[35]

The pope's decision was based upon the distinction between "proved" opinions which could be *held*, and "probable" opinions, which could be *defended, taught,* and *discussed,* but not *held*. This distinction rested upon epistemological presuppositions which partly reflected the educational practice of the period. The executors of the injunction abused the pope's decision by distorting the distinctions upon which it had been based and introducing a new one: a distinction between *to defend* and *to*

35. Ibid., p. 342: "Noi Roberto Cardinale Bellarmino, havendo inteso che il Sig.ʳ Galileo Galilei sia calunniato o imputato di havere abiurato in mano nostra, et anco di essere stato per ciò penitenziato di penitenzie salutari, et essendo ricercati della verità diciamo che il suddetto S. Galileo non ha abiurato in mano nostra nè di altri qua in Roma, nè meno in altro luogo che noi sappiamo, alcuna sua opinione o dottrina, nè manco ha ricevuto penitenzie salutari nè d'altra sorte, ma solo gl'è stata denunziata la dichiarazione fatta da Nostro Signore e publicata dalla Sacra Congregazione dell'Indice, nella quale si contiene che la dottrina attribuita al Copernico che la terra si muova intorno al sole e che il sole stia nel centro del mondo senza muoversi da oriente ad occidente, sia contraria alle Sacre Scritture, e però non si possa difendere nè tenere. Et in fede di ciò habbiamo scritta e sottoscritta la presente di nostra propria mano, questo dì 26 di Maggio 1616. Il. med.ᵐᵒ di sopra Roberto Card.ˡᵉ Bellarmino". Finocchiaro, p. 153.

teach on the one hand, and *to discuss* on the other. Such a distortion, it has been claimed, made sense only within a certain intellectual climate, in which theological constraints upon the world of knowledge were common and well understood. This intellectual climate may be associated with a group of Dominican theologians and officials of the Inquisition.

Bellarmine's policy was likewise accompanied by a new distinction: that between *holding* (*tenere*) and *defending* (*difendere*) an opinion, which he stated were forbidden for Copernicanism, and *teaching and discussing* it, which were not expressly forbidden, and hence were left open as a free option for Galileo. It was Bellarmine's certificate, however, which introduced the greatest ambiguity into the process of interpretation. What could have been meant by forbidding to *defend,* while allowing, "ex silentio" to *teach?* What, in other words, was the difference between *teach* and *defend* in Bellarmine's mind, and what did the process of teaching imply? It is obvious that by excluding the act of teaching from the interdiction, Bellarmine expressed his concern with the enterprise of education, which had been a major preoccupation both for himself personally and for the Jesuit order in general. My hypothesis is that there was a connection between the educational interests of the Jesuits, their special brand of Thomistic theology defended by them in the *De auxiliis* debate, and their interpretation of the status of astronomical hypotheses in general and the Copernican and Galilean claims in particular. Likewise, there was a connection between the traditional interests of the Dominicans as defenders of the purity of faith, their claim to authority as "true" Thomists, and their position vis-à-vis the Copernican and Galilean innovations. An attempt to understand these connections is at the heart of my story of Galileo and the church.

3

1633

The events of 1633 demonstrated the split between two factions in the church in a much more explicit way than those of 1616. This has already been claimed by G. de Santillana in his masterwork, *The Crime of Galileo*. Here, in a nutshell, is Santillana's assessment of the situation:

> The search for the point of indictment had led to a long-drawn-out fumble between different conceptions; and as the first scene comes to a close, we see those conceptions embodied in two factions which are still far from agreeing on a common line. The Dominicans of the Inquisition, no longer the ruthless ones of a generation before, were still trying to handle the affair on a restrictive legalistic basis; but they had against them the will of the Pope and the plans of a curial group allied with the Jesuits, of whom men like Inchofer were the spearhead, who were pressing for judicial slaughter.[1]

Santillana's analysis of the Inquisitorial documents of 1633, informed by numerous letters of the main actors in the drama, is detailed, thorough, and enriched by a fluent narrative which weaves all the available evidence into a lively story. Yet, it is an incomplete story, whose lacunae still allow for its accommodation within a traditional binary narrative, where the church is portrayed as an embodiment of the power of authority – ruthless, arbitrary, and mostly incomprehensible – while cognitive interests are represented by the scientist alone. Santillana does not wait for others to draw these conclusions from his story. Rather, he performs the accommodation himself, taking care to eliminate unnecessary ambiguities from his text:

> One thing can be concluded for certain: that an extremely capable outfit of "hypocrites without Nature and without God," as Micanzio calls them, was operating effectively inside the hierarchy. Some of them were undoubtedly fanatics like Inchofer; others, mere power politicians. Some were even scientific renegades like Scheiner (at least if we believe Father Kircher), who had chosen to commit the Church to what they knew to be ultimate discredit for the sake of personal revenge and of currying favor with the authorities. The rest is lost in the

1. Santillana, *The Crime* . . . , p. 249.

mists of time. Were the events even contemporary, it would be hard to reconstruct them. The working of great administrations is mainly the result of a vast mass of routine, petty malice, self-interest, carelessness, and sheer mistake. Only a residual fraction is thought. To try to look into them is as unprofitable as to stare at the wall of Plato's cave.[2]

This leaves us with the traditional picture of the church as an enormous power-mechanism bound to clash with science. Indeed, Santillana manages to show that the church was not only that; that there were other forces, less cynical, more faithful to some legalistic tradition which, although Inquisitorial, was not completely arbitrary. But his explanation of the church's clash with Galileo is sheer violent repression, motivated by fanaticism, personal revenge, and ignorance. This may be only part of the story, however. Starting with the hypothesis about the two different interpretations given to the censure of the Copernican opinion by the Jesuit Bellarmine on the one hand, and by the decree of 1620 on the other, it is now possible to add Santillana's conclusion about the two factions in 1633, and attempt to use it as a point of departure for a different kind of narrative about science and the church. The two factions – the Jesuits and the Dominicans – had cognitive interests represented by their different interpretations of the Copernican hypothesis, which they took upon themselves to defend. These cognitive interests were embodied in concrete institutions which implied a structure of power-relationships between the two groups. Their rivalry was also implied in their different theologies, which probably played different social roles in the constitution of their respective world views. Galileo, the documents could disclose, also had a consistent intellectual-political position, which was emerging as a third option in this triple struggle for cultural hegemony. In other words, the new narrative beginning to crystallize conceives of the church as a knowledge–power structure whose consensual basis for cultural hegemony had already been shaken long before Galileo, perhaps with the emergence of the Jesuits as an alternative cultural elite to the traditional Dominicans. The science of Galileo only upset some precarious balance and polarized the existing forces.

The first step in establishing such a story is to show that the documents lend themselves to a reading which "saves the hypothesis". This is my aim in further examining the documents of 1633. First I shall give a hearing to three voices – the Dominicans, the Jesuits, Galileo – in the course of which both consistencies and inconsistencies between the positions of 1616 and 1633 will be pointed out. This will be done by examining some key documents for each of these positions: Riccardi's letter [document 1] for the Dominican position; Inchofer's report [document 2] for the Jesuit

2. Ibid., p. 290.

position; and the three hearings of the trial and the plea for self-defence [documents 3, 4, 5, and 6] for Galileo's position. A second group of documents will be analysed in order to detect clues of a lack of consensus and expose attempts at covering up discrepancies in the official policy of the church. This second group consists of three documents: the report of the Special Commission [document 7]; the "chiusura d'istruttoria" final report to the pope [document 8]; and the sentence [document 9].

The Dominican Voice

(1) The Dominican position in 1633, as represented in a letter of Niccolò Riccardi, the Master of the Holy Palace and the chief authority with regard to licensing, seems to be consistent with the order's values and interests as expressed in the documents of 1616. At this point, the reader should be reminded that Riccardi, also called familiarly "Padre Mostro" because of his immense girth and erudition,[3] was the Dominican father who held the responsibility for the imprimatur of the *Dialogue,* and whose hesitations brought about long procrastinations which lasted about eighteen months, in the granting of the desired license.[4] The essence of Riccardi's views of the matter is expressed in a letter of 24 May, which contains his instructions to the Florentine Inquisitor on whom he had – unwillingly, it should be recorded – conferred the responsibility of issuing the imprimatur. In the original file of the Holy Office, the letter is attached to the memorandum of the Special Commission [document 7] nominated by the pope as soon as the book was first circulated in Rome:

> Mr. Galilei is thinking of publishing there a work of his, formerly entitled *On the Ebb and Flow of the Sea,* in which he discusses in a probable fashion the Copernican system and motion of the earth, and by means of this supposition, corroborating it in turn because of this usefulness . . . However, I want to remind you that Our Master thinks that the title and subject should not focus on the ebb and flow but absolutely on the mathematical examination of the Copernican position on the earth's motion, with the aim of proving that, if we remove divine revelation and sacred doctrine, the appearances could be saved with this supposition; one would thus be answering all the contrary indications which may be put forth by experience and by Peripatetic philosophy, so that one would never be admitting the absolute truth of this opinion, but only its hypothetical truth without the benefit of Scripture.[5]

3. Ibid., p. 170.
4. See above, ch. 1.
5. *Opere*, XIX, pp. 327; trans. by Finocchiaro, 1989, p. 212.

Riccardi displays neither ambiguity nor hesitation where the correct interpretation of the status of the Copernican theory is concerned. Basing himself on the pope's authority, Riccardi, in fact, repeats the interpretation suggested by the decree of 1620,[6] according to which it was not because of a temporary lack of scientific evidence that Copernicanism had been assigned a hypothetical status; rather, Copernicanism was in principle a hypothetical theory, not dangerous since the absolute truth should *never* be conceded to this opinion. When understood as an abstract construction of the mind, incapable of scientific proof, it is perfectly suitable and acceptable.

Traditionally, it should be noted, the pope's epistemological argument has been considered a device which later permitted the influential group of Jesuits most interested in Galileo's judicial incrimination to convince the pope that he had been duped. Beyond the immediate political considerations, however, it is far more significant to realize the complete continuity and consistency which existed between the pope's theological argument and the Dominican interpretation, expressed both in the decree of 1620 in the injunction of 26 February and in Riccardi's words. If indeed this Dominican interpretation reflects the climate of opinion of one faction of the church establishment, a climate of opinion conditioned by a specific theology and a whole range of educational interests, then, not only the short-range politics but also the structural conditions which made possible the political manipulations of the Jesuits are exposed. Santillana's contention that all this belongs to the realm of historical contingencies which can never be reconstructed[7] thus loses its weight and history discloses one more irony. For it then becomes necessary to account for the Galilean affair not in terms of the short-range action of one group with strong vested interests, but in terms of the dialectics of rival groups, with interwoven cognitive-political interests within the church establishment. The facade of authoritative uniformity which the church had always attempted to present thus crumbles in front of the bitter factionalism which existed despite all attempts to reconstruct and rationalize the traditional structure of authority.

The Jesuit Voice

(2) Although the Jesuit position in 1633 is not consistent with Bellarmine's stance in 1616, there is a kind of symmetry between the two in the

6. See above, ch. 2.
7. See above, this chapter.

difficulty which they present to the interpreter. Ambiguity seems to be an inherent feature of the Jesuit position. Some conclusions, however, can nevertheless be drawn both from the text of Melchior Inchofer, the Jesuit consultor of the Inquisition, who wrote the most detailed report on the *Dialogue,* and from his peculiar status within the society. Inchofer's text[8] contained an unambiguous statement of his understanding of the hypothetical status of the Copernican theory:

> Because Copernicus, satisfied with a simple system, merely explained celestial phenomena in terms of this hypothesis by what, so he thought, was an easier method; whereas Galileo, having searched for many additional arguments, confirms Copernicus's discoveries and introduces new ones, and this is a twofold defense.[9]

Unlike Bellarmine,[10] who had never prohibited an attempt to confirm and establish a mathematical hypothesis as possible truth, and in fact admitted that if a proof were found it would be necessary to change the interpretation of Scripture, Inchofer followed, almost verbatim, the words of the decree of 1620, which prohibited one to construe the Copernican hypothesis as entirely true ("ut verissima adstruere"), while still allowing its use as a computational device. On the crucial issue, therefore, the Jesuit position, unlike the Dominican, seems to have changed between 1616 and 1633, and the Jesuit voice of 1633 differed substantially from the Jesuit voice of 1616. In other words, there is no way Inchofer's view can be reduced to Bellarmine's position of 1616. Accordingly, Inchofer piled up accusations against Galileo condemning him for teaching, defending, and holding the Copernican theory ("I am of the opinion not only that Galileo teaches and defends the view of Pythagoras and Copernicus but also, if we consider the manner of proceeding, of reasoning, and then of expression, that he is vehemently suspected of firmly adhering to it, and indeed that he holds it"), as well as for transgressing the boundaries between mathematics and philosophy ("Galileo promises to proceed in the manner of a mathematical hypothesis, but a mathematical hypothesis is not established by physical and necessary conclusions".)[11] Thus, in his accusations, Inchofer expressed both Bellarmine's sensibilities concerning teaching, education, and the organization of knowledge, and the Dominican theological sensibilities, allowing only the narrowest possible use of the Copernican theory as a computational device.

Inchofer's text, when read alongside Riccardi's, creates an illusion of

---

8. *Opere,* XIX, pp. 349–356.
9. Ibid., p. 351; Finocchiaro, p. 265.
10. See above, ch. 2.
11. *Opere,* XIX, pp. 350, 352; Finocchiaro, pp. 264, 266–267.

uniformity in the church position vis-a-vis Galileo. This uniformity, I contend, is contingent, not structural: and it is precisely this contingency which made it possible to define the conflict of Galileo and the church in terms of a conflict of science and religion by constructing a historical story of a binary oppositional character. A hint of the illusory nature of the supposed uniformity of the church can be found in the very choice of Inchofer in the role of an Inquisitorial consultor. For it was none other than Inchofer himself who had been censored by the society, and then examined and proscribed by the Congregation of the Index "until corrected".[12] This seemingly minor biographical fact is quite typical of the ambiguous nature of the Jesuit position in general: it was the Jesuits, indeed, who expressed the harshest criticism of the *Dialogue*: however, this criticism could always be claimed to have been made by a person hardly representative of Jesuit opinion, since he himself had been subject to the internal censure of the society, and had even been condemned by the Inquisition.

Galileo's Voice

The documents of the three depositions of the trial of Galileo and his plea in self-defence reveal fluctuations and recurrent shifts in the judicial strategy of the Inquisition, as against sustained consistency on the part of the accused. Galileo had to defend himself against two different charges: that he had intentionally concealed the injunction ("praeceptum") imposed upon him by the Commissary of the Inquisition on 26 February 1616, in Cardinal Bellarmine's palace and in the cardinal's presence; and that in his book he had blurred the distinction between the absolute and the hypothetical, treating the Copernican opinion as true rather than a hypothetical or probable one. The first charge, if proved, was a clear disciplinary offence against the "political" authority of the church. The second charge could be, and indeed was, interpreted in two different ways: it could be attributed to carelessness or even arrogance in writing, but still be seen as a mistake to be corrected by the author; or, it could be understood as an indication of Galileo's true conviction – namely, his secret adherence to Copernicanism. This meant that he had transgressed

12. W. R. Shea, "Melchior Inchofer's 'Tractus Syllepticus'; A Consultor of the Holy Office Answers Galileo", in: P. Galuzzi (ed.), *Novità celesti e crisi del sapere. Atti del Convegno Internazionale di Sudi Galileiani*, Florence 1984, p. 287.

not only the Inquisition's injunction ("praeceptum") but also Bellarmine's admonition ("monitum") not to *hold* the Copernican opinion, and hence that the *Dialogue* was written with an evil intention, challenging the church's intellectual authority.

(3) The first deposition on 12 April 1633[13] focussed its attention on the first charge, namely Galileo's disciplinary offence. The Inquisition was seeking to make Galileo confess to having disobeyed the injunction ("praeceptum") which he had not mentioned when applying for the imprimatur. Such a confession was intended to confirm the suspicion that he had obtained the imprimatur by means not entirely honest, so that the proper punishment could be found for that offence.

But the investigation did not end with a full confession as one might have expected to happen in the harsh circumstances of imprisonment in the hands of the Inquisition. Galileo did not unequivocally acknowledge the severe form of the prohibition placed upon him, even if he did not entirely deny it. He acknowledged the admonition he had received in Bellarmine's residence in the presence of the Commissary of the Inquisition, a notary and two witnesses, but he failed to acknowledge the severe form of the prohibition, claiming that he did not recall it. In a sense, this was a kind of nonacknowledgement. In his defence, he produced Bellarmine's certificate of May 1616 in which he was forbidden to *defend* the Copernican doctrine, but not forbidden to *teach* it.[14] In order to justify the way in which the imprimatur had been obtained, Galileo pointed out that his work had been read by four censors of the Inquisition and had been found acceptable.

Thus, the investigation reached an impasse. As against the document of 26 February 1633 containing the injunction ("praeceptum"), which had been found unsigned in the file of 1616, Galileo produced a document signed by the Cardinal of the Congregation of the Holy Office which confirmed his claim that such an injunction had not been imposed on him. In this situation, it was hard to substantiate the claim of a disciplinary offence without questioning the legal and social validity of granting the imprimatur.

(4) The document of 30 April 1633[15] does not in fact contain an investigation, but a confession by Galileo in which he recognized one of the accusations made against him. A letter of the Commissary of the Inquisition of that period, V. Maculano, to Cardinal F. Barberini, hinted at unex-

13. *Opere,* XIX, pp. 336–342.
14. See above, ch. 2.
15. *Opere,* XIX, pp. 342–343.

pected difficulties which the court had encountered after the first investigation.[16] Maculano succeeded in convincing the pope to adopt a different policy in dealing with the case in view of the "difficulties" to which he alludes. As a result, he received special permission to attempt to obtain from Galileo a "confession" of guilt. In Maculano's opinion, the confession would "redeem" the honour of the law courts of the Inquisition, which would then be able to manifest its "clemency" towards the accused.[17]

Maculano's letter provides a unique testimony outside the files of the Inquisition which enables us to reconstruct something of what was transpiring behind the scenes. There is no doubt that Maculano interpreted the outcome of the first investigation as a defeat for the position of the Inquisition, which had sought to base the entire trial on Galileo's disciplinary offence. When it became clear that this charge could not be proven, Maculano saw the need for a basic change of legal strategy. As a result, he concentrated on what seemed to him the less serious charge of a failure to treat the Copernican doctrine as a hypothesis, and sought to make Galileo admit to this charge by means of a promise of an alleviation of the sentence. And indeed, the document of 30 April made no mention at all of the charge on which the investigation of 12 April had centered. Instead, Galileo acknowledged that on rereading his book he himself had become aware of the forcefulness of his pleading on behalf of the Copernican hypothesis. Such pleading, he admitted, could give the reader the impression that he believed in this doctrine in an absolute sense.[18] Galileo claimed that the error which he recognized in his confession had not been committed intentionally or knowingly, but as a result of a tendency to floridity of expression, caused by excessive ambition, ignorance, and lack of caution.[19]

(5) The plea in self-defence was submitted on 10 May, written and

16. V. Maculano to F. Barberini, 28 April 1622, *Opere,* XV, p. 106: ". . . hanno dell'altro canto considerate varie difficoltà quanto al modo di proseguir la causa . . ."

17. Ibid., p. 107: ". . . Il Tribunale sarà nella sua reputatione, co'l reo si potrà usare benignità, e in ogni modo che si spedisca, conoscerà la gratia che li sarà fatta con tutte l'altre consequenze di sodisfatione che in ciò si desiderano . . ."

18. *Opere,* XIX, p. 343: ". . . avalorati all 'orechie del lettore più di quello che pareva convenirsi ad uno che li tenesse per inconcludenti e che li volesse confutare . . ."

19. Ibid., ". . . È stato dunque l'error mio, e lo confesso, di una vana ambitione e di una pura ignoranza et inavertenza . . ."

signed by Galileo himself.[20] Just as the confession of 30 April had deviated from the main charge on which the investigation of 12 April had centered, so this document deviated from the confession with regard to the blurring of the boundary line between the hypothetical and the absolute, and returned to an attempt, on the part of Galileo, to defend himself against the charge of a disciplinary offence against the injunction ("praeceptum"). Here, Galileo used two arguments. In connection with Bellarmine's certificate, which he had kept for all those years, Galileo said that he had entirely forgotten the stringent injunction of the Inquisition of 26 February 1616. All that he was aware of was Bellarmine's admonition that he should not *hold* ("tenere") or *defend* ("difendere") the Copernican doctrine. This admonition he had interpreted as a repetition of the decree of the Index of 1616,[21] which left it unclear whether the Copernican doctrine had the status of a "possible" truth, or merely of a fictional, arbitrary invention. By mentioning Bellarmine's certificate, as well as the decree of the Index of 1616, Galileo was insinuating that the document on which the charge of a disciplinary offence was based was contradicted by the two other documents, but he did not insist on the legal significance of this fact. He only claimed that, if he had erred, it had been in good faith and not with evil intent. Galileo's other argument was that in requesting an imprimatur he had taken all the necessary steps and submitted his book to four successive examinations by the Inquisition. These two arguments repeated his positions put forward in the first hearing. In that document, too, Galileo confessed his errors in the formulation of the arguments of his book, but he regarded them as errors of style which could be corrected. These errors, he claimed, derived from his desire to appear witty and clever, but did not constitute a conscious offence against the rules laid down by the Inquisition.[22]

(6) The final deposition, on 21 June,[23] revealed a further change in legal strategy. This investigation was based on the church's decision to focus the discussion on the question of whether Galileo upheld the Copernican

20. Ibid., pp. 345–347.
21. Ibid., p. 346: ". . . non resta punto da dubitare che il comandamento fatto in essa sia l'istesso precetto che il fatto nel decreto della S.ʳᵃ Congregazione dell'Indice . . ."
22. Ibid., p. 347: ". . . quei mancamenti che nel mio libro si veggono sparsi, non da palliata e men che sincera intenzione siano stati artifiziosamente introdotti, ma solo per vana ambizione e compiacimento di comparire arguto . . ."
23. Ibid., pp. 361–362.

theory[24] – that is to say, whether he had not only transgressed against the more stringent interpretation of the injunction ("praeceptum"), but had also disregarded Bellarmine's admonition ("monitum"), which could be understood as an implied permission to discuss Copernicanism as a "possible" truth. In other words, at that stage Galileo was accused of possessing an absolute faith in the Copernican doctrine and of attempting to conceal this fact from the Inquisition. Galileo never admitted to this charge. To the repeated questions asking whether he upheld the Copernican doctrine, he answered negatively, even under the threat of torture.[25]

The full complexity of the Galilean position seems to be recorded in these documents, irreducible either to an arrogant challenge or to complete submission. Although Galileo conceded full political authority to the church, he at the same time was well aware of the delicate foundations of that authority, consisting of a subtle combination of power relationships and intellectual claims to knowledge. Politically, he was attempting an extremely dangerous game by playing two sources of authority against each other: the Inquisitorial injunction ("praeceptum") on the one hand and Bellarmine's admonition ("monitum") and his signed certificate on the other. This move, however, was backed by the firm belief that he had not transgressed Bellarmine's intellectual boundaries between the hypothetical and the absolute with evil intent. Galileo's refusal to ever admit that he had believed in the absolute truth of the Copernican theory amounted to a conviction that his concept of scientific truth and Bellarmine's were basically – although perhaps not rhetorically – in agreement, or that at least they did not exclude each other. Basing himself upon what seemed to him to be an intellectual consensus, he then dared to play a political game, consciously or unconsciously becoming, by that very move, a significant factor in the restitution of a balance of power within the church.

24. Ibid., p. 361: "An teneat vel tenuerit, et a quanto tempore citra, solem esse centrum mundi, et terram non esse centrum mundi et moveri etiam motu diurno . . ."

25. Ibid., pp. 361–362: Concludo dunque dentro di me medesimo, nè tenere nè haver tenuto dopo la determinazione delli superiori la dannata opinione . . . ; Io non tengo nè ho tenuta questa opinione del Copernico, dopo che mi fu intimato con precetto che io dovessi lasciarla; . . . . : Et ei dicto, quod dicat veritatem, alias devenietur ad torturam; Io son qua per far l'obedienza; et non ho tenuta questa opinione dopo la determinatione fatta, come ho detto. . ."

Factionalism within the Church

Before focussing our gaze on the three last documents [7, 8, and 9], the reader should be reminded of the Dominican and Jesuit positions in 1633, reconstructed above. The Dominicans, represented in this case by Niccolò Riccardi, had already suggested in the decree of 1620 a most stringent interpretation of the Copernican hypothesis, as an abstract structure of the mind, useful for computational purposes but not open to further confirmation as physical truth. This interpretation was in accordance with the Inquisition's practical decision to warn Galileo against holding, teaching, or defending Copernicanism "in any way whatsoever, verbally or in writing". In this way, Galileo was not entirely silenced, for he could still discuss ("tractare") Copernicanism, even under this injunction, as a computational device. Riccardi's letter is entirely consistent with this position, and testifies to the pope's sharing it as well.

The Jesuit position of 1633, represented by Melchior Inchofer, deviated substantially from Bellarmine's interpretation of the status of Copernicanism, which he had expressed in several documents, including the certificate and the letter to Foscarini. Whereas Bellarmine had not forbidden, in principle, attempts to confirm and establish Copernicanism as a physical truth, Inchofer explicitly rejected any such attempt, for he assumed Copernicanism to be merely an arbitrary invention bearing no relation to reality. Thus the very fact of Galileo's writing of the *Dialogue* was presented by him as transgressing Bellarmine's admonition not to *hold* or *defend* the Copernican hypothesis, although such an interpretation badly distorted the meaning of all the documents related to Bellarmine's name.[26] Following the documents of the three hearings presented above, it is now possible to show how these respective positions were conditioned and, in their turn, also shaped the reality which brought about Galileo's fall, contrary to the expectations of many contemporaries.[27]

The first stages of the investigation exposed the weakness of the Dominican position. For it was the Dominicans who controlled the *imprimatur*, and it was they who had finally made possible the publication of a book whose author could perhaps claim that he did not "hold" the Copernican position but was certainly unable to maintain that he had treated it as a mere computational device. The discovery of the injunction of 1616 was a way out for the Dominicans, for it enabled them to blame and

---

26. See above, ch. 2.
27. See Santillana, *The Crime* . . . , pp. 199–201.

punish Galileo not for the actual contents of his book, which they themselves had read and approved, but for the disciplinary offence of concealing the injunction and obtaining the *imprimatur* by means which were devious and not altogether honest. In these circumstances, it is quite clear that the Dominicans had vested interests in emphasizing the disciplinary offence and minimizing the significance of transgressing the boundaries between the hypothetical and the absolute. The judicial strategy practiced at the first hearing of the trial probably reflected the interests of the Dominicans, as it concentrated wholly on the disciplinary offence, completely ignoring the deviation from the hypothetical. As mentioned before, however, Galileo's incomplete acknowledgement of the injunction, coupled with the discovery of Bellarmine's certificate, completely upset the original judicial strategy of the Inquisition. At the same time, it also exposed the Jesuits to acute danger. For if the Dominicans had enabled the publication of the *Dialogue* by granting the *imprimatur*, it was no doubt Bellarmine's certificate which had enabled Galileo to attempt a physical interpretation of Copernicanism – something which was clearly forbidden by the Dominicans – and yet to claim that he had, in fact, treated it only as a possibility, a probable opinion, without transgressing Bellarmine's admonition. Bellarmine's certificate incriminated the Jesuits, instead of the Dominicans, for creating the conditions which made possible the appearance of the dangerous book. Thus, it became the particular interest of the Jesuits to show that Galileo had transgressed the boundary between the hypothetical and the absolute, actually held the Copernican opinion, and wrote the *Dialogue* with evil intention. Hence, he could be accused not only of challenging the injunction, but of knowingly ignoring Bellarmine's admonition. It was at this point, when both sides had grown well aware of their respective vulnerabilities, that the rivalry became a real power struggle.

At first, attempts were still made by each side to save its own interests at the expense of the other. The story is well known, only reinterpreted here in the light of my basic hypothesis. The Commissary General of the Holy Office, Vincenzo Maculano da Firenzuola, acted in accordance with Dominican sensibilities in his attitude towards the legal procedures of the Inquisition. Realizing the impossibility of basing the case upon the injunction from 1616 after the disclosure of Bellarmine's certificate, he attempted an arrangement out of court based on Galileo's confession to the lesser offence (from the point of view of the Dominicans), namely deviation from the hypothetical. From the point of view of the Jesuits, that confession only helped to strengthen their claim that Galileo had actually transgressed Bellarmine's admonition, not only the injunction. It was this move which impelled some of them, according to Santillana, to hope to have "the *Dialogue* burned as the work of a confirmed heretic, and Galileo confined for the rest of his days in the holds of Castel Sant' Angelo".[28] As is well known, this was not to happen.

Although Galileo's enemies may have taken advantage of the confession, the trial ended with a compromise, for it seems that no group could really outweigh the other in this struggle of power. The most severe measures were not taken, nor was the arrangement outside the court carried out. The compromise achieved was a trial "de vehementi": Galileo was found to be strongly suspected of heresy, and an abjuration, which meant public humiliation, was not to be avoided.[29]

Three documents still remain to be examined: the memorandum of the Special Commission [document 7]; the "chiusura d'istruttoria" – a final report to the pope – [document 8]; and the sentence [document 9]. Santillana, in his account of the trial, which is still the most authoritative, regarded all three as bizarre documents, partly incomprehensible, and not meeting the generally high juridical standards of the Inquisition. Thus, the memorandum appears to him as "strangely inclusive";[30] the "chiusura" is condemned as "a judicial skullduggery" manifesting "a strange uniformity",[31] and the sentence is said to be based "on a kind of three-valued logic whereby Galileo was brought to trial as though the injunction were true and was then sentenced as though it were not very serious."[32] Once the split within the church is admitted, however, and the need for compromise is also realized, these documents acquire a particular logic of their own. The strange inconclusiveness of the memorandum is then understood as the result of a draw between the factions; the "chiusura" and the sentence represent an act of covering up the split, rendered necessary by the "no-win" situation, and the joining of forces at the end of the day. Let us now look more closely at these documents.

The memorandum of the Special Commission from September 1632 [document 7][33] was the first occasion where the two different charges against Galileo were invoked. According to Finocchiaro[34] the Special Commission consisted of three members: Riccardi, the papal theologian Agostino Oreggi, and Inchofer. The commission did not produce a single

---

28. Ibid., p. 286.
29. On the juridical and social meaning of the abjuration see: Mereu, *Storia del'intollerenza . . .*, ch. 7.
30. Santillana, *The Crime . . .*, p. 212.
31. Ibid., p. 281.
32. Ibid., p. 314.
33. *Opere*, XIX, pp. 324–327.
34. Finocchiaro, *The Galileo Affair*, p. 355, n. 64.

report; neither did it come up with three individual ones. The document consists of two distinct parts, differing both in contents and in form.

The first part related to Galileo's transgression of the boundary between the hypothetical and the absolute. It also touched upon the disciplinary offence, whereby Galileo was charged with having been "deceitfully ['fraudolentemente'] silent about an injunction given him by the Holy Office". This part of the document concludes with the statement: "One must now consider how to proceed, both against the person and concerning the printed book".[35]

The second part was concerned with deviating from the hypothetical. There followed a comment, however, which made it clear that this transgression could be amended by the author "if the book were judged to have some utility which would warrant such a favor".[36] This comment was in perfect accordance with the Dominican attitude – expressed in Maculano's letter – which tended to interpret deviation from the hypothetical as a minor transgression. Only at the end was the injunction of 1616 mentioned. It seems reasonable to suppose, then, that the second part was written by Riccardi, whereas the first part, which piled up accusations without discrimination, conforms to the spirit of Inchofer's report written later in 1633. Thus read, the document indeed testifies to different orientations within the church at this early stage of Galileo's second encounter with the church.

The "chiusura" [document 8][37] represented a gross attempt at covering up the differences between the factions, after the impossibility of a one-sided victory had become obvious. The clue to the act of covering-up lay in its ignoring the difference between Bellarmine's admonition and the Commissary's injunction of 1616. Instead, both steps were attributed to Bellarmine, who was claimed to forbid *holding* the Copernican opinion and even *discussing* it:

> Consequently, on 25 February 1616 His Holiness ordered the Lord Cardinal Bellarmine to summon Galileo and give him the injunction that he must abandon and not discuss in any way the above-mentioned opinion of the immobility of the sun and the motion of the earth.
>
> On the 26th the same Cardinal, in the presence of the Father Commissary of the Holy Office, notary, and witnesses, gave him the said injunction, which he promised to obey. Its tenor is that "he should abandon completely the said opinion, and indeed that he should not

35. Ibid., p. 325; trans. by Finocchiaro, p. 219.
36. Ibid., p. 327; Finocchiaro, p. 222.
37. Ibid., pp. 293–297.

hold, teach, or defend it in any way whatever; otherwise the Holy
Office would start proceedings against him".[38]

Indeed the document demonstrated uniformity within the church, as Santillana rightly observed; however, this uniformity was neither "strange"
nor the result of a "slick job", as he claimed. Rather, it was the natural
outcome of a well-coordinated campaign in which Galileo was made to
appear as though transgressing Bellarmine's admonition – and not just the
Inquisition injunction – so that he could then be accused of wrong intentions and strongly suspected of heresy.

There is little to be added to Santillana's subtle analysis of the sentence,[39] the devious inner workings of which are described by him in
terms of "the acrobatics of the text". His analysis will therefore be quoted
at some length:

> The sentence shows the hand of a competent jurist. He has discarded
> . . . [the final report (i.e., "chiusura")] and worked over the original
> sources. . . . The judge goes on . . . piling up grounds for crime of
> intention assiduously. . . . The famous personal prohibition of 1616,
> the cynosure of the case, is not glossed over. It is revealed at last to
> the world, and the proceedings connected with it also disclosed, just
> enough to imply, even at a little risk, that Galileo had always had bad
> intentions. It has to be there, because it is the only thing that can get
> the judge past the licenses . . . but the judge is obviously not comfort-
> able about it. Instead of dwelling on it as the basic point of incrimina-
> tion . . . he manages to shift the center of gravity quickly to the Bellar-
> mine certificate, where he feels on firmer ground. . . . The shift of
> ground is effected in the curious "but" section: "And all this you
> urged not by way of excuse for your error but that it might be set
> down to a vainglorious ambition rather than to malice. But this cer-
> tificate produced by you in your defense has merely aggravated your
> delinquency . . ."
>
> How could the certificate "aggravate" what is already declared in
> the previous sentence to be malice . . . ? The ominous roll of the big
> drum is obviously there to divert attention from the concluding sen-
> tence, which has to mention a command; and it is hoped that the
> reader, carried along by the whole sixteen lines preceding, will take
> the personal prohibition as an integral part of Bellarmine's notifica-
> tion – as, indeed, had been not hinted but brazenly affirmed in all
> previous informal statements. The ambiguity had to be kept whirling
> on a pinpoint. But now the judge has reestablished the ground under
> his feet. "What is really damning" he thunders, "is this paper which

38. Ibid., p. 294; Finocchiaro, p. 282.
39. Ibid., pp. 403–406.

you brought here in your defense" . . . From this point on, in fact, the injunction is lost sight of. The sentence goes downhill in a formal mumble about "seeing and considering all that has to be seen and considered" and comes to a conclusion based on the theological points alone, which are juridically the weakest.[40]

There is one last point to be made: the acrobatics of this text reflects the acrobatics of different cultural orientations, different institutional interests, different theologies within the church establishment, all culminating in a short power struggle whose victim turned out to be Galileo Galilei. The injunction had to be mentioned, to justify the granting of the *imprimatur* by the Dominicans. The verdict had to be based upon the theological arguments, however, in order to make the crime appear as a transgression of Bellarmine's admonition; this was the only way in which the Jesuits could conceal their continuous dialogue with science which might otherwise have been considered the factor which made the publication of the book possible.

Power and Knowledge in Galileo's Trials

With these last remarks on the logic informing the textual strategies of the sentence, the task of rereading the documents of the "trials" is completed and the rudiments of a story complementary to the traditional drama of Galileo and the church are beginning to emerge. The radical hypothesis at the heart of the hermeneutic process attempted here seems to be partially vindicated by the plausibility of the story when checked against the original documents. At this stage, it seems reasonable to suppose that the church was not monolithic in its relation to scientific knowledge in general and Copernican astronomy in particular; that the different cultural orientations of groups within the church had dictated different attitudes towards Galilean science, and that Galilean science merely upset a precarious balance and polarized positions within an institution whose consensual basis for cultural hegemony had already been shaken long before his time.

The Dominicans and the Jesuits represented two alternative Counter-Reformation intellectual elites whose claim to authority, and hence to political influence, was intimately connected with their claim to knowledge – scientific and theological alike. The Dominicans were a well-established order whose interests were bound up with the particular conception of its members as defenders of the purity of faith both through

40. Santillana, *The Crime* . . . , pp. 313–316.

the education of the people in the right doctrine and through Inquisitorial means of control. The Dominicans' claim to authority was also rooted in their pretension to the status of "true Thomists". The Jesuits, a newly founded order, had their power based in education. Their widespread influence was primarily rooted in their educational programme, but culminated in a special brand of Thomistic theology which they defended in the controversy *De auxiliis*.

In the following pages I shall examine the historical process – stretching between the Council of Trent and the first trial of Galileo – by which the cultural field of the Counter Reformation was structured. An analysis of the decrees of the Council of Trent will provide a clue to the theological and disciplinarian constraints imposed by reformed Catholicism on that field, as well as the new possibilities which were opened. A preliminary account of the educational policy of the Dominicans, and a fuller one of the Jesuits', will throw further light on the practices which shaped the positions of the rivaling elites in the cultural field. The culmination of that rivalry in the sphere of theology will complete the background underlying their respective attitudes towards the new astronomy.

The challenge presented by Galileo's Copernicanism, however, will be interpreted through the analysis of one concrete scientific exchange between Galileo and the Jesuit astronomer Christopher Scheiner. It will thus be shown that the Jesuit position crystallized through a dialogue with Galileo's science, and not simply through its rejection. However, the specific limits of that dialogue will also be exposed. Those specific limits rather than the global rejection of the new science or contingent personal motivations shaped the Jesuits' position in the trial of 1633.

PART II

The Cultural Field of the Counter Reformation

4

The Council of Trent

The Doctrinaire Phase of the Counter Reformation

The Counter Reformation[1] is generally imagined to have been a repressive period in which the laws of the Inquisition were imposed on all spheres of life, especially in the intellectual domain. D. Cantimori, one of the

1. The historiography of the Counter Reformation may be divided into three main currents. The origins of the first current are rooted in the image of the Counter Reformation created by the Enlightenment and idealist philosophy. This current emphasizes the authoritarian and reactionary aspects of the Counter Reformation. The scholars who have most substantially contributed to research in this direction are: B. Croce, "La vita morale in Italia nel '600", *La Critica* (1928), pp. 162–164; idem, *Storia dell'età barocca*, Bari 1929; F. Chabod, *Per la storia religiosa dello Stato di Milano durante il dominio di Carlo V,* Bologna 1938; D. Cantimori, "Riforma cattolica", *Società, II* (1946); V. de Caprariis, "L'Italia nella Controriforma", in: N. Valeri (ed.), *Storia d'Italia,* II, 1965, pp. 383–474. The second current in the historiography of the Counter Reformation has been engaged in sharp polemics with the traditional image of this period, which had previously been considered mainly as a reaction to the Protestant Reformation. This current tends to emphasize the impulse for change and innovation which autonomously brought about the Catholic reform. Its most prominent representatives are: P. Brezzi, *Le riforme cattoliche dei secc. XV e XVI,* Rome 1945; H. Jedin, *Storia del Concilio di Trento,* trans. by G. Lecchi and O. Niccoli, Bresia 1973–1982, 3 vols.; Evennett, *The Spirit.* . . . The third current tends to emphasize the characteristic tendencies for reform which were common both to the Protestant Reformation and to the Catholic Counter Reformation, and grew out of the theological problems of late scholasticism: J. Lortz and B. Ullianich, *Storia della Chiesa,* trans. by B. Ulianich, Rome 1960; M. Bendiscioli, *La riforma cattolica,* Rome 1958; P. Prodi, *La crisi religiosa del sec. XVI: Riforma cattolica e controriforma,* Bologna 1964. On the problem of education during the Counter Reformation, which is of particular interest for my following arguments, see: L. Volpicelli (ed.), *Il pensiero pedagogico della Controriforma,* Florence 1960; D. Balani and M. Roggero, *La Scuola in Italia: Dalla Controriforma al secolo dei Lumi,* Turin 1976.

major historians of the period, expressed a common opinion when he wrote:

> This repression was based on the most rigorous dogmatic orthodoxy and the systematic suspicion of all who failed to show profound subservience to the clergy. . . . From the time of Paul IV onwards, the Spanish type of severe and grave ecclesiastic predominated at Rome; the humanist was replaced by the theologian, austere and upright, harsh and uncompromising towards laity and heretics alike.[2]

Repression, monolithic dogmatism, and cultural reactionism have been particularly emphasized by intellectual historians, who chose to organize their account of the emergence of the modern world around a series of confrontations between the church and a number of lay intellectuals, whose rebellion was conceived as a necessary precondition for progress. For these historians, repression was exemplified in the restoration of the Roman Inquisition in 1542, and the creation of the Index of forbidden books in 1587. Monolithic dogmatism resulted in establishing the unambiguous authority of the church institutions, and cultural reactionism was expressed in a preference for a strict theological formation, rejecting the late medieval and Renaissance plurality of cultural orientations in favour of the Thomism of the thirteenth century.

In the linguistic arsenal of the twentieth century, repression, authoritarian dogmatism, and reactionism are all pejorative terms used by the secular winners of cultural hegemony to describe the inherent faults of their historical predecessors. This linguistic perspective, however, is somewhat alien to the set of problems and issues which preoccupied both the architects of the Catholic reform in the Council of Trent (1545–1563) and the executors of the cultural policies which were later derived from Tridentine principles. The problem of the Counter Reformation, I contend, is not exhausted, or even fully understood in terms of repression, monolithic dogmatism, and reactionism. In fact, I shall strive to show that its understanding requires an alternative set of concepts such as education rather than repression, balance of power between rival dogmatic positions rather than uniformity, and a subtle control of cultural orientation rather than reaction.

These alternative categories are crucial to my attempt to reconstruct the conditions of the dialogue between clerical intellectuals and laymen at the dawn of the Scientific Revolution. In order to revive this long-forgotten cultural dynamic I suggest we go back to the doctrinaire phase of the Counter-Reformation era, in which a set of doctrinal and practical

---

2.   Cantimori, "Italy and the Papacy," *The New Cambridge Modern History,* II, Cambridge 1975, p. 272.

positions was formulated in the decrees of the Council of Trent (1545–1563). The thread linking the various subjects dealt with in those decrees leads one to think of the Counter Reformation in terms of a reconceptualization of the relation between the "transcendental" and the "mundane". As a result of this reconceptualization the role of education in the Catholic milieu became much more prominent than ever before. The emphasis on education, however, also entailed a need to control the contents of instruction. The orientation of the Council of Trent towards Thomism signaled an attempt to develop the Thomistic principles of the organization of knowledge into a subtle means of control. The decrees of the Council of Trent, however, failed to display the monolithic dogmatism to be expected from modern stereotypes of the Counter-Reformation era. On the contrary, the formulation of the official Catholic position on all major issues discussed by the council was ambiguous and prone to at least two rival interpretations.

The doctrinaire phase of the Counter Reformation created the conditions which could enhance dialogue between science and religion, especially by emphasizing the rationalistic interpretation of the religious world view suggested by Thomism, and the concern for education as a means of raising the religious consciousness among Catholics. This same doctrinaire phase, however, also shaped the constraints upon any such dialogue by revealing the need for institutionalized means of control over the contents of education, and by turning the traditional plurality of positions within the Catholic Church into a rivalry between various institutionalized cultural orientations.

My account of the doctrinaire phase of the Counter Reformation, and its intellectual and institutional significance is based on two different sources. First, I shall scrutinize the decrees of the Council of Trent in search of a major dogmatic and disciplinarian issue which it faced and the type of solutions it promoted. Secondly, I shall recover from Thomas's texts those principles which allowed for the integration of a secular world of knowledge into a religious world view. These principles were adopted by the Counter-Reformation church as a subtle means of control over the contents of Catholic education.

The Decrees of the Council of Trent (1545–1563)

For analytical purposes I have divided the decrees of the Council of Trent into two main groups: the doctrinal decrees concerned with the elaboration of dogma and the disciplinary decrees delineating the implications of dogma for religious forms of life.[3] Underlying both kinds of decree was the need to reestablish, in dogmatic as well as institutional terms, the recognition of the Catholic Church as the sole channel of grace and salvation vis-à-vis the Protestant view of the church as a mere congregation of believers chosen by predestination. Consequently, a reconceptualization of the relation between "the mundane" and "the transcendental" world was legally sanctioned and the different possibilities of bridging them were restructured.

The point of departure of my argument is the reformulation of dogma. While analysing the four main issues encapsuled in the doctrinal decrees, namely, the Scriptures and their interpretation, original sin, the doctrine of justification, and the doctrine of the sacraments, I shall expose the peculiar Catholic emphasis upon the potentiality of humanity to play a significant role in the path to salvation. Thus, the recognition of interpretation as an inherent part of the sacred message pointed out the role of human reason embodied in human institutions in the attainment of revealed truth; the conception of human nature as partially redeemed from original sin after baptism represented a limit to the abyss between grace and this-worldliness; the insistence on the role of free will in the doctrine of justification disclosed yet another aspect of the bridge between the sinful individual and the infinite goodness of God; and the conceptualization of the operation of sacraments in terms of real physical processes understood by human science also established a link between the natural-human order and the order of grace. The doctrinal approach represented by Catholicism stood in marked contrast to the Protestant insistence on salvation as a wholly transcendent act depending on the incomprehensible will of God alone. It is therefore all the more surprising to discover, through the analysis of the disciplinary decrees, that not unlike in Protestantism the reconceptualization of the bridge between the "mundane" and the "transcendental" was translated into practice in terms of an innovative approach to education.

The Doctrinal Decrees

The Interpretation of the Scriptures (decrees formulated at the fourth session on 8 April 1546). The decrees concerning the Scriptures dealt with

---

3. All citations from the decrees are taken from: M. Marcocchi, *La Riforma Cattolica: Documenti e testimonianze,* Brescia 1970, 2 vols., henceforth Marcocchi, and from Joannes Dominicus Mansi, *Sacrorum Conciliorum nova et amplissima collectio,* Graz 1961, Vol. 33, hencefore *SC.*

the preservation, interpretation, and transmission of the Christian message. The first decree stated that the church was responsible for preserving the purity of the message of the Gospels,[4] thus confirming the church's institutional right to exclusive authority of mediation between revelation and the body of believers.

The recognition of a privileged body of people with the authority to interpret the Scriptures was expressed in the statement that the source of revelation was the Scriptures *and* the "oral traditions" of the church.[5] Thus, in contrast to the Protestant principle of *sola scriptura* (scripture alone) which recognized only a single transcendental source of ultimate truth, the Catholic Church also assigned a place to human interpretation, the product of human reason, in the path leading to revealed truth. However, it was precisely this recognition of the intrinsic value of human reason which, from the point of view of the church, necessitated a strict control of interpretation, the process of transmitting the Gospel, and any form of expression concerning these subjects. The prerogative of interpretation, stated the second decree, belonged solely to the church, which had the right to increase or restrict all freedom of expression or action with regard to the matters in which it had a definite monopoly.[6] Stretched to its limits, this decree could be taken to mean that lay Catholics were not really supposed to read the Scriptures on their own, upon the assumption that all such reading might involve an unauthorized interpretation.

Both the final formulations of the decrees and the exhaustive discussions which took place during the Council of Trent reveal the complexity of the church's position concerning the problem of interpretation. The idea that revealed truth is not only to be found in the Scriptures but also partly in the church's traditions of interpretation represents a nonfundamentalist approach to the sacred text which assigns a significant role to the interpretive power of human reason. Yet, the church wished to exercise control over the process of adapting the Gospel to changing needs and situations.

4. Marcocchi, II, p. 572, Sess. IV, April 1546: ". . . ut sublatis erroribus puritas ipsa evangelii in ecclesia conservetur . . ."
5. Ibid.: ". . . hanc veritatem et disciplinam contineri in libris scriptis et sine scripto traditionibus . . ."
6. Ibid., p. 575: ". . . ut nemo, suae prudentiae innixus, in rebus fidei et morum, ad aedificationem doctrinae christianae pertinentium, sacram scripturam ad suos sensus controquens, contra eum sensum, quem tenuit et tenet sancta mater ecclesia, cuius est iudicare de vero sensu et interpretatione scripturarum sanctarum, aut etiam contra unanimem consensum patrum ipsam scripturam sacram interpretari audeat, etiamsi huiusmodi interpretationes nullo unquam tempore in lucem edendae forent . . ."

From the description of the discussions at Trent given by the historian H. Jedin,[7] it appears that the Catholic world was divided at that period between two opposing approaches to the problem of interpretation. One school of thought, led by the church humanists who were close to the Protestant and fundamentalist position, regarded the Scriptures as a sufficient source of religious truth in themselves. The other school, consisting mostly of the more traditional theologians, adopted the traditional Catholic viewpoint that the consensus of opinion of the church fathers was also to be considered a source of revelation. This school proposed a specific formulation of that position, stating that religious truth was to be found *partly* in the Scriptures and *partly* in nonwritten traditions (*Partim in libris scriptis partim sine scripto traditionibus*). This formulation was not accepted, however. Instead, a much more ambiguous version was endorsed, which left the official policy undecided, i.e., In *libris scriptis* et *sine scripto traditionibus*. This formulation failed to clarify the relationship between the two kinds of sources. In fact, this version could legitimately be interpreted in two different ways. According to one, the oral traditions were incorporated in the written traditions: this version corresponded to the fundamentalist position. According to the other, the oral traditions constituted an additional source of religious truth.

The compromise reached at Trent is characterized by an inherent ambiguity, due to a failure to achieve consensus. It is this opacity rather than clear dogmatism which is fundamental to the understanding of the mentality of the Counter-Reformation church.

Original Sin and the Doctrine of Justification (decrees formulated at the fifth session on 17 June 1546, and on 13 February 1547, respectively). The decree on the Scriptures and their interpretation allowed for building a bridge between the "transcendental" and the "mundane" through the recognition of the interpretative capacities of human reason. Later on, the doctrine of justification would establish a link between the two orders through an accentuation of the validity of human will. The doctrine of original sin, placed in between the reevaluation of reason and will, represented an act of setting the limits to human capacities, both intellectually and morally. Although man, owing to his guilt, had lost his original righteousness and his nature was defiled, the taint of sin transmitted from generation to generation was to some degree removed by the sacrament of baptism, even if subsequently man still continued to retain a certain tendency to sin.[8] This decree encapsuled in a nutshell the Catholic dogma

---

7. Jedin, *Storia* . . . , II, pp. 67–118.
8. Marcocchi, II, pp. 586–587, Sess. V. 17 June 1546: "Manere autem in baptizatis cuncupiscentiam vel fomitem, haec sancta synodus fatetur et sentit;

of original sin. Upon this cornerstone the doctrine of justification, the very heart of Catholic moral theology, was constructed. A Catholic believer passes from a condition of inclination to sin to a condition of grace.[9] Only as a result of justification can one attain faith, hope for redemption and charity. These are the conditions for a perfect union with Christ, and for becoming a lively member of his body.[10] Finally, only those who are justified are offered eternal life.[11] This is in fact the crown of God's justice.[12] The decree concerning justification describes the process of achieving a condition of grace in the following manner: justification originates solely from God and is an act of God,[13] despite the fact that it is called "our" justification because it exists within us and is bestowed upon us by God on account of Jesus.[14] Its causes are all rooted in God. Its final cause is the praise of God and his son Jesus in eternal life; its efficient cause is the mercy of God who confers justification upon us in his compassion and not on account of man's innocence; the material cause is Jesus' sinlessness transmitted to mankind through baptism; and the formal cause, the righteousness of God.[15] Human will or even the Old Testament are insufficient in themselves to liberate one from sin.[16]

At the same time, however, the path leading to salvation includes stages in which a serious effort of human cooperation is required. When a man

---

quae cum ad agonem relicta sit, nocere non consentientibus et viriliter per Christi Iesu gratiam repugnantibus non valet. . . . Hanc concupiscentiam, quam aliquando Apostolus peccatum appellat, sancta synodus declarat, ecclesiam catholicam nunquam intellexisse, peccatum appellari, quod vere et proprie in renatis peccatum sit, sed quia ex peccato est et ad peccatum inclinat . . ."

9. Ibid., p. 590, Sess. VI, 13 January 1547: ". . . ut sit translatio ab eo statu, in quo homo nascitur filius primi Adae, in statum gratiae . . ."

10. Ibid., p. 594: "Unde in ipsa iustificatione cum remissione peccatorum haec omnia simul infusa accipit homo per Iesum Christum cui inseritur: fidem, spem et charitatem. Nam fides, nisi ad eam spes accedat et charitas, neque unit perfecte cum Christo, neque corporis eius vivum membrum efficit".

11. Ibid., p. 604: "Atque ideo bene operantibus usque in finem et in Deo sperantibus proponenda est vita aeterna . . ."

12. Ibid.: ". . . Haec est enim illa corona iustitiae . . ."

13. Ibid., p. 605: "Ita neque propria nostra iustitia tamquam ex nobis propria statuitur, neque ignoratur aut repudiatur iustitia Dei . . . illa eadem Dei est . . ."

14. Ibid.: ". . . quia a Deo nobis infunditur per Christi meritum . . ."

15. Ibid., pp. 593–595.

16. Ibid., p. 588: ". . . ut non modo gentes per vim naturae, sed ne Iudaei quidem per ipsam etiam litteram legis Moysi inde liberari aut surgere possent . . ."

is first summoned by God's grace, and while in the course of this process, he must respond of his own free will.[17] The response of the will to God expresses itself as a belief in the truth of revelation,[18] a recognition of sin and of the mercy of God, and a capacity to hope; its consequence is the recognition of the redeeming God as the source of all righteousness, and the beginning of a new life.[19] Subsequent sections of the decree attempt to explain the interaction of the divine and human factors in the process leading to salvation. Justification is entirely divine, since it is conferred in consequence of grace and not as a result of faith or actions.[20] Persistence in following the path strewn with obstacles also depends entirely on God, as the section on "persistence" points out.[21] Nevertheless, there can be no salvation without the cooperation of the human will by means of chastity and good works.[22] Nobody is exempted from performing the divine commandments, however much of a believer he may be.[23] According to the divine law, not only unbelievers but also sinners are excluded from the sphere of grace.[24] Moreover, in his path towards salvation, the Catho-

17. Ibid., p. 591: "... ipsius iustificationis exordium, in adultis a Dei per Christum Iesum praeveniente gratia sumendum esse, hoc est, ab eius vocatione, ... eidem gratiae libere assentiendo et cooperando ..."
18. Ibid., p. 591–592: "... fidem ex auditu concipientes, libere moventur in Deum, credentes, vera esse, quae divinitus revelata et promissa sunt ..."
19. Ibid.: "... et dum, peccatores se esse intelligentes, a divinae iustitiae timore, quo utiliter concutiuntur, ad considerandam Dei misericordiam se convertendo, in spem eriguntur, fidentes, Deum sibi propter Christum propitium fore, illumque tamquam omnis iustitiae fontem ... ; denique dum proponunt suscipere baptismum, inchoare novam vitam et servare divina mandata ..."
20. · Ibid., p. 595: "... gratis autem iustificari ideo dicamur, quia nihil eorum, quae iustificationem praecedunt, sive fides, sive opera, ipsam iustificationis gratiam promereretur ..."
21. Ibid., p. 600: "Similiter de perseverantiae munere ... nemo sibi certi aliquid absoluta certitudine polliceatur, tametsi in Dei auxilio firmissimam spem collocare et reponere omnes debent ..."
22. Ibid., p. 597: "... mortificando membra carnis suae et exhibendo ea arma iustitiae in sanctificationem per observationem mandatorum Dei et ecclesiae: in ipsa iustitia per Christi gratiam accepta, cooperante fide bonis operibus, crescunt atque magis iustificantur ..."
23. Ibid.: "Nemo autem, quantumvis iustificatus, liberum se esse ab observatione mandatorum putare debet ..."
24. Ibid., p. 603: "... divinae legis doctrinam defendendo, quae a regno Dei non solum infideles excludit, sed et fideles quoque fornicarios, adulteros, molles, masculorum concubitores, fures, avaros, ebriosos, maledicos, rapaces, ceterosque omnes, qui letalia committunt peccata ..."

lic not only requires faith but must also participate actively in the sufferings of Jesus[25] and must utilize the channels through which grace is provided such as the sacraments, confession, and prayer in which churchmen are actively involved.[26]

The unusual length and circularity of the description of justification attests to the problem confronting the writers. It consisted in finding a middle way between the Reformist concepts of divine grace as the sole source of all faith, and of justification by faith alone, and the Pelagian heresy which assigned too prominent a role to the natural ability of man to recognize his sins and attain revealed truth, faith, and the love of God.

On this question also the council was torn between two conflicting views. One came close to the Protestant position through a doctrine of double justification. This doctrine conceived of two kinds of justification: one was human and natural, consisting in human effort, understanding, and good works but incomplete and insufficient; the other was entirely of God, and only by means of the latter could salvation be attained.[27] An alternative view of justification sought to adhere to the traditional Catholic doctrine, maintaining that justification was rooted in God but given to man in the course of the cooperation of his free will. In this view, however, the nature of the relationship between the divine will manifested in grace and the role of human free will failed to be explained unequivocally. The problem remained unsolved, and different interpretations of the workings of grace and the process of responding with the will were elaborated. These interpretations, cast in scholastic terminology, involved such philosophical concepts as causality (was the relation between grace and free will like the relation between a physical cause and its effect?), contingency (was the status of human actions in relation to God's predestinatory decree contingent or just hypothetical?), and certainty (was God's foreknowledge of the future acts of men certain?). The same concepts, how-

25. Ibid., p. 598: "Itaque nemo sibi in sola fide blandiri debet, putans, fide sola se haeredem esse constitutum haeriditatemque consecuturum, etiamsi Christo non compatiatur . . ."

26. Ibid., pp. 601–602: "Christus Iesus sacramentum instituit poenitentiae. . . . Unde docendum est, christiani hominis poenitentiam post lapsum multo aliam esse a baptismali, eaque contineri non modo cessationem a peccatis et eorum detestationem . . . , verum etiam et eorumdem sacramentalem confessionem, saltem in voto et suo tempore faciendam, et sacerdotalem absolutionem, itemque satisfactionem per ieiunium, eleemosynas, orationes et alia pia spiritualis vitae exercitia . . ."

27. Ibid., p. 606; Jedin, *Storia* . . . , II, chs. 5 and 8. The representative of this doctrine in the Council of Trent was Girolamo Seripando, see: H. Jedin, *Girolamo Seripando*, Würzburg, 1937.

ever, were also in the center of scientific debates on the certainty of mathematics, on the hypothetical nature of mathematical entities, on the status of astronomical constructions as physical causes, and the like. This created the framework for a mediating ground between science and religion.

The Sacraments (decrees formulated at the seventh session on 3 March 1547). At the heart of this decree was the concept of the sacrament as understood by scholastic theology, i.e., as containing (*continere*) grace and conferring (*conferre*) it on anyone who does not create an obstacle to its reception (*non ponentibus obicem*).[28] As against the Protestant understanding of the sacrament as a God-given sign of the promise of grace, the effectiveness of which depends on faith in the divine promise,[29] Catholicism stressed the reality of the sacrament as a channel for grace. Grace is bestowed not only on account of faith, but also on account of the operation of the sacrament[30]; still the actual operation of the sacrament nevertheless demands the cooperation of free will in the form of not "raising an obstacle" to its action. The mysterious yet physical actuality of the operation of the sacrament requires that the power to bestow the sacrament is not given to all believers,[31] but is associated with an act of sanctification; namely, with the sacrament of ordination, which turns the priest into a member of the privileged group of mediators of grace.

The decree concerning the sacraments, like all the other dogmatic decrees, conceptualized the connection between the physical, mundane world and the transcendental one. The sacrament conceived as a real act of transmission of grace and the emphasis on the cooperation between the human will and grace embody this connection.

The formulation of the decree by means of a specific rejection of Protestant concepts[32] suggests an attempt to prevent a further elaboration of the

28. *SC*, Sess. VII, 3 March 1547: "Si quis dixerit, sacramenta novae legis non continere gratiam, quam significant; aut gratiam ipsam non ponentibus obicem non conferre . . . anathema sit".
29. Ibid. ". . . quasi signa tantum externa sint acceptae per fidem gratiae vel justitiae . . ."
30. Ibid. "Si quis dixerit, per ipsa novae legis sacramenta ex opere operato non conferri gratiam, sed solam fidem divinae promissionis ad gratiam consequendam sufficere, anathema sit".
31. Ibid. "Si quis dixerit, Christianos omnes in verbo & omnibus sacramentis administrandis habere potestatem, anathema sit".
32. As the former citations plainly show, the rhetorical strategy of the decree focuses on Luther's doctrine of "justification by faith alone", condemning it step by step.

subject. In fact, the council deliberately decided not to debate the nature of grace. How does grace bring about an increase in faith or the cooperation of man's will? Is its action truly physical or only moral? These questions were not resolved in the theological discussions at Trent, and they were connected with a whole series of assumptions about God, man, and nature concerning which there flourished differences of opinion in the period between the Council of Trent and the Galilean episode.

The Disciplinary Decrees

The council's concern with disciplinary problems represented a complementary aspect of its concern with doctrinal problems. The view that the saving of souls was the chief object of the Catholic Church led to a preoccupation with the image of the priesthood, its way of life, its intellectual training, and the teachings relevant to salvation which it passed on to the believers. Most relevant for my own particular purposes, however, is the way in which the disciplinary concerns gradually focussed on the problem of education, through which the new sense of Catholicism as a way of mediating between the "mundane" and the "transcendental" was to be implemented.

The council called for a tightening of discipline and concern for appropriate behaviour.[33] Moral aberrations in the life of the secular priesthood were explicitly condemned and strict observation of the canonical law was ordered.[34] In particular, the council sought to root out the fault most prevalent in the Catholic hierarchy: namely, absenteeism (*absentia*),[35] the prolonged absence of churchmen from their seats which enabled them to receive several benefits (*beneficia*) simultaneously.[36] The council sought to

33. Marcocchi, I, p. 517, Sess. XIV, 25 November 1551: "Cum proprie episcoporum munus sit, subditorum omnium vitia redarguere, hoc illis praecipue cavendum erit, ne clerici, praesertim ad animarum curam constituti, criminosi sint, neve inhonestam vitam ipsis conniventibus ducant . . ."

34. Ibid., p. 507, Sess. VI, 13 January 1547: "Ecclesiarum praelati ad corrigendum subditorum excessus prudenter et diligenter intendant . . ."

35. Ibid., p. 505: "Romano pontifici denuntiare teneatur, qui in ipsos absentes prout cuiusque maior aut minor contumacia exegerit, suae supremae sedis auctoritate animadvertere et ecclesiis ipsis de pastoribus utilioribus providere poterit, sicut in Domino noverit salubriter expedire".

36. Ibid., p. 510, Sess. VII, 3 March 1547: "Quicumque de cetero plura curata aut alias incompatibilia beneficia ecclesiastica, sive per viam unionis ad vitam, seu commendae perpetuae, aut alio quocumque nomine et titulo contra formam sacrorum canonum et praesertim constitutionis Innocentii III, quae incipit *De multa,* recipere ac simul retinere praesumpserit, beneficiis ipsis

impose norms of simplicity and modesty on the members of the secular clergy[37] and the observance of discipline on those who were subject to a rule.[38] In addition to the concern for the life-style of the priesthood, the decrees also contained measures advocated by the Counter Reformation for the religious and intellectual formation of the clergy and for the improvement of the Catholic education of the people. The importance given to these two aspects of educational policy is remarkable. The council ordered the creation of a chair of Scripture in the proximity of every church and public school[39] as well as every monastery.[40] It also decreed the appointment of a teacher who, without remuneration, would instruct poor priests and laymen in the foundations of grammar,[41] in order that they be able to proceed to properly organised studies of the Scriptures. In addition, the Council ordered the creation of seminaries next to all cathedrals

iuxta ipsius constitutionis dispositionem ipso iure, etiam praesentis canonis vigore, privatus exsistat".

37. Ibid., p. 574, Sess. XXV, 4 December 1563: "In primis vero ita mores suos omnes componant, ut reliqui ab eis frugalitatis, modestiae, continentiae ac, quae nos tantopere commendat Deo, sanctae humilitatis exempla petere possint . . ."

38. Ibid., p. 564: "Quoniam non ignorat sancta synodus, quantum ex monasteriis, pie institutis et recte administratis, in ecclesia Dei splendoris atque utilitatis oriatur: necessarium esse censuit, quo facilius ac maturius, ubi collapsa est, vetus et regularis disciplina instauretur, et constantius, ubi conservata est perseveret . . ."

39. Ibid., p. 495, Sess. V, 17 June 1546: "Eadem sacrosancta synodus . . . statuit ac decrevit, quod in illis ecclessiis, in quibus praebenda aut praestimonium seu aliud quovis nomine nuncupatum stipendium pro lectoribus sacrae theologiae deputatum reperitur, episcopi, archiepiscopi, primates et alii locorum ordinarii eos, qui praebendam aut praestimonium seu stipendium huiusmodi obtinent, ad ipsius sacrae scripturae expositionem et interpretationem per se ipsos, si idonei fuerint, alioqui per idoneum substitutum . . ."

40. Ibid., pp. 497–498: "In monasteriis quoque monachorum ubi commode fieri queat, etiam lectio sacrae scripturae habeatur. . . . In conventibus vero aliorum regularium, in quibus studia commode vigere possunt, scarae scripturae lectio similiter habeatur . . ."

41. Ibid., p. 497: "Ecclesiae vero, quarum annui proventus tenues fuerint, vel ubi tam exigua est cleri et populi multitudo, ut theologiae lectio in eis commode haberi non possit, saltem magistrum habeant, ab episcopo cum consilio capituli eligendum, qui clericos aliosque scholares pauperes grammaticam gratis doceat, ut deinceps ad ipsa sacrae scripturae studia (annuente Deo) transire possint . . ."

and large churches for the training of suitable young people, and especially those who were poor, for the priesthood.[42]

The main educational questions deliberated by the Tridentine Church made their appearance in the course of discussions on the cultural and intellectual equipment required by the priesthood in order to carry out its task. One school of thought, whose influence was particularly marked in the early stages of the council (1545–1547), was influenced by the humanistic current of the Catholic movement of spiritual rebirth and sought to strengthen the connection with the original source of faith, namely, the Scriptures.[43] Moreover, in reaction to the speculative tendencies predominant in scholastic theology, this school of thought proposed the preparation of catechisms in the spoken vernacular which would provide a systematic exposition of Catholic dogma in terms which were simple, direct, and comprehensible to the people.[44] The decrees ordering the establishment of chairs of Scripture next to every church and cathedral represented a victory for the orientation rooted in Christian humanism.

The intellectual and institutional implications of these moves, however, were not lost on those members of the council who belonged to the traditional teaching elite of the Catholic Church, the mendicants in particular,

---

42. Ibid., p. 549, Sess. XXIII, 15 July 1563: ". . . sancta synodus statuit, ut singulae cathetrales, metropolitanae atque his maiores ecclesiae pro modo facultatum et dioecesis amplitudine certum puerorum ipsius civitatis et dioecesis, vel eius provinciae, si ibi non reperiantur, numerum in collegio, ad hoc prope ipsas ecclesias vel alio in loco convenienti, ab episcopo eligendo, alere ac religiose educare et ecclesiasticis disciplinis instituere teneantur . . . "

43. Jedin, *Storia* . . . , II, pp. 119–146.

44. This suggestion was accepted in the last period of the council. See Marcocchi, I, pp. 560–561, Sess. XXIV, 11 November 1563: "Ut fidelis populus ad suscipienda sacramenta maiore cum reverentia atque animi devotione accedat: praecipit sancta synodus episcopis omnibus, ut non solum, cum haec per se ipsos erunt populo administranda, prius illorum vim et usum pro suscipientium captu explicent, sed etiam idem a singulis parochis pie prudenterque, etiam lingua vernacula, si opus sit et commode fieri poterit, sevari studeant, iuxta formam a sancta synodo in catechesi singulis sacramentis praescribendam, quam episcopi in vulgarem linguam fideliter verti atque a parochis omnibus populo exponi curabunt; necnon ut inter missarum solemnia aut divinorum celebrationem sacra eloquia et salutis monita eadem vernacula lingua singulis diebus festis vel solemnibus explanent, eademque in omnium cordibus (postpositis inutilibus quaestionibus) inserere, atque eos in lege Domini erudire studeant".

who feared an erosion of their position as a result of the victory of the humanists. The establishment of chairs for teaching the Scriptures meant transferring part of the responsibility for education to the bishops. The humanistic reformers conceived of the role of a bishop in the tradition of the church fathers like Paul, Ambrose, and Augustine, who were preachers and spiritual mentors as well as ecclesiastical administrators. And, indeed, in the course of the discussions at the council, there was a very far-reaching proposal for transferring the task of preaching on Sundays and festivals to the bishops and parish priests.[45] This was a direct blow to the mendicant orders which, from the thirteenth century onwards, had possessed a monopoly in preaching activities, and had received as a result many privileges from the popes which exempted them from the jurisdiction of the diocesan bishops. Furthermore, the emphasis on the teaching of the Scriptures and the proposal to prepare catechisms of the principles of faith also raised the question of the intellectual training suitable for traditional preachers. One of the specific proposals put forward was to oblige the religious orders and especially the Dominicans to give preference to the study of the Scriptures over scholastic theology.[46]

As a result of these discussions, two cultural orientations emerged which, to a great extent, also corresponded to institutional allegiances. On the one side was the secular clergy and especially some reformist bishops with humanist tendencies who insisted that a preferential status should be given to the Scriptures both in the subject-matter of sermons and in the education of preachers. And, in order to reinforce this demand and give it reality, they asked for the right to supervise the preachers, proposing that the sermons of the mendicants both in monastery churches and in parish churches should be made subject to the authorisation of a bishop. The bishops also demanded the authority to take action against heretical preachers.[47] On the other side were the mendicants who sought to preserve the ancient privileges which had been conferred upon them by the papacy and which assured them their independence. The intellectual leaders of the mendicants, and especially the Dominican Domingo de Soto, did not deny the need for spreading knowledge of the Scriptures. They refused, however, to emphasize the Scriptures at the expense of scholastic theology which they saw as Catholicism's most effective weapon in the struggle against the Protestants, who in those years detested scholasticism.[48] In the first period of the council, the humanistic

45. Jedin, *Storia* . . . , II, p. 127.
46. Ibid.
47. Ibid., p. 130.
48. Ibid., p. 140.

orientation won a partial victory, reflected in the demand that chairs of Scripture should be set up in the monastery schools and in the requirement that preachers appear in front of their superiors before giving a sermon.[49] In the third period of the council, the scholastic theologians won a more definite victory with the decision to establish seminaries for the training of the priesthood. The council, however, did not stipulate in detail the nature of the instruction to be given in these seminaries. This remained open for debate among executors of the cultural policies of the church, which were only delineated in broad terms and left much to be determined by rival cultural orientations of the future.

The Tridentine reconceptualization of the relationship between the "natural" and the "transcendental" is reflected in the doctrinal as well as the disciplinary decrees of the council. The Catholic Church wished to bring its identity into focus by emphasizing its function as an institution mediating between the two worlds. It therefore stressed its exclusive rights in the sphere of Scriptural interpretation, in the instruction of the people in the true doctrine, and in the ministration of the sacraments. Consequently, however, the justification for the perpetuation of the authority of the Catholic Church lay in the recognition of the relevancy of human capacity to create a bridge between the two worlds. The validity of an interpretation of the Scriptures by means of human reason, the belief in a residue of natural morality remaining after original sin, the insistence on the need for the cooperation of the human will in the process of justification and on the effectiveness of the ministration of the sacraments by the representatives of the Church: all these expressed an acknowledgement of the adequacy of the natural human potential to bridge the immense gap between the transcendental and the mundane.

The practical consequence of this acknowledgement was the insistence of the disciplinary decrees on the necessity for a proper training in the understanding of the Scriptures, for guiding the will by means of sermons, and for a thorough education of the priesthood which would give the bestowal of the sacraments the quality of sanctity. Within the framework of this policy, the Church's intellectual elite – traditionally the mendicants, and especially the Dominicans, who were soon to be joined by the Jesuits combining humanistic and scholastic tendencies – was given the crucial task of deciding what was to be taught, reconciling it with orthodox doctrine and disseminating it. At the same time, the long-drawn-out discussions which took place at Trent before the decrees appeared, and the cau-

49. Ibid., p. 145.

tious wording of the decrees themselves, reveal difficulties in arriving at a cultural orientation based on doctrinal unity. This, it appears, remained more of an aspiration than a reality. Moreover, it was soon to be discovered that the intellectual elite of the church – especially the Dominicans on the one hand and the Jesuits on the other – was not united in its practical interests. Differences of opinion deriving from clashes of interest therefore continued to manifest themselves behind the scenes after the doctrinaire phase of the Counter Reformation was over.

The Thomism of the Counter Reformation

Despite the difficulties in arriving at a unified cultural orientation, there is much evidence of the attachment of the council to Thomist ideas. The historian Marcocchi has expressed his judgement of the matter by saying:

> A comparison between the doctrinal decrees and the ideas expressed by the senior and junior theologians during the special and general congregations reveals that the Thomist current had a very profound effect on the whole work of the Council and on the formulation of the decrees themselves.[50]

Other scholars have also shared this view.[51]

Thomism could offer an approach in which the link between human reason and the process leading to salvation was much stronger than anything which Christianity had previously known. Thus, the Thomist system was suited to the spirit and needs of Catholicism in its attempt to crystallize its identity vis-à-vis the reform movement by emphasizing moral seriousness mingled with Christian humanism. Moreover, the bridge which Thomism constructed between rational knowledge and the doctrine of salvation involved principles of the organisation of knowledge which could serve as means of controlling the educational programmes required by the Council of Trent, ensuring their relevancy to religious aims and guaranteeing their orthodoxy. An analysis of the Thomist concept of the relationship of natural and transcendental truth and of the principles of the organisation of knowledge which it involved is therefore essential for understanding the doctrinarian phase of the Counter Reformation.

50. Marcocchi, *La Riforma* . . . , II, p. 565.
51. See M. Dupront, "Du Concile de Trente: réflexions autour d'un IVème centenaire", *Revue historique*, 1951, pp. 262 ff.

Thomas's spiritual world, like Augustine's before him, centered on the concept of an ultimate truth flowing from a divine source[52] which was considered to be beyond and above human reason.[53] In this world, God's truth could be disclosed through revelation and transmitted through the Holy Scriptures and their exegesis.[54] However, it could never be fully attained in the "here and now". Ultimately, this truth was only available to those chosen for salvation,[55] and salvation was utterly dependent on God's predestination, which was not subject to human understanding or will.[56] Thus, with regard to revelation as the source of ultimate truth and the attainability of that truth in the state of salvation, Thomist thought was strictly orthodox, wholly rooted in Christian doctrine as formulated in the writings of the church fathers and in particular those of Augustine.

Nonetheless, the originality of Thomas's system lay in his greater emphasis on the role of rational human thought, both in this world as well as on the path toward salvation. Thomas's notion of contemplation illustrates the spirit of Thomist attitudes in these matters. On the one hand, Thomas considered contemplation as the crown of all knowledge. Although he accepted the Aristotelian premise that the process of knowledge started from data supplied by the senses (*Nihil est in intellectu nisi prius fuerit in sensu*), in his opinion real knowledge consisted of the abstraction of the essence of things by the intellect. The notion, of Platonic origin, that that which was immaterial and spiritual was logically more real than that which was material led to the view that knowledge of the immaterial was more real than knowledge of the material. Thus, the philosophical concept, which abstracted the essence from a group of individual particulars, came to represent a connection between the realistic structures of the world and the rationalistic structures of the mind.[57] Aided by his senses and his intellect, man could ascend from an individual's confused cognition to a true knowledge of the essence of things and the highest knowledge of the principles of thought, order, and causality.[58] This last stage was identified by Thomas as the contemplation of things, for in

52. Thomas Aquinas, *Sum. theol.*, II-II, q. 1, a. 1 c; q. 1, a. 10 c; q. 17, a. 6 c.
53. Ibid., I, q. 32, a. 1 c; q. 32, a, 1 ad. 3.
54. Ibid., II-II, q. 1, a. 9 ad. 1.
55. Ibid., I-II, q. 3, a. 5; Thomas, *Scriptum super libros sententiarum*, IV, prooem., q. 1, a. 1 c.
56. Thomas, *Sum. theol.*, III, q. 48, a. 6 c.
57. Ibid., I, q. 84, a. 1c; q. 84, a 6; q. 85, a 1; q. 85, a 6c; q. 8, a 1c.
58. *In II Phys.* lec. 1.

its highest state, which required a long training of the intellect in the various disciplines of learning, human reason could become an adequate tool for grasping the fundamental order of things which ultimately sprang from the divine intellect. However, Thomas did not view contemplation as merely the crown of all knowledge. It was also conceived as a bridge to salvation. Thomas said:

> It is plain, . . . that in the vita contemplativa man has a part with those above him, . . . God and the Angels. . . . But in those matters which belong to the vita activa, other animals, however imperfectly, have somehow a part with him. And so the final and perfect beatitude, which is looked for in the life to come, in principle consists primarily in contemplation, and secondly in the true operation of the practical intellect directing human actions and passions.[59]

Salvation, then, was perceived more in terms of intellectual contemplation than in terms of actions, which were what man had in common with the animals and not with God. In preparing the mind for the contemplation of things, and in realizing that part of himself which is closest to God, man was also preparing himself for the contemplation of God, which was only possible in the world to come for those chosen to be aided by the endowment of grace. In contradistinction to Augustine, Thomas believed that human reason unaided by grace was an adequate tool for understanding God's created world, which is the only way of knowing God himself, albeit indirectly. Therefore, knowledge of mundane things acquired legitimation not only in instrumental terms, as it had done for Augustine, but in religious terms as well.

Thomas's notion of contemplation shaped and coloured his conception of the relation between mundane and ultimate truth. For Thomas, the source of all truth was one: God's intellect, which was inseparable from his will. Although it contained regions inaccessible to human reason, other parts were subject to natural investigation.[60] "Grace does not destroy nature; it perfects her", said Thomas, suggesting the impossibility of opposition between natural and divine truth. This obviously implied that divine truth was embedded in mundane reality. The investigation of mundane reality complemented the attainment of ultimate truth through revelation and grace.

The basic assumption which enabled Thomas to integrate the whole body of Aristotelian philosophy into the framework of Christian thought was closely related to his conception of the nature of divine and mundane truth and the relationship between them. If truth is one, and its source

59. *Sum. theol.*, I-II, q. 3, a. 5.
60. *Scriptum super libros sententiarum, prooem.*, q. 1, a. 1c.

divine, the transcendental world of God's intellect (which is inseparable from his will) and mundane reality (his creation) are united by the same laws of causality and reason. These, for Thomas, were obviously based on the Aristotelian concept of necessity. Necessity, therefore, became for Thomas the main organizational principle by which various types of knowledge were related to each other in a hierarchical order.

Real knowledge (*scientia*) was a knowledge of causes – that is, the necessary connection between the effect and its reason. The realm of true knowledge excluded a practical cognition of objects of art which were forever immersed in the senses, and were not capable of yielding necessary knowledge.[61] It ascended from a knowledge of natural, material substances acquired through an abstraction from sense data and the application of deductive reasoning for establishing the causal connections between them (physics) to a knowledge of pure, immaterial substances which supplied the principles governing the world of natural substances (metaphysics), and to those truths of theology which can be rationally explained and understood.[62] On the other hand, the possibility of acquiring a necessary knowledge of the world of nature and of God was based on the assumption that in fact (although not of necessity) God operates in accordance with the same laws of necessary causality.

In this scheme, the status of different bodies of knowledge was determined by their relevance to the aim of a preparation for salvation. Such a preparation, according to Thomas, consisted in contemplation which led from the temporal, physical reality, via the metaphysical essence of any possible reality, to the ultimate reality of God. Mathematics, the third part of philosophy according to Aristotle, had only a limited significance on this path, due to the ambiguous status of its entities in relation to physical reality. Following Aristotle's line of thought, Thomas regarded mathematical entities as "quasi-substances", devoid of properties of change (place or time), and therefore not wholly pertaining to physical reality.[63] Mathematics, hence, was assigned a middle position between physics and metaphysics. Its main value was to provide a model for correct reasoning to be applied by the physicist (the natural philosopher) in his own branch of real knowledge.

Thomas, however, believed that there were fields of inquiry in which mathematics seemed to pertain more directly to a knowledge of reality. Mechanics, optics, astronomy, all sciences of observed facts interpreted in a theoretical manner by mathematical methods, were defined by

61. *De verit.*, q. 3, a. 3c.
62. *Sum. theol.*, I, q. 1, a 8 ad 2.
63. *In Boeth. de Trin.*, lect. 2, q. 1, a. 3.

Thomas as "mixed sciences".[64] Since they used mathematical methods, they were included in the category of mathematical sciences and were therefore considered "subalternated", devoid of autonomy. From this Thomas derived the status of the theories which attempt to explain observed phenomena by mathematical principles. While Ptolemaic astronomy used epicycles and eccentrics to explain physical phenomena, Thomas argued that these were merely hypotheses and not necessarily true in nature.[65]

The following Thomist assumptions were used to justify this contention: since mathematical entities were not considered real substances, they could not assure the necessity of physical causes. In addition, since the connection between observed phenomena and an astronomical hypothesis (e.g., the eccentrics and epicycles) was established by a dialectical, logically weak syllogism (from effect to cause), it did not result in a real scientific demonstration. Thus, astronomical theories failed to acquire the status of real physical causes in the Thomistic system. They were considered probable, capable of "saving the phenomena", but not necessarily true. They therefore remained marginal to the main axis of the organization of knowledge suggested by Thomas.

Two features of Thomas's world view are relevant to understanding the church's cultural policies during the Counter Reformation. First, there was the legitimation in terms of salvation which the intellectual activities could gain within the framework of the Thomist world view. Second, there was the significance of the Thomist principles of the organization of knowledge for the integration of profane bodies of knowledge at the same time as retaining the orthodox core of traditional Christianity. Undoubtedly, any attempt to change these principles implied a need to modify the whole structure of the relationship between rational and ultimate truths. Nonetheless, such attempts became more likely at a period which necessitated a rethinking of Catholic education and in which the involvement in the scientific world was a direct result of the general educational concern.

64. *Ibid.,* q. 5, a 3; In *II phys.* lect. 3 nn. 6–9; *Sum. theol.,* I-II, q. 35, a. 8; II-II, q. 9, a. 2 ad 3.
65. *Sum theol.,* I, q. 32, a 1; In II *De caelo,* lect. 17.

.

5

The Dominicans

A Traditional Intellectual Elite of the Catholic Church

The Council of Trent required the Catholic establishment to prepare the priesthood for an educational mission and to guide education in the Thomist direction. Who was, however, to put the "ideology" of the reform into practice?

Traditionally, it was the Dominicans who saw themselves as responsible for instructing the people in the true doctrine and for preserving the purity of Catholic dogma. As a result they became outstanding among the mendicant orders for their intellectual orientation.[1] From the thirteenth century, the Dominicans ran an organised and well-ordered educational system which not only included schools in the individual monasteries, but also *studia generalia* – regional universities – qualified to confer academic degrees in theology. Moreover, the Dominicans were the first to adopt Thomism as a comprehensive religious and intellectual framework, regarding it as most suited to the orthodox requirements of the Catholic faith.[2] Significantly enough, it was also the Dominicans who mainly dominated the Inquisition,[3] the most visible instrument of control of the Catholic Church.

The Dominicans' self-perception as instructors, their Thomist orientation, and their predominant role in the Inquisition made them the most prominent intellectual elite of the Catholic Church. Nevertheless, to the best of my knowledge, there is no description in academic literature of the intellectual and institutional framework in which the Dominicans op-

---

1. A. Koperska, *Die Stellung der religiosen Orden zu den Profanwissenschaften im 12. und 13. Jahrhundert*, Freiburg 1914; C. Douais, *Essai sur l'organisation des études dans l'ordre des frères prêcheurs*, Paris 1884.
2. P. O. Kristeller, "The Contribution of Religious Orders to Renaissance Thought and Learning", *Medieval Aspects of Renaissance Learning*, Durham 1974.
3. *The Oxford Dictionary of the Christian Church*; A. G. Gardew, *A Short History of the Inquisition*, London 1933, pp. 26–27.

erated in the sixteenth and seventeenth centuries and no monograph has been written about Dominican education.

These lacunae in historical information account for the very preliminary nature of my study of the Dominican intellectual and institutional milieu, which is largely based upon original documents from a number of sources. One major source is the *Constitutiones,* printed during the Counter Reformation in three editions: 1566, 1607, and 1650.[4] Each edition contained the original form of the rule adopted by the order – i.e., the Rule of St. Augustine – and a body of constitutions divided into three categories: a few basic constitutions which could never be changed; others which could be changed in exceptional circumstances; and the rest which could be changed if agreed by the body of representatives which governed the order on three consecutive occasions. The combination of a rule and different kinds of constitutions ensured the original way of life of the Dominicans as compared to the other orders. It institutionalized well-defined ways of introducing change at the same time as preventing it from being achieved too easily or quickly, or without a broad consensus.[5] The *Constitutiones* contained a special section pertaining to all matters of study, entitled "De studentibus". The survival of three different versions of *Constitutiones* allows one to follow changes introduced into the "De studentibus" section in each of these versions. Such a comparison is most illuminating for revealing subtle changes in the cultural policy of the Dominicans.

A second and more elaborate source for the study of the educational problems of the Dominicans in the Counter-Reformation era and their solutions are the *Acta* of the General Chapters.[6] The Dominicans, as is well known, had the most developed constitutional system among the religious orders. It consisted of two kinds of committees of elected officials, and "diffinitors" endowed with the authority to consider policy, to legislate, and to punish. These "committees" were known as provincial and General Chapters, and convened alternately every year.[7] The *Acta* of the General Chapters in the years 1540–1650 usually began with the opening addresses to the rather solemn meetings of representatives from

4. *Regula beati Augustini et constitutiones Ordinis Fratrum Praedicatorum,* Rome 1566; Venice 1607; Rome 1650. My study refers to ch. XIV ("De studentibus"). Henceforth *Constit.,* place and date of publication.
5. See R. B. Brooke (ed.), *The Coming of the Friars,* London 1975, p. 97 ff.
6. B. M. Reichert (ed.), *Acta Capitulorum Generalium Ordinis Praedicatorum (1558–1628), Monumenta Ordinum Praedicatorum,* Rome 1901, vols. X-XI, henceforth *Acta,* year of capitulum.
7. Brooke, *The Coming . . . ,* pp. 98, 193–196.

all provinces of the order. In addition, they also contained whole sections concerned with educational matters, including deliberations and decisions only part of which were later incorporated in the *Constitutiones*. The *Acta* therefore constitute a mine of precious and often buried information about Dominican intellectual problems and institutional forms of life.

A third and last source of information are early histories of the order. The most useful among them is the *Histoire de Maîtres Généreaux de l'Ordre des Frères Prêcheurs*, which is mainly based upon the recollections of the Masters General of the order.[8] In spite of the rather narrow, personal perspective of this kind of history, it nevertheless provides a narrative framework and temporal continuity to the vicissitudes of the order which the other legal sources obviously lack.

My study of the Dominican educational activities after the Council of Trent begins with an account of the role of learning in the order's ideals of life: learning as contemplation, as a means of instructing the right doctrine to others, and as a practice of self-negation and isolation. This account is based upon a reading of the *Constitutiones* and a comparison of the different versions mentioned above. There follows a description of the atmosphere of cultural crisis expressed in the opening addresses to the General Chapters. The broad lines of cultural policy laid down by the chapters as a result of the crisis and the implications of this policy for the contents and organization of studies complete the picture of the intellectual orientation peculiar to the Dominicans in the post-Tridentine era.

The Duty to Teach and Study

The section "De studentibus" of the *Constitutiones*[9] begins with justification of study as the most suitable means of achieving the Dominican ideal of life.[10] Three arguments supported this assertion. First, study was regarded as essential for those who aspire to the *vita contemplativa*. Contemplation of the things of God is bound up with study, which alone can

---

8. R. P. Mortier, OP, *Histoire des Maîtres Généraux de l'Ordre des Frères Prêcheurs*, Paris 1913, vol. VI.
9. See note 4.
10. *Constit.*, Rome 1650, p. 241: ". . . studium [sacrarum scientiarum] Religioni nostrae quam maxime congruit . . ." The edition of 1566 does not specify which kind of studies are most adequate, but states generally that "studies" are adequate.

purify it from error, while at the same time ignorance (of the Scriptures) can falsify the nature of contemplation.[11] Second, study was regarded as necessary because the order was founded to instruct others in the true doctrine. Instruction of others cannot be achieved without study and contemplation of the Scriptures, for ignorance is the mother of error, which would be disseminated precisely by those who have taken it upon themselves to instruct the people.[12] Third, study was to guard the soul against the temptations of the flesh.[13]

The justification of study as a means to the contemplative life, and hence to salvation, illustrates the Dominican concept of the relationship between intellectual activities and the quest for revealed truth. The *Constitutiones* seek to stress the idea that intellectual activity can be a means of approaching God. The need to establish revealed truth through the power of reason receives a certain sanction in the *Constitutiones*. The ambivalent nature of this very Catholic argument finds expression in the numerous restrictions which are placed on study. In its earliest, original form, the Dominican Rule states that one should be cautious in granting people permission to study, whether it is a matter of attending lectures or of copying. An appointed brother ought to be nominated; in order to exercise control over the students and correct whatever needs correction.[14] Many constraints were imposed on pursuing secular studies, and the language used to describe them is harsh and uncompromising: The students should not study the books of the heathens or philosophers or occupy themselves with the secular sciences or liberal arts without special permission from the Master of the order, the General Chapter, or the pro-

11. Ibid.: "Imprimis, propter illud quod vitae contemplativae proprium est; nam contemplatio ordinatur ad rerum Divinarum considerationem, quae recte studio dirigitur. Etenim studium liberat contemplantem a periculis, quae sunt errores, in quibus saepe versantur ii, qui contemplantur, cum tamen scripturas sacras ignorent . . ."

12. Ibid.: "Secundo, studium Religioni nostrae necessarium est, quoniam ipsa est instituta ad tradendum aliis contemplata per doctrinam; quae sane res fieri non potest sine studio scripturarum eamque contemplatione. . . . Ignorantia quae mater est cunctorum errorum, a cunctis vitanda est, et maxime ab his qui officium docendi in populo susceperunt . . ."

13. Ibid.: "Tertio, literarum studium Religioni nostrae convenit, quoad illud, quod ei cum omnibus Religionibus est commune: valet enim primo ad prostigandam carnis lasciviam . . ."

14. *Constit.*, Rome 1650, p. 240: ". . . quoniam circa studentes *a.* diligens est adhibenda cautela, aliquem specialem fratrem habeant, sine cuius licentia non scribant quaternos, nec audiant lectiones; et quae circa eos in studiis corrigenda viderit corrigat, et si vires eius excedant, Praelato propanat . . ."

vincial prior.[15] Later on, however, in subsequent versions, the rule was reinterpreted to mean there should be room for secular studies as an auxiliary aid to the contemplative life. The study of the secular sciences was permitted as long as it did not become an end in itself; the student should not devote all his time to it, to the detriment of sacred studies and the Scriptures.[16] Finally, secular studies were permitted to the degree that they directed one to theology, which was regarded as the end of contemplation, leading one to God. One should devote oneself to them day and night, both in the house and outside.[17]

The Dominican *Constitutiones* offer an elaborate justification of learning in terms of the contemplative life, in which the tension between the speculative fruits of the intellect, on the one hand, and knowledge derived from Scriptural revelation, on the other, is never hidden. The balance is achieved by stressing that both learning and the Scriptures are essential foundations of the contemplative life.

The defence of learning as a constitutive element of the contemplative life is related to the internal vocation of the Dominicans. Externally, however, study was conceived as necessary for the task of instructing others in the right doctrine, a vocation denoted by the term "officium docendi" – the duty to teach. A constitution from 1228, among the earliest ever recorded,[18] stated that to avoid a gulf between the spiritual life of the brothers and the intellectual life of the period, and to prevent a degeneration of the internal schools of the order, these schools should be open to the public. Very rapidly, then, the justification of study for the purpose of instructing others had actually involved the Dominicans in broader educational practice, thus turning them into the teaching elite of the Catholic Church.

15. Ibid., p. 240: "In libris Gentilium et Philosophorum non studeant, et si ad horam inspiciant, *b.* Saeculares sententias non addiscant, neque artes quas liberales vocant, nisi aliquando circa aliquos Magister Ordinis, vel Capitulum Generale vel Prior provincialis, vel Capitulum provinciale, voluerit aliter dispensare . . ."

16. Ibid., p. 241: "Delcaramus quod quamvis liceat fratribus nostris studere scientiis saecularibus, non tamen diu in illis versari debent, et omnes aetatis suae tempus consumere, sed potius scripturarum studio, atque eis, quibus consulere possunt saluti animarum, debent se assidue, ac solicite exercere, a curiositate, et inani gloria cavendo atque ex illis non sibi solunt, sed etiam aliis ad bene . . ."

17. Ibid., p. 240: ". . . *c.* sed tantum libros theologicos, tam iuvenes quam alii legant. Ipsi vero in studio taliter sint intenti, ut de die, de nocte, in domo, in itinere legant aliquid, vel meditentur . . ."

18. Koperska, *Die Stellung . . .*, pp. 108–129.

Last is the recommendation of studies as a means of guarding the soul against the temptations of the flesh. A peculiarly medieval state of mind was here manifested, which tended to see intellectual activities as a kind of retreat from the world. Learning was not conceived of as a social activity; on the contrary, learning was culturally constructed as a means of separation and isolation, whose function was asocial and affected the private rather than the public domain.

The three justifications of study provided in the original *Constitutiones* of the Dominicans suggest a rich field of possibilities within which the educational vocation could develop: studies were conceived as a way to God, a way to one's fellow men, and a way to oneself. Nevertheless, the *Constitutiones* imposed a number of constraints upon that field. Although intellectual activities received a religious sanction and secular studies were allowed or even encouraged, profane knowledge had always to be oriented towards a religious goal and especially towards the field of theology. Moreover, although the Dominicans never ignored the public significance of their role as teachers and actually opened their schools to the laity, they nevertheless tended to see their vocation in terms of a retreat from society and from public life in general.

If the Dominicans had to take upon themselves the task of reforming Catholic education along the lines suggested by the Tridentine decisions, their practice was nevertheless constrained by the long tradition of teaching and learning within which they operated. Traces of these constraints are amply found in Dominican texts of the sixteenth century which reveal an attempt to reform education in response to the needs of the contemporary world.

Dominican Educational Policy

The opening addresses to the General Chapters of the Dominican order express the mood prevailing in those years among those friars who felt the need to act. The decline of the Dominican religion; the diminuation of zeal, learning, and virtues; the passing of a vanished golden age are recurring themes in many of the addresses.[19] A sense of crisis also perme-

19. *Acta*, 1558, p. 4: "O deplorandam calamitatem, o casum miserabilem, o religionem diminicanam tanto fervore erectam, tanto studio auctam, tanto sanctissimarum virtutum censu locupletatam! Quo te redactam conspicimus. . . . O flebilis et lacrimosa mutatio! Commutatum est aurum in argentum, argentum in aes, aes in ferrum, ferrum in lutum et coenum, ita ut pristinae illius religionis ne vestigium quidem relictum videatur . . ."

ates the words of the Master of the order who opened the chapter of
1561:

> Where, then, are all the exceedingly learned and pious fathers of that
> religion who were a treasure not only for us, but for the whole of
> Europe? Where is the large number of universities which once ex-
> isted? Where are all the schools and the debates [in the places of
> instruction] where the most highly regarded sacred and profane stud-
> ies once flourished? Where is the pious observance of our laws and
> customs, whereby so often the souls of our children were induced to
> sweetly embrace the Christian virtues? Our churches are destroyed
> or depleted or used for secular purposes; the schools are no more
> than mediocre, studies have declined, talents have decreased. Our
> teachers are not respected and thus our lights have grown dim.[20]

True, the speaker chooses a much used rhetorical topos ("ubi
sunt . . .") to express anxiety. Still, the sense of danger should not be un-
derstated. Not only a sense of crisis, but the sentiment that a return to
the original Dominican ideal of life is the solution to the tremendous
problems facing the order is also recurrent in those addresses. It was
through study and the exercise of their talents that the Dominicans could
extricate themselves from their present troubles.[21] Moreover, accentuat-
ing traditional values would also help to solve the problems of the church
which itself was experiencing great difficulties. The Dominicans have the
duty to be the first in the field, the pioneers who would treat this sickness
and overcome it.[22]

The opening addresses said a great deal about the decline of the Domin-

---

20. Ibid., 1561, p. 28: "Ubi enim doctissimi et religiosissimi illarum religionum
    patres, qui non nobis duntaxat, sed toti Europae ornamento erant? Ubi tan-
    tus studiorum numerus? Ubi tot ludi et scolae in quibus humanarum et
    divinarum scientiarum studia quam gloriosissima florebant? Ubi legum et
    sanctionum nostrarum observantia, quibus tanquam vehiculis filiorum nos-
    trorum animi ad dulces virtutum amplexus quam saepissime deducebantur!
    Templa nostra aut destructa sunt aut nuda remanserunt vel in prophanos
    usus deputata. Scolae solo aequatae, studia perierunt, emarcuerunt in-
    genia, sanctiones nostrae despectae et lumina denique omnia extincta sunt
    . . ."
21. Ibid., 1564, p. 50: "Quo circa longe magis opportunum esse puto, ab his
    deplorandis revocato animo, nostra ingenia, vires et studia in hoc unum
    enixissime transferere . . ."
22. Ibid.: "Quis enim ignorat, universalis etiam ecclesiae praesentes fluctus,
    quos cernimus, suam hinc originem ac radicem deduxisse, et si natura com-
    paratum est, eadem via recessum fieri, qua fuerit accessus factus, intelligere
    iam potestis, unde nobis exordiendum sit, quo principio, qua fronte mor-
    bum hunc diluere conveniat . . ."

ican "religion", warned against deviations from the Rule, and deplored the present condition of studies. There were even hints that the crisis was causing an erosion of the order's position within the church. Thus, for example, the ruling from 1615 that in every province one or two houses should be set aside for purposes of study was mentioned in connection with the privileges which the Dominicans enjoyed on account of their educational activities.[23] The ruling reflects a need to justify these privileges by realising their vocation in practice. The cure, however, was always conceived of in terms of a restoration, a returning to the glorious past. And the explicit criticism was entirely channeled within. The Dominican vocation in the outside world was still conceived of in terms of protecting the church by preserving the purity of doctrine and persecuting its external enemies, the heretics. This was the task transmitted to them by their forefather, the founder of the holy Dominican order. Never would they renounce the glory of this calling, which would perpetually be reconfirmed.[24]

The history of the order, chiefly written from the point of view of its leaders and largely based on their notes,[25] may provide a broader perspective for interpreting the sense of crisis manifested in the opening addresses to the General Chapters. The second half of the sixteenth century was a period in which a serious attempt to reform the Dominican convents failed to achieve practical results. By the end of the century study was

23. Ibid., 1615, p. 243: "Quia multa privilegia nostro ordini eiusque provinciis a summis pontificibus conceduntur et singularia pro iisdem provinciis [sic] a magistris ordinis per necessaria praescribuntur, ut huiusmodi litterae vel apostolicae vel magistrorum ordinis perpetuo conserventur, mandamus omnibus pp. provincialibus, quatenus de concilio patrum suarum provinciarum, prout necessitas exposcet, unum vel duos conventus primarios pro huiusmodi litteris conservandis designent, in quibus conventibus omnes provinciales, finem sui officiis imponentes de scitu patrum provinciae tales litteras reponere tenebuntur . . ."

24. Ibid., p. 239: "Non est abbreviata manus Domini neque miraculorum continuatione fatigata, sed ut mirabilia audivimus facta ab initio in institutione ordinis et patres nostri annuntiaverunt nobis, cum sanctissimum patriarca noster s. Domenicus exigua manu primorum patrum, haereses debellandas suscepit et turbato orbi et ecclesiae pacem miraculis et praedicatione confirmavit, ita etiam non paucos his temporibus ordinis nostri patres et fratres esse voluit, in quibus ob eximiam, quam profitentur et vitae et doctrinae puritatem, ordo noster nihil plane sui splendoris fluxu temporum amisisse, sed plurimum sibi de novo conciliasse comprobetur . . ."

25. R. P. Mortier, *Histoire* . . .

confined to a small elite within the order, an elite whose privileges had traditionally been derived from its educational task, and who had vested interests in studies. This elite was generally reluctant to introduce reforms into the Dominican convents. At the same time, education on a primary level was largely neglected. As the role of educating the Catholic masses was gradually transferred to the Jesuits, who constituted a new, vigorous teaching elite, the Dominicans themselves tended to send their novices to Jesuit schools,[26] which provided up-to-date pedagogical methods and higher standards than any local Dominican school.

Many of the *Acta* of the General Chapters with regard to education gain significance in this broader context. In 1558, a ruling confirmed that in every convent in which there were young people there would be a lecturer to instruct them in grammar and the liberal arts, and also a teacher of casuistry.[27] In 1569, the specific roles of various teachers and educators were carefully redefined.[28] The General Chapters repeatedly stressed the need to provide a decent education before granting a degree.[29] In 1558, 1564, and 1569, the provincial priors were placed under the obligation of visiting convents, observing the lectures and exercises, and punishing those who were neglectful of their duties and seeking to improve their moral character.[30] In 1569, concern was expressed not only about the intellectual level of holders of academic degrees but also about that of preachers. It was decided that anyone who had not studied theology for

---

26. Ibid., p. 325.
27. *Acta*, 1558, p. 8: "Item confirmamus ordinationes infrascriptas factas in praecedenti capitulo, ut conformiter ad constitutiones nostras in omni conventu, ubi sunt iuvenes, sit aliquis lector, qui eos doceat grammaticam vel artes iuxta eorum capacitatem. Qui lectores ita tractentur, ut eidem officio incumbere possint. Sint etiam in conventibus lectores casuum conscientiae, et assidue de ipsis collationes fiant . . ."
28. Ibid., 1569, pp. 86–87.
29. Ibid., 1561, p. 29: "Inchoamus hanc, quod in capitulo quartodecimo de studentibus distinctione secunda ubi dicitur: nec aliquis magistretur in theologia, nisi prius actus exercuerit pro forma et gradu praedicti magisterii per quattuor annos ad minus in aliqua universitate, addatur: vel legitima exercitia fecerit in lectura et aliis actibus, quae in eadem universitate in qua promoveri debet, de iure fieri consueverunt . . ."
30. Ibid., 1569, p. 91: "Ordinamus et mandamus in meritum sanctae obedientiae reverendis provincialibus et visitatoribus, ut in eorum visitationibus diligentiam adhibeant ad inquirendum de solicitudine lectorum tam in legendo quam in aliis exercitiis literariis exercitatione, et quos negligentes repererint, corrigant et puniant, secundum quod iudicaverint eorum demerita . . ."

at least two years would not be permitted to preach, and that anyone disobeying this rule would be severely punished.[31] An attempt to institutionalize these tendencies is reflected in the ordinance of 1571, in a text entitled "De studiorum promotione ordinationes". The necessity to select suitable candidates for study was officially confirmed by a ruling specifying the proper method of selection. It was to be carried out by the provincial prior or the prior of the convent in consultation with the fathers. These had the right to vote on the decision of accepting students or on the dismissal of unsuitable ones.[32] The ordinances stated with an unusual insistence that the practice of giving certain candidates "preference" for reasons other than their aptitude for study was totally unacceptable. Similarly, those with tasks in education – regents, magisters, and lecturers – were not to be chosen by "preference" but solely on account of their talents.[33]

The profusion of ordinances concerning standards in teaching and study manifest a set of attitudes which affected the internal and external policies of the order. Internally, there was a tendency to close ranks, to focus the group around a common body of doctrine, and to distrust other groups. One of the most striking examples of this phenomenon is to be found in a ruling from 1556 forbidding studies in any universities which had not received the explicit approval of the master of the order. A relatively short list of universities throughout Europe was drawn up; studies in any other places or certificates from other institutions were not considered valid and did not give its possessors the right to serve as educators

---

31. Ibid., p. 93: "Ordinamus ut nullus nostri ordinis frater ad verbum Dei pro sacra concione publice ad populum tractandum admittatur, neque etiam in oppidis, qui theologiam per duos ad minus annos non audierit; cuius oppositum facientes poena gravioris culpae puniantur . . ."

32. Ibid., 1571, p. 132: "Ordinamus imprimis ut iuvenes ad studia mittendi non praeferantur iure antiquitatis, sed eligantur potius a priore provinciali vel conventuali de consilio patrum et illi prae caeteris assumantur, qui probatis moribus fuerint et qui praevio examine magis idonei fuerint inventi. Item quod in quolibet conventu, in quo viget studium generale, omnes magistri, etiam actu non regentes, vocem habeant in admissione vel studentium formalium vel quoruncunque aliorum fratrum, qui ad studia mittuntur, et simile suffragium habere volumus in emissione studentium, cum ad proprius conventus reverti oportuerit . . ."

33. Ibid., p. 133: "Item regens, baccalaurei, magistri studentium alii que lectores non iure antiquitatis assumantur nec ad determinatum tempus assignentur, sed a reverendis provincialibus cum diffinitoribus constituantur, qui magis idonei fuerint ad discipulos instituendos . . ."

in the order.[34] This list was not included in the *Constitutiones* of 1556, but it appeared in those of 1607 and was expanded in that of 1650.[35]

At the same time, the awareness of an increasing discrepancy between ideals and practice only sharpened an extremism towards the outside world. Many resolutions were concerned with the need to fight heretics and to persecute them everywhere.[36] This "cleansing of the ranks" began first of all at home. If someone was found in possession of a heretical book, he had to make an abjuration and was suspended from all his functions for at least a year. Anyone who was suspected of heresy, of friendship with heretics, or of a readiness to talk to them, was forbidden to preach and had to be delivered to the provincial priors until cleared of the accusation. If the suspicion was found to be justified, he was suspended for a long period, and if guilty of heresy, he was never forgiven.[37]

The Dominicans also adopted an extreme position in the war against the heretics outside the walls of the convent. The ordinances recommended an attitude of outright hostility towards anyone suspected of compromise. Moreover, anyone who seemed ready to compromise was regarded as a sinner and was punished by being dismissed from his task.[38]

---

34. Ibid., 1558, p. 7: "Item confirmamus determinationem conventuum et universitatum studii iam per duo capitula praecedentia approbatam, extra quas universitates nolumus aliquem fratrem nostri ordinis cursum facere seu graduari posse; et si secus factum fuerit, totum sit irritum et inane . . ."
35. *Constit.*, Venice 1607, p. 123; Rome 1650, pp. 257–258.
36. *Acta*, 1561, p. 34; 1569, p. 82; 1571, p. 120; 1650, p. 240.
37. Ibid., 1561, p. 34: "Ordinamus et inviolabiliter observari volumus, quod apud quemcumque inventus fuerit liber haereticus, omni excusatione explosa compellatur ad abiurandum haeresim et per annum ad minus suspendatur ab officio praedicandi et omni executione sacrorum ordinum. Et suspectis de haeresi sive ex familiaritate cum notatis sive ex stilo dicendi et loquendi sive ex omissione interdicitur praedicandi et celebrandi potestas, donec ad arbitrium provincialium se purgaverint vel satisfecerint. In eandem vero suspitione relapsi non possint restitui nisi per duo capitula provincialia; tertio vero relapsis in eadem suspicione nulla fiat perpetuo relaxatio . . ."
38. Ibid., "Ordinamus et praecipimus singulis praedicatoribus, ut aperte et animose pro fide loquantur contra haereses, servata tamen modestia, et teneantur proponere ac asserere dogmata fidei et ecclesiae Romanae, quoties locus inciderit, praecipue per quadragesimam. Quod si quis comprobetur omissione notabili peccasse, pro hac sola culpa privetur officio celebrandi et praedicandi . . ."

The Organisation of Studies

Traces of the effects of the policy laid down by the General Chapters upon the contents and organisation of Dominican education can be seen in the *Constitutiones* of 1607 and 1650, in comparison with those of 1566. The modifications taking place from one version to another testify to subtle changes in orientation which may be interpreted as the peculiarly Dominican reaction to the demands posed by the era of the Counter Reformation.

The *Constitutiones* of 1566, quoting the *Acta* of the General Chapter of 1305, reasserted that the curriculum of the Dominican schools of higher education (the *studia generalia*) consisted of two cycles, philosophical and theological, each lasting from four to five years. The course of philosophy was divided into two main parts: logic and Aristotelian natural philosophy. The following ruling of 1566 provides a clue to the irrelevancy of these studies to the intellectual problems of the sixteenth century: nobody, it was stated, could be accepted as a teacher of logic if he had not first attended a two-year course in the "new logic".[39] A preoccupation with thirteenth-century developments in the field of logic is further evidence of a long period of resistance to change, lasting until the mid-sixteenth century. The disappearance of this ruling from subsequent versions of the *Constitutiones* reveals the process of adaptation which the Dominican educational system underwent between 1566 and 1607.

According to the *Constitutiones* of 1566, the theological course was to be based on the *Sentences* of Peter the Lombard. Nevertheless, reference was made to repeated *Acta* of the General Chapters (1309, 1313, 1329, 1342) which recommended the reading of Thomas Aquinas's theological works. Moreover, these *Acta* also required a solution along Thomist lines to any question which arose in academic disputation.[40] The *Constitutiones* of 1607 marked a change in the Thomist policy of the order. They incorporated the *Acta* reconfirmed in the General Chapter of 1558, and replaced Lombard's *Sentences* with Thomas's *Summa theologiae* as the

39. *Constit.*, Rome 1566, p. 83: "... quod nullus ad legendum logicalia admittatur nisi prius audierit logicam (novam) duobus annis et duobus aliis annis naturalia ..."

40. Ibid.: "Declaramus et districte ingiungimus lectoribus et sublectoribus universis quod doctrinam sancti Thomae de Aquino quod ceteris sanior est et communior et toti mundo ac ordini nostro utilis prosicua et honorabilis legant et secundum ipsam determinent et in eadem scholares suos informent et studentes in ea diligenter studeant ..."

basic text in theology.[41] In addition, this version of the *Constitutiones* also stated that it was not enough to recommend Thomist doctrine, but that anyone deviating from it in writing or by word of mouth was to be punished by dismissal from the post of lecturer and a suspension of the privileges connected with that position.[42]

The Dominican educational reforms, however, were not limited to the buttressing of Thomas's authority. In addition, the *Constitutiones* of 1607 and 1650, and the *Acta* of the General Chapters in those years also testified to a number of systematic steps towards a reorganization of the educational system in the second half of the sixteenth century and the first two decades of the seventeenth century. These steps were institutionalized in two sets of ordinances: the "Ordinationes de studiorum promotione" of 1571,[43] and the "Pro reformatione studiorum" of 1651.[44]

A desire for spiritual renewal was strongly manifested in the ordinances of 1571. With regard to the means most likely to bring about that renewal, however, there was an oscillation between two different tendencies. One reflected the pressures to adapt to intellectual change, and involved the selective introduction of humanistic teachings, especially in the field of ancient languages. The other, opposite tendency, was a wish to preserve the uniqueness of the Dominican tradition through a stress on theological studies.[45]

The ambivalent attitude towards humanistic literature was revealed in the novel interpretation given to that section of the *Constitutiones* which forbids studying the books of the heathens ("In libris gentilium . . ."). In this interpretation, an initial suspiciousness towards the works of the humanists, judged to be full of wickedness and moral corruption, was mingled with a recognition of the necessity of learning ancient languages. The reading of the works of Erasmus was expressly prohibited, and the

41. *Constit.*, Venice 1607, p. 119: "Item in cap. Salamantim, 1551 ordinatum fuit quod in Theologia legatur totus articulus S. Thomae . . ."
42. *Acta*, 1558, p. 8: "Advertant etiam diligentissime iuxta ordinationes tot capitulorum generalium lectores et fratres, ne a solida sancti Thomae doctrina recedant, novitatem sectantes. Qui vero contra praefati doctoris doctrinam verbo vel scripto temere aliquid dixerit, ab officio lectoratus depositus sit ipso facto eoque perpetuo sit privatus, super quo provinciales diligentissime invigilent . . ."
43. Ibid., 1571, pp. 132–133.
44. Ibid., 1615, pp. 245–249.
45. Ibid., 1564, p. 63: "Item ordinamus, ut in promovendis studiis nostris sollicitos se praebeant reverendi provinciales et in eis reformandis sedulo invigilent provideantque in primis, ut temporis ad studendum praefixi maior pars theologiae quam artibus impendatur . . ."

argument that they contained useful knowledge was rejected. On the other hand, a selective study of Greek and Latin texts was allowed, on receipt of special permission from the priors.[46] At the same time, a desire was also expressed to renew the splendid Dominican tradition of theological studies and to adapt them to the needs of the period. Thus, a proposal was made to divide the theological course into two branches of study: speculative theology, based on the writings of Thomas, and moral theology, based partly on Thomas and partly on Peter the Lombard.[47] The introduction of a distinction between the two branches signaled the emergence of a new discipline, moral theology. In spite of the initial decision to base the teaching of the new discipline on authoritative texts, already in 1615 there was added to these texts a rather modern work by Thomas de Vio Cajétan (1468–1534), who added to his commentary on Thomas eighty-two questions on practical matters relating to Catholic morality.[48] The typically humanistic interest in the practical aspects of life thus succeeded in impressing itself on the most traditional and authoritative field of knowledge, namely theology.

A decision was also made to shorten the philosophical course to a period of three years. This decision may be taken to represent a response to criticism, originating during the Council of Trent, of an exclusive emphasis on speculative knowledge as a preparation for scholastic theology. The course was now divided into three parts, each of which was devoted to one year of study. The first year was dedicated to the study of "dialectics", including not only formal logic but also modes of thought which could be applied to matters connected with practical experience. The relevancy of this part of logic to casuistry and moral theology is obvious. The second year was dedicated to natural philosophy and the third year to metaphysics.[49] Although the pressures to assimilate humanistic contents did

46. Ibid., 1569, p. 92: ". . . exequutioni mandetur, ne praetextu bonarum, quas vocant, literarum et politioris linguae Erasmi libros ad consimilium habeant et legant, unde mala dogmata moresve pravos imbibant. Graecam autem linguam autem Haebraicam non addiscant sine peculiari sui provincialis licentia in scriptis habenda, in qua fiat mentio expressa praeceptoris, a quo talem linguam addiscere debeant . . ."

47. Ibid., 1571, p. 133: "Sacrae vero theologiae quadriennii, nisi aliter reverendo provinciali de consilio magistrorum et officialium studii visum fuerit; quo tempore ita distinguantur lectiones, ut altera sit speculativa ex prima parte vel prima secundae divi Thomae vel etiam de incarnatione, altera vero moralis de secunda secundae vel IV. sententiarum . . ."

48. Thomas de Vio Cajétan (1468–1534); see: *Dictionnaire de Théologie Catholique,* Paris 1910.

49. *Acta,* 1571, p. 133: "Tempus triennii dialecticae, phylosophiae et metaphysicae deputetur . . ."

not find full institutional expression in 1571, a constitution was nevertheless passed requiring all students to engage in regular "literary exercises"
in order to preserve the quality of their Latin, which remained the language of preaching.[50]

The ordinances of 1615 confirmed all the tendencies to reform mentioned above. From the institutional point of view, they reflected a concern for the quality of teachers and pupils and contained certain measures
for controlling the standard of studies. In order to be accepted for a
course of study, for instance, one needed to have a written recommendation.[51] Before students were promoted, they were subjected to a double
examination – of their quality of life and observance of religious duties,
and of the quality of their learning.[52] There was no real change in the
intellectual contents of the courses, but teachers were instructed to abandon the various ideas of different philosophical schools and to interpret
the Aristotelian text according to the interpretation accepted by the fathers of the order, and were threatened with dismissal if they failed to
comply.[53]

It would be hard to exaggerate the significance of this brief ruling. In
fact, it was symptomatic of the crystallisation of a body of doctrinal
knowledge, not only in theology but also in philosophy. It also expressed

50. Ibid.: "Item ordinamus, quod omnes studentes et studii officiales non nisi
      latino sermone loquantur et scribant etiam extra litteraria exercitia et li-
      teras, quas conscribere eos contigerit, non vulgari sermone, sed latino
      deinceps conscribant . . ."

51. Ibid., 1615, p. 247: ". . . neque studentes tam artium quam eiusdem sacrae
      theologiae in aliis externis provinciis ad lectiones admittantur, nisi a suis
      lectoribus, quorum primo erant discipuli, de moribus et qualitate studiorum
      in scripto testimonium habeant . . ."

52. Ibid., p. 248: "Deinde vero, quoniam in diffinitorio generalis capituli pluri-
      mae quaestiones scrupulosae circa requisita promovendorum ad gradum
      baccalaureatus et magisterii oriuntur, ita ut etiam et proponentes diffinitores
      capituli generalis, et admittentes in generali diffinitorio, merito pavore vel
      quaerelarum, si propositi non promoveantur, vel conscientiae, si indigni
      praemiis donarentur, perhorrescant, idcirco huiusmodi difficultates ampu-
      tare volentes, praecipimus patribus provincialibus et diffinitoribus capitu-
      lorum provincialium, ut in suis petitionibus omnium promovendorum aeta-
      tem, mores religiosos, eruditionem, lecturae et meritorum qualitatem et
      quantitatem temporis lecturae ac studiorum exprimant, ut saepius ordina-
      tum est . . ."

53. Ibid., p. 246: ". . . omnibus lectoribus et studiorum moderatoribus sub
      praedictis poenis mandamus, ut varietate auctorum relicta, in philosophicis
      quidem Aristotelem, eius textum commentariis patrum nostri ordinis expli-
      cando . . ."

a desire to attach the intellectuals of the order to the Thomistic world-picture in its entirety, at a period when in the outside world a serious questioning of its chief philosophical principles was taking place. This form of Thomism, officially imposed by the order upon teachers and students, may be defined as "doctrinal" Thomism, to be distinguished from other forms of Thomism which also grew out of the endeavours of the Tridentine Church to base its cultural policies upon a Thomistic orientation. The decision to commission certain fathers to compose readers of scholastic theology, moral theology, and the Scriptures for the use of the teachers of the order[54] further demonstrates the tendency of the Dominicans to isolate themselves from other theological currents. Also, the failure of the ordinances of 1615 to reflect any significant assimilation of humanistic tendencies suggests that it must have been a peak period of suspiciousness and self-isolation.

The Dominicans, however, were not completely impervious to educational renewal. A short while afterwards, in 1628, a constitution was passed to set up in every *studium generale* of philosophy and theology a chair of ancient languages, especially Greek. A person was put in charge of these studies, rules were established, and it was stated that teachers and pupils were to be encouraged by a rapid advance in academic status.[55]

Despite the sense of crisis in the sphere of education, the Dominicans of the sixteenth century clearly attached a supreme importance to intellectual activities. They had inherited a splendid tradition which regarded the

54. Ibid., p. 249: "Mandamus omnibus provincialibus, praecipue in regnis Hispaniarum, ut selectis aliquibus patribus doctioribus committant, quatenus libros theologiae scholasticae, moralis et sacrae scripturae scribant . . ."

55. Ibid., 1628, p. 357: "Denuntiamus, ex decreto sanctae congregationis de Propaganda Fide habitae die 11 aprilis 1625 instituenda esse studia linguarum a sacro Viennensi concilio et a fel. record. Pauli V. constitutione praescripta et aliarum etiam, quae magis ex usu esse videntur, ut Illyricae et Graecae vulgaris. Quare ex eiusdem decreti praescripto districte mandamus reverendis patribus provincialibus sub poena absolutionis ab officio, ut intra quatuor menses ab harum notitia unusquisque ipsorum in aliquo seu in aliquibus conventibus, in quibus artium et theologiae studia vigent, certum religiosorum magis idoneorum numerum constituant et linguarum praedictarum studiis addicant sub aliquo religioso, ubi studii praefectus non extiterit, qui studiorum huiusmodi curam et solicitudinem habeat; qui ut muneri suo serius incumbat, iisdem privilegiis, quibus magistri et studiosi, fruatur. Certae autem regulae et constitutiones, quibus numerus et species linguarum, tempora lectionum, studii repetitionumque ac alia huiusmodi a

life of study as the means to fulfill the individual and social vocation of members of the order. Facing the need both to improve the level of studies in their schools and to meet the demands of the church in educating the people, they oscillated between two trends. On the one hand, they expressed the wish to return to the golden age of the Dominican past by concentrating on sacred fields of study, metaphysics, and theology in particular. On the other hand, pressure was also felt to regenerate spiritual life by the gradual absorption and adaptation of new contents to religious concerns. Consequently, the Dominicans responded by allowing selective reading of humanistic literature and by introducing the study of Greek and Hebrew into the curriculum.

Basically, however, the Dominican cultural orientation remained anchored in the medieval outlook and its emphasis on the *vita contemplativa*. Institutionally, they favoured isolation from the secular world and its influences. Religious fanaticism mounted as they ordered aggressive tactics against anyone suspected of heresy. Intellectually, studies remained oriented towards the contemplative life. Freedom of opinion was curtailed as they imposed strict adherence to Thomistic theology and under severe penalty prohibited any deviation from Thomistic philosophy.

The Dominicans transformed Thomism from a religious-intellectual system into a binding doctrine, and in so doing conferred on it a status which it had never enjoyed in Catholicism before the Counter Reformation. Seen against this background, it is easier to understand the success of an alternative intellectual elite, the Jesuits, who were engaged in creating a much more open religious-intellectual establishment in the same years during which the Dominicans sought to renew their spiritual life while jealously clinging to the closed forms of monasticism.

---

magistris et studiosis diligenter observanda praescribantur; quarum constitutionum observandarum cura praedicto religioso vel studii praefecto committatur; qui si suo deerit officio, pro modo culpae puniatur. Decernimus autem, ut praedicti magistri et studiosi respective ad honores prae ceteris promoveantur, et si in suo munere negligentes fuerint, iis, quibus gaudent honoribus, priventur et ad alios via et spes praecludatur.

6

The Jesuits

An Alternative Intellectual Elite

The need to create an active elite, capable of transposing doctrinal principles into a mode of action, shines through the language of the bull *Regimini militantis,* which announced the founding of the Society of Jesus in 1540.[1] This need may be partly explained by the relative inability of the Dominicans to adapt themselves to the demands of the post-Reformation era. At the same time, it also forms the relevant context for understanding the Jesuits' original way of coping with the same demands through the invention of entirely new strategies.

My account of the Jesuit cultural orientation begins with distinguishing their originally apostolic and practical justification of studies from the traditional "intellectualist" and instrumentalist arguments of the Dominican tradition. To an even greater extent, however, the "modernity" of the Jesuits was expressed in the development of an educational "discourse" which exceeded the limits of practical needs or the justification of studies in practical terms. The changes in the terminology purporting to the educational vocation testify to the ideological importance attached to the society's increased concern with education as its main field of action. A further attempt to justify the educational activities by presenting knowledge as a bridge to salvation involved the adoption of a Thomist framework of thought but led to a special interpretation of Thomism.

The Ideology of Educational Mission

The small group which gathered around Ignatius Loyola (1491–1556) in the 1530s was notable for its mixture of men of action and intellectuals who placed apostolic principles of Christian living above theory and study. "The mission of this order", reads the papal bull:

1. Paul III, "Regimini militantis ecclesiae . . .", 27 September 1540, *Bullarium Romanum,* Torino 1860.

exactly like that of the apostles, extends throughout the entire world, to all peoples, to Christians as well as unbelievers. Moreover, its mission is also universal in that it embraces all apostolic missions, both those that belong specifically to the priesthood and those which are concerned with works of charity.[2]

Ignatius himself did not see education and learning as the proper goal of the order, which he thought should be chiefly directed to more urgent tasks such as preaching, practical charity, and missionary activities in Europe and beyond. He therefore found it necessary to state in the first version of the Constitutions of the order that no studies would be carried on or lectures given within the framework of the society.[3] Nevertheless, a short time after the society was founded the educational question arose in connection with the need to train confessors, missionaries, and preachers. Despite his suspicion of intellectualism as a way of life, Ignatius was conscious of the necessity for a solid educational grounding for novices. In a letter of 1542, he expressed the hope that members of the order would increase their knowledge of the Latin language and the liberal arts.[4] The realisation of this hope depended on sending selected groups of priests to the various European universities. A first group was already sent in 1540 to the University of Paris. A year later, a group was sent to the University of Padua, and two other groups were sent to Louvain and Coimbra in 1542.

Very soon, however, the Jesuits discovered that the universities did not provide an education suited to the needs of the society and the aims which it had set itself. An early letter of Polanco, one of the founding fathers later to become the secretary of the society, expressed doubts concerning the philosophical and theological teaching provided at the University of Padua and questioned its effectiveness.[5]

Only a few years later, however, the Jesuit *studium* was already organising its teaching activities. Constitutions were written,[6] referring to exercises and disputations as well as to the rules of Christian behaviour which the students were expected to observe.

2. Paul III, "Regimini militantis ecclesiae . . .", 27 September 1540, in L. Lukács (ed.), *Monumenta Paedagogica Societatis Iesu*, Rome 1974, II, Introductio, p. 1. Other volumes of the *Monumenta Paedagogica* used in this book are: I, Rome 1965; III, Rome 1974; IV, Rome 1981, henceforth *MP*.
3. *MP*, I, Introductio, p. 6, n. 23: ". . . No estudios ni lectiones en la compañia . . ."
4. *MP*, I, p. 7, n. 4: ". . . mucho deseo los yzi ésedes fondar mucho bien en latin, y despúes en sus cursos de artes enteramente, sin azer quiebra alguna . . ."
5. *MP*, I, p. 358, I. A. de Polanco to O. I. Laínez, Padua, 18 May 1542.
6. *MP*, I, pp. 3–17, Constitutiones scholasticorum S. I., Padua 1546.

The *studium* in Padua was established within the confines of the Jesuit house, where teachers and pupils lived together, and lessons were occasionally provided. This was a departure from the original idea according to which Jesuit students were to live in their own separate residence, bind themselves to the triple monastic vow (poverty, chastity, and obedience), but would leave their residence to go and study at a university. The Constitutions of 1546 attempted to explain the reasons why teaching activities were carried on within the precincts of the common residence despite the initial rejection of this practice. It was claimed that the teaching was necessary to ensure the acquisition of knowledge needed for the practice of confession and preaching.[7] Nevertheless, the Constitutions recoiled from allowing a full engagement in teaching activities and insisted that, as a rule, lessons, whether public or private, were not to be given in the Jesuit residence, and that this was only to be done occasionally and with special permission.[8]

Similar doubts and apprehensions concerning the place of intellectual concerns in an apostolic vocation were expressed in the texts of the founding fathers in the first years of the society. Ignatius, in a letter of 1548, warned that one should guard against the arrogance and ambition that can result from frequenting universities.[9] Other documents state that study should not be made into an end in itself.[10] They stress that one should learn only what is relevant to the purposes of the society.[11] Criticism of the difficult and obscure aspects of scholasticism so typical of the humanists is also present in these texts.[12] All these comments seem to imply a reaction against the intellectualistic bias which dominated the

7. Ibid., p. 6: ". . . per bene imparare tutto ciò che è necessario per confessare esshortare et predicare la parola di Dio per la salute spirituale di molte anime . . ."
8. Ibid.: ". . . in la casa non si legga lettione alcuna in publico, nè in secreto; tamen alcuna volta rara per ricuperare alcuna lettione persa, con licentia del superiore, potrebbe leggersi alcuna lettione per ricuperatione della persa, non tamen che si perdi alcuni di fora".
9. Ibid., p. 376, Ignatius de Loyola to A. de Araoz, 3 April 1548.
10. *MP*, II, p. 49, I. Laínez, Regulae scholarium externorum: ". . . non ut scientiam solum aut opes et honores adipiscantur, sed ut veritatis cognitione sibi et aliis ad honorem et gloriam Dei opitulentur . . ."
11. *MP*, I, p. 48, I. Laínez, De modo studenti philosophiae tractatus: "Quanto ale lettioni odite no ho da dirvi altro, se non che pilliate quello trovarete più a proposito . . ."
12. Ibid., p. 368, I. A. de Polanco to I. Laínez, Rome, 21 May 1547: ". . . y es ser regalados, usar y habituarse a no entender sino en cosas fáciles y sabrosas; y así facilmente los spanta o enjoa el tratar cosas, en que se hallan las

thought and actions of the traditional elite. Repeated warnings of the need to preserve one's purity of motive while studying appeared in the constitutions of all Jesuit colleges. They reveal an acute awareness of the dangers inherent in intellectual activity, liable to distract the student's attention from his main goal, the glorification of God and the concern for the salvation of souls.[13]

Yet, despite some hesitations and misgivings concerning the centrality of teaching and study, the 1550s and 1560s saw a tremendous expansion of Jesuit educational activities.

In 1551 the Roman College was established in Rome.[14] It originated with a group of fifteen students and three teachers who were inspired by the humanist idea of a trilingual school aiming to provide a thorough grounding in the ancient languages as the basis for all other learning. They were imbued no less with a desire to return to the original Christian sources and to revive the Catholic faith by reading them, explaining them, and disseminating the doctrine derived from them. In a few years the Roman College became one of the best institutions of higher learning in Europe. On the basis of the courses in languages and the humanities, an arts faculty was created which provided a philosophic-scientific education, and, above it, a faculty of theology.[15] The Roman College thus came to be acknowledged as the Jesuit university. It competed successfully with the University of Rome[16] and became a source of pride for the entire Catholic world. In 1554, Polanco wrote to F. Borgia:

> The importance I attach to these institutions, and especially the college, not only for the whole society but also for the entire church, is such that I do not know of any others like them in the entire Christian world.[17]

Jesuit colleges were created by the dozens in the 1550s and the first half of the 1560s, and their numbers grew to hundreds by the end of the

qualidades contrarias, de dificultad, y desambrimiento, como vemos en las artes y theología scholástica . . ."

13. MP, I, p. 66, Regula rectoris Collegii Romani: "Tutti si sforzino di haver la intentzione recta non cercando le lettere per altro fine che della divina gloria et aggiuto delle anime; et così si risolvino de studiare da vero con grande diligentia per conseguire il fine detto, et satisfare alla santa obedientia, et nelle sue orationi dimanderanno gratia di aggiutarsi e fare profetto nelle lettere . . ."

14. R. G. Villoslada, SJ, *Storia del Collegio Romano*, Rome 1954.

15. The privilege of endowing academic titles was given to Ignatius of Loyola by Pope Paul 4th on 17 January 1556; see Villoslada, *Storia* . . . , p. 34.

16. Villoslada, *Storia* . . . , pp. 39ff.

17. Ibid., p. 15.

century.[18] In most of them, the studies remained along the lines of the humanistic education which was provided in the trilingual colleges. The Jesuits were among the first educators in Europe to adopt the idea of dividing classes according to ages and subjects, and were responsible for many other pedagogical innovations such as the use of physical exercises, the theatre, etc.[19] A few of the colleges became genuine universities with a faculty of arts (philosophy) and a faculty of theology. Quite apart from the establishment of Jesuit universities, however, the members of the order succeeded in penetrating the theological faculties of the existing universities and exerting much influence there. This happened first at Ingolstadt, and subsequently in Vienna and other places.[20]

The educational activities of the Jesuits soon extended beyond the boundaries of the various local colleges and assumed a more comprehensive character. In 1552, Ignatius Loyola, together with Cardinal Giovanni Morone,[21] had the idea of founding a college in Rome for the training of cadres of priests from the German provinces and the countries close to Germany (Poland, Bohemia, Hungary), with the special aim of combatting heresy. Thus, the German College came into being – an institution which, for teaching purposes, was dependent on the Roman College. It was a boarding school where the students lived in a separate building, were drilled by Jesuit instructors in exercises and repetition of the material, and were trained according to the moral and spiritual norms of the Jesuits. The institution, which soon became a centre of education for the priesthood from every part of the Catholic world, received the blessing of Pope Julius III. It was supported financially by the pope and the College of Cardinals, who contributed money for the purpose, each according to his capacities or as he saw fit.[22]

The success of the Jesuits in the sphere of education also attracted many lay students to their institutions. Pupils who lived in Rome and attended the Roman College studied there for free. Already in 1558, moreover,

18. L. Lukács, "De origine collegiorum externorum deque controversiis circa errorum paupertatem obortis (1539–1608)", AHSI 29 (1960), pp. 189–245; 30 (1961), pp. 3–89.
19. J. B. Herman, SJ, La Pédagogie des Jésuites au XVIe siècle. Ses sources, ses caractéristiques, Paris 1917; A. P. Farrell, SJ, The Jesuit Code of Liberal Education, Milwaukee 1938; G. Codina Mir, S.I. Aux sources de la pédagogie des Jésuites. Le "Modus parisiensis", Rome 1968: D. de Dainville, SJ, L'éducation des jésuites au XVIe et XVIIe siècles, Paris 1978.
20. See, for example, MP, I, pp. 518–519; 400; 533; 593.
21. Villoslada, Storia . . . , p. 24; G. E. Ganss, SJ, Saint Ignatius's Idea of a Jesuit University, Milwaukee 1956, p. 23.
22. Villoslada, p. 25.

secular boarding students who did not see themselves as destined for priesthood were accepted into the German College. In that same year, the rule was adopted in the Jesuit college at Coimbra that these students were to live separately from other students.[23] These, unlike the cadres of novices for the priesthood whose expenses were entirely paid by the college, paid for their tuition, food, and all other expenses.

In the 1560s, the Jesuits took yet another kind of educational institution under their wing. These were the local seminaries which, to a large extent under the influence of the Jesuit delegation to the Council of Trent, had been decreed by the Tridentine legislation.[24] The first one was the Roman seminary founded by the pope who, after consultation with the cardinals, decided to put it under the control of the Jesuits. This institution also relied for its teaching on the Roman College. Its pupils lived separately in their own residence in accordance with rules carefully formulated by the Jesuits,[25] who devoted much thought and energy to their application.

The attitude of the earliest members of the society towards teaching and study and the way in which Jesuit educational activity developed indicate that it did not stem from any emphasis on the intellectual aspect of faith. On the contrary, it was in the name of the *vita activa* that the Jesuits proceeded to train confessors and preachers. Their tremendous success in that field then led them to develop institutions for the education of ever-wider circles in the Catholic world, ranging from schools for young people, via universities which trained a local Jesuit priesthood together with Catholic laymen who studied there, to the central seminary for priests from the provinces and the local seminaries for training the secular priesthood.

The importance which educational activities had assumed for the Society of Jesus is demonstrated by the need to differentiate between two kinds of Jesuit houses. One was called the *domus professae* and involved all those who had completed their studies and were directly engaged in the saving of souls. Then there were colleges made up of teachers and students working together in the all-important educational vocation.[26] In the hierarchy of the order, the *domus professae* was regarded as the higher

23. *MP,* II, Introductio, p. 23.
24. J. A. Donhoe, *Tridentine Seminary Legislation: Its Sources and Formulation,* Louvain 1957.
25. *MP,* II, p. 379 ff., Constitutiones seminarii romani.
26. *MP,* I, pp. 395–396, Informatio de collegiis Societatis Iesu: "... c'è differentia fra li collegii et case della Compagnia, perchè li collegii sono ordinati propriamente per gli scholari, acciò imparino le lettere, che sono necessarie

institution and had the task of supervising the colleges. The quantitative relationship between these two kinds of institutions, however, explains the preponderant influence of the colleges in the formation of the society's policies. The following figures illustrate this point: between 1540 and 1556 – that is, in the first sixteen years of the society's existence – only two *domi professae* were founded, as against forty-six colleges. This proportion continued to be maintained in the following years. In 1579, the society had 10 *domus professae* and 144 colleges, and, in 1600, 16 *domus* and 245 colleges.[27]

In order to justify this extensive educational activity in terms of the original aims of the society, an educational discourse developed during the first thirty years of its existence which exemplified the Jesuit outlook, mentality, and mode of thought.

Already in the 1540s one could find attempts at a justification of educational activities, differentiating them from the main vocation of the society but seeing them as a task imposed by the requirements of the situation. Father Jay, who had been sent to look into conditions in the German province, came to the conclusion that at least half of the members of the society ought to devote themselves to education. In a letter to Salmerone he wrote:

> For my part, I am of the opinion that although our vocation is not directed towards fulfilling the duty of ordinary university professors or lecturers, nevertheless, in view of the [situation] of extreme need which exists in this unfortunate country, it would be an excellent thing if a few of the members of our society whom God has provided with suitable talents for this duty [*ufficio*], if asked to do so, would respond [to the request] to serve for nothing – that is, without a salary – in this duty [*ufficio*] . . . in this way, it may be possible through training to attract scholars [knowing] the German language to the spirit of holiness so that, when the time comes, one may hope that God, by means of learned, exemplary men, will bring the country back to the church in the same way as Satan swept it into error through the wicked apostates.[28]

---

al'instituto della Compagnia, et hanno intrate per sustentare detti scholari, non si potendo in altra cosa adoperare dette intrate, se no in beneficio delli stessi collegii et scholari. Le case sono per li professi della Compagnia, li quali con le lettere prima imparare attendono ad aggiutar l'anime, et servano povertà in particulare et in commune, non potendo havere alcune intrate nè possessioni. Et questi tali professi, overo il loro preposito generale, hanno la superintendentia delli collegi quanto al modo de vivere, et instructione delli scholari, et quanto al metere et cavare le persone, secondo che si reputa conveniente per il divino servitio . . ."

27. Lukács, "*De origine collegiorum* . . . "
28. *MP,* I, p. 364–365; C. Jay to A. Salmerón, Worms, 21 January 1545.

In Jay's thinking, there could not be a total identification of teaching and study with the main vocation of the society, which was essentially apostolic and moral, not intellectual. It was not knowledge but the religious spirit that was most lacking in Germany, he claimed; it may have even been through the emphasis on knowledge rather than spirit that the country had reached its present unfortunate situation. Yet, for all that, it seemed to him that teaching and education were the best means of drawing people into a cultured environment in which they could imbibe the Catholic spirit which was necessary in order to prevent a slide into heresy.[29]

Two texts written by Polanco at the end of the 1540s and the beginning of the 1550s provide more arguments in favour of educational activities. Polanco, unlike Jay, did not speak of the "duty [*ufficio*] of teaching". He put forward a series of arguments in favour of teaching activities, most of them based on a consideration of the "usefulness" of the knowledge gained with the Jesuits and the education they provided. First of all, there was the usefulness of the knowledge of the ancient languages for an understanding of the Scriptures. Polanco also believed that learning languages was the best means to a systematic development of the intellectual faculties, as these studies are neither too easy nor too difficult, and because they require a rational and concrete approach. Furthermore, this knowledge formed the basis of many other subjects which were needed both by the Jesuits themselves and by others who received their education from them.[30]

In a text of 1551, Polanco distinguished between the usefulness of learning to the Jesuits and the necessity of the knowledge they provided to outsiders. The members of the society could on the basis of this knowledge teach others, serve as preachers and fulfill the tasks of the priesthood, and undertake any task incumbent on "the workers in Jesus's vineyard". As for the external pupils, he claimed, a Jesuit education not only brought about spiritual advance and helped in attaining salvation, but it also served as a basis for every kind of career, within the church or outside it, in the arts, in government, or in the law.[31]

But a need to develop a complete ideology of the educational mission,

---

29. Ibid., p. 363: ". . . perchè questa povra patria più ha bisogno de spirito et mortificazione che de molta scientia; anzi la scientia senza spirito li ha redotti a questa lor disgratia. Tamen per tirarli allo spirito, mi pare seria bello mezo che se potesse nelle università havere persone docte et spirituale . . ."

30. *MP*, I, pp. 366 ff., I. A. de Polanco to I. Laínez, Rome, 21 May 1547.

31. *MP*, I, p. 417, I. A. de Polanco to A. de Araoz, Rome, 1 December 1551: "Que de los que solamente son al presente studiantes saldrán con tiempo diversos, quién para predicar y tener cura de las ánimas quién para el govierno de la tierra y administratión de la justitia, quién para otros cargos . . ."

based on stronger claims than the "usefulness" of knowledge and its practical advantages, clearly arose when voices were heard in opposition to an extension of this activity at the end of the 1550s and during the 1560s. The objections which the members of the society raised were generally directed at a single target: the boarding schools for lay students, which required a tremendous investment of energy in the enforcement of discipline and in the teaching of students whose basic knowledge was very limited. The internal debate among the Jesuits on this subject reached its peak between the years 1565 and 1570, and the position adopted determined to a great extent the subsequent educational style of the society.

Just as the considerable involvement of the Jesuits in teaching originally derived from practical requirements (the need to train confessors and preachers, the need to combat heretics, the need to attract people into a cultured Catholic environment), so was the opposition to this involvement originally based on practical reasons. L. Maggio,[32] the rector of the German College, listed a series of arguments against such involvement in a text he wrote in 1559. He claimed that the novices of the society did not wish to concern themselves with drilling children, an occupation which disturbed them in their own studies. The pupils, for their part, hardly knew anything, and required very basic teaching. The novices were inexperienced and could not cope with disciplinary problems. The result was that the children bore the novices a grudge about things which the latter regarded as a matter of Christian righteousness. Consequently, the society did not gain any cadres from those boarders; quite the contrary: even those who had originally thought of joining the society became disinclined to do so. On the other hand, the society was hurt by the disturbance which this teaching represented to the peace of the novices.[33]

Maggio's attitude to involvement with teaching and education was not entirely negative, however. He did not deny what he called the "universal importance" of Jesuit education in its inculcation of Christian doctrine and a Christian life-style in accordance with ecclesiastical tradition, with regular prayers and attendance at masses, sermons, and confession. Moreover, teaching boarders gained the society the support of the pupils' parents, and thus wider social support. Also, it enabled the society to train a great many members in the sphere of "governo", the art of government and administration.[34]

By the middle of the 1560s, the movement against the teaching of

32. *MP*, II, pp. 799 ff., L. Maggio, Circa collegii germanici convictores relatio.
33. Ibid., pp. 800–802.
34. Ibid., pp. 802–803.

boarders had become an opposition of principle. In a letter of 1564 to General Laínez, a prominent theologian in Portugal, named de Torres, summed up the complaints of his colleagues against the "educational mission" (*ministerio*) which, he believed, had undermined the original purposes of the society and was to the detriment of other matters more directly connected with its original aims.[35] This letter echoed similar feelings among the Spanish Jesuits, including the provincial, A. Roman.[36]

The series of documents from the forties, fifties, and sixties referring to the educational policy of the society testifies to the modification in the approach to teaching activities reflected in the terminology used by supporters and opposers alike. In Jay's letter of 1545,[37] education was described as an *ufficio* – one duty among many others. In the same way, in the texts of the 1560s, much was said about study as an activity, a form of work undertaken on behalf of the society: *opera litteris*.[38] The terminology used by Torres, however, showed a qualitative change in the understanding of the place of education. Now it was no longer regarded as one duty among others or considered simply a form of work. Torres referred to education as a *ministerio* – a mission. This change in terminology bears living witness to the process whereby the educational mission came to be identified almost entirely as the society's main purpose.

One last major text, written by M. Lauretano, head of the German College, summed up the debate over the significance of the teaching voca-

---

35. *MP*, III, pp. 362–365, M. de Torres to I. Laínez, Lisbon, 12 October 1564: ". . . por tener grande temor que vademos en seco en este ministerio, con mucha infamia y descrédito de la Compañia; . . . la esperiencia muestra muchos y, a mi juizio, muy grandes inconvenientes que piden remedio, o por mudança o por otra orden, estrechando más el ministerio; porque, por entender en éste tan exactamente, se debilita el primario y principal instituto de la Compañia que es attender a la salvación de las almas por la predicación, administración de sacramentos y otros ministerios más immediatos a ella que éste . . ."

36. Ibid., p. 389, A. Roman to F. de Borgia, Valencia, 28 January 1566: ". . . y si se entiende (como parece) que ha de ser casa de convictores, y de que haya de tener cura la Compañia, a todos nos ha parecido mucho convenir que en ninguna manera se accepte; porque es una introductión que no sólo no accredita la Compañia, pero parece modo de codicia y de acoger gentes por ganancia; y es negocio muy embaraçoso, y que a todos non parece bien tractar tanto con mochachos . . ."

37. See p. 116 in this chapter.

38. *MP*, II, p. 49, Regulae scholarium externorum; I, p. 531, Excerpta eorum quae Monumenta Paedagogica illustrant in chronico patris Polanci.

tion and the crystallisation of the discourse of an "educational mission" in the 1570s.[39] Providing a summary of the critics' arguments at the beginning, it then proceeded to refute them with ad-hoc, commonsensical counter arguments, and to suggest the long-run political implications of the broadening educational enterprise.

The arguments of the opponents, Lauretano contended, were dominated by the fear that a secular existence in the colleges of boarders was incompatible with the monastic idea of "otherworldliness", of separation from the world: "That religious people like us should become educators of secular youth seems inappropriate", he wrote. "It requires us to leave our residence, to eat and live with them, which is not without a certain danger, or at least a suspicion of the people of ['this'] world."[40]

Instead of concentrating on the education of laymen, it was argued, one could occupy oneself with things closely connected to the service of God.[41] In the process of working with secular people, the Jesuits might become accustomed to "free" behaviour and might risk losing some of their members to the secular world.[42] Moreover, it was feared that living continually in the company of the members of the order might make the pupils aware of the human weaknesses of their educators, which could change them from possible recruits to the order into its enemies.[43] Furthermore, it was said that, in engaging in their secular educational activities, the Jesuits were forced to take lay interests into account. For instance, the high social standing of parents sometimes prevented children from being sent away from the boarding schools, even when they were unsuitable and unworthy of remaining in those institutions.[44] And finally, it was argued that the educational mission of the Jesuits was only a matter of chance, a circumstantial involvement with things of no substance, an accidental concern with activities of doubtful value.[45]

The critics' arguments, as presented by Lauretano, betrayed the profound change in forms of life entailed by the educational vocation of the Jesuits. Life in the *saeculum* involved the educators in the practical concerns of this world and forced them to structure differently the boundaries between themselves and their extern students. Lauretano vehe-

39. *MP*, II, pp. 995–1004, M. Lauretano, Utrum convictus iuvenum nobilium in collegio germanico conservandus sit? Rome 1572.
40. Ibid., p. 1001.
41. Ibid., p. 1002.
42. Ibid., p. 1003.
43. Ibid.
44. Ibid., p. 1004.
45. Ibid., p. 1002.

mently denied, however, the presumed negative consequences of this change. His denial was mainly based on the evidence of experience. Experience had shown the fears of the opponents to be unjustified. The society had not lost any members. And even if there were any risks in dealing with this-worldly matters, the society had to assume them, relying on the spiritual fortitude, faith, and fidelity of its members.[46]

The peculiar mixture in Lauretano's rhetoric of practical, commonsensical arguments and a reliance upon a strong and binding value system recurs in his positive message. It is the political aspect of this vision which determines the ideological tone of the whole text. The educational work which the society engaged in he called "an undertaking which is most important for the whole Christian republic."[47] The Council of Trent, he recalled, had ordered the setting up of seminaries for the training of priests and people who held office in the church. For this reason, the German College was founded not only as a Jesuit seminary but a universal seminary for the whole Catholic world.[48] In that difficult period, moreover, it was not only the priesthood which needed to be educated but the entire Catholic population. The Society of Jesus set about this task first of all through the education of the aristocracy.[49] The German College was universal, not only because it served the needs of the Catholic Church throughout Europe, but because it trained generations of future rulers.[50] In the hands of these young noblemen whom the Jesuits educated in a Christian spirit would be placed the responsibility for whole families, for entire cities.[51] This was the way to deeply influence large sections of the Catholic world.[52] The conditions of the period necessitated an education in Christian values, especially among the secular population. The Italian universities were full of idlers who were interested only in amusements, and this was in addition to the arrogance they gained from the knowledge they acquired, which led to atheism.[53]

In the eyes of Lauretano, then, the Jesuit educational system was conceived as a practical solution to the most pressing problems of Catholic society, both internal and external. Education had to function as a barrier against "atheism" which threatened to spread in secular

46. Ibid.
47. Ibid., p. 995.
48. Ibid.
49. Ibid.
50. Ibid., pp. 995–996.
51. Ibid., p. 996.
52. Ibid.
53. Ibid., pp. 996–997.

society. At the same time, education was also a medium through which the church could exert its influence upon the heretics in an attempt to attract them back to Catholicism. Interestingly enough, it was by reflecting upon the methods of the heretics themselves that Lauretano became aware of the need to counterbalance their effectiveness by engaging in education.[54]

Lauretano's justification of the educational enterprise in terms of a universal mission, catering to the needs of the Catholic world, shows how far the Jesuits had moved from their point of departure. Originally, the Jesuit involvement with education was associated with the internal needs of the society, especially the need to train cadres of the Jesuit priesthood. In the 1550s and 1560s, however, increasingly broad areas of needs were taken into consideration until they came to include – in addition to the needs of the Society of Jesus – the needs of the Catholic Church as an institution, and finally the needs of the corpus of believers constituting the church: i.e., the needs of society in Catholic countries as a whole.

However, in order to ensure the infiltration of Catholic values from the Jesuit educators into lay society, and to prevent the exertion of influence in the opposite direction, Lauretano also emphasized the need to institutionalize certain means of control over the educated population. This meant, first of all, sifting the candidates for entry into the Jesuit colleges with an eye to accepting only those most likely to fulfill the hopes of teachers and spiritual mentors.[55] With great psychological insight, Lauretano explained how a spiritual training in accordance with Jesuit norms could not fail to leave a permanent mark on the pupils:

> Even if a few cast away the instructions they received from us and lead a life which is not particularly praiseworthy, nevertheless our work with them was not in vain, for the things they learnt are always before them, and with the help of the directing hand of the benevolent God, sometimes by this means and sometimes by that, He causes them to return to the lessons of the past, and these remain with them in such a way that they decide to return to the true path, to remember the religious piety of their life in the college, the prayers they prayed morning and evening. They remember the frequency of the holy sacraments, the discussions connected with salvation concerning things whose matter is God. They remember the fear and abhorrence they felt of every kind of sin. They remember the devoutness and the taste which the Holy Spirit bestowed upon them in accordance with their capacities. They remember the piety with which they preserved the

54. Ibid., p. 998.
55. Ibid., p. 1004.

images of the saints in their places which with a holy curiosity adorned their little houses of prayer. And the memory of these things and of other, similar things produces a wonderful impression in their hearts, and thus the good training in some way provides them with considerable benefits.[56]

Finally, Lauretano mentioned confession as a very effective means of control, through which one could ascertain the results of a Jesuit education.[57]

Tracing the development of Jesuit activities and rhetoric in the second half of the sixteenth century, one sees that the educational involvement grew out of constraining circumstances: the need to train confessors and preachers, the struggle against heresy, especially in the German frontier provinces, and the aim of apostolic activity, of which a basic education for the poor was considered a part. Education was felt to be a powerful instrument capable not only of recruiting people to the society or training the priesthood, but also of moulding the dominant elite of Catholic Europe. Education was a means of influencing the world. Its effectiveness derived first of all from the applicability of the knowledge provided by the society in a wide range of spheres: ecclesiastical as well as secular professional careers, the art of government and administration, and even the general development of the intellectual faculties. The cultural atmosphere of Jesuit education, imbued with spiritual and moral values, was considered a second source of influence upon the world. The obligatory confession and examination of consciences, the close contact with the spiritual directors – all these were conceived as capable of inculcating the society's life-style among the population of students.

The Jesuits' justification of studies by means of practical arguments, framed in the terminology of the active life, encapsulates their deviation from the intellectual traditions and practices of the old elite of the Catholic Church. The Jesuit orientation was directed towards the world, whereas the Dominican orientation aimed at preparing the soul for a separation from the world. This accounts for the intellectual openness which characterized the Jesuits and their preference for useful knowledge in the areas of humanistic studies, practical mathematics, and administration, all of which were considered unsuitable or lacking in value by the traditional intellectual elite.

At the same time, however, the Jesuits also sought to develop a more fundamental approach to the relationship between intellectual activity and the ultimate aim of the society, defined in terms of the quest for salvation. The fourth part of the Constitutions of 1558 demonstrates this point:

56. Ibid., pp. 997–998.
57. Ibid., p. 999.

Since the objective of the society and of its studies is to help one's neighbour towards a knowledge and love of God and towards the salvation of his soul, . . .[58]

The necessity of coping with the teaching vocation in this-worldly, practical, and political terms determined the secular facet of the Jesuit educational discourse. The reconciliation of such secular aspects with a religious vision was no less of a challenge, however. And with the same zeal and originality that they had shown in producing an up-to-date intellectual programme, the Jesuits rose to this challenge, developing and modifying their ideas about the soteriological function of knowledge.

Knowledge as a Bridge to Salvation

Among the earliest expressions of the Jesuit concern with the relationship between study and salvation is a letter of P. Favre[59] addressed to the first group of Jesuit students in the University of Paris. The students, wrote Favre, were primarily confronted with the problem of finding the connection between their studies and their chief and ultimate goal, namely the love of God which leads to salvation.[60] Renewing the connection with the Christian gospel and re-creating the link between faith and reason through the right intention: these were the two issues Favre was concerned with. In the present, he felt, the pendulum was swinging towards an emphasis on sentiment, willpower, and devotion, and consequently the principle of understanding needed strengthening.[61] A confirmation of the importance of understanding could be found in the words of Jesus himself,[62] and all the great figures of Catholicism in the past, especially Thomas Aquinas.[63] If

58. *MP*, I, p. 281, Capita de studiis constitutionibus Societatis Iesu: "Cum Societatis atque studiorum scopus sit proximos ad cognitionem et amorem Dei et salutem suarum animarum iuvare . . ."
59. Ibid., pp. 355–357, P. Favre, Sodalibus Societatis Parisiis studentibus, Ratisbon, 12 May 1541.
60. Ibid., p. 355: "Nuestro Redemptor, Jesu Christo os dé a todos cumplida gracia, para que de tal manera podáis llevar vuestros studios al preconcepto scopo vuestro, sin afloxar el arco de las intentiones, que al cabo os podáis gozar en el Señor del triunfo que reportaréis, si con el spiritu del saber no apagardes el spiritu del santo sentir . . ."
61. Ibid., p. 356: ". . . quiere que tengamos su spiritu, no solamente para el sentir de la voluntad y coraçon, mas etiam para el saber del intendimiento . . ."
62. Ibid.: ". . . 'quisquis aliquid scit, nondum novit, quomodo oporteat ipsum scire' . . ."
63. Ibid.

one was conscious from the beginning of the ultimate aim and adhered to it closely, even when involved in study, the passage from knowledge to the truth and to God could be quite straightforward.[64]

Favre's ideas were further developed in the "Constitutions which must be kept in order that the studies in the Jesuit colleges will be to the praise of God and his honour", written by Polanco between 1548 and 1550.[65] Polanco regarded study not simply as a means to the ultimate goal, but as a form of worship. Just as prayer and contemplation serve God if suffused with a spirit of charity, so does study if done with a pure intention.[66] What distinguished Polanco's perspective from similar traditional justifications was the separation of studies from contemplation. This marked the attainment of a certain degree of autonomy for studies. Simultaneously, a high degree of legitimation was achieved for noncontemplative intellectual activity as a path to God. This tendency culminated in the "Rules for the scholars of the society" drawn up by H. Nadal in 1563.[67]

Nadal explicitly differentiated between two ways of achieving the aims of the society. One, the way called "divine", consisted of offering the sacraments, prayer, and all the ceremonial practices. The other was the path of study:

> Let everybody know that the society has two means by which it strives for its end: the one is a certain force, spiritual and divine, which is acquired through the sacraments, prayer, and the religious exercise of all virtues and which is granted by the special grace of God; the other force exists in the faculty which is ordinarily acquired through studies.[68]

Intellectual activities, study, and the sciences were to be regarded as spiritual weapons which had received a divine sanction, transferred to those

---

64. Ibid., p. 357: ". . . necessario es, que etiam in medii veritate, id est, en vuestros studios, también os reposéis, trabajando in illo ipso eodem mediatore nostro, qui est veritas, quae a Patre exivit et ad Patrem rediit per lineam rectissimam . . ."

65. Ibid., pp. 38–45; "Constitutiones que en los collegios de la Compañia de Jesú se deven observar para el bien proceder dellos a honor y gloria divina".

66. Ibid., p. 39: ". . . y que como en la oratión y contemplatión Dios se sirve por la charidad donde proceden, así se sirve mucho del studiar en qualquiera facultad, quando puramente por amor suyo el studio se toma . . ."

67. *MP*, II, pp. 114–121, H. Nadal, Regulae pro scholaribus Societatis.

68. Ibid., p. 116: "Sciant autem omnes duo esse Societatis media, quibus ad suum finem contendit: alterum est vis quaedam spiritualis ac divina, quae acquiritur usu sacramentorum, orationis, virtutum omnium exercitio religioso, ex peculiari gratia a Deo accepta; alterum est positum in facultate quae ex studiis comparari solet . . ."

who were preoccupied with them. Nadal found legitimation for this point of view in the words of Jesus himself, who said that errors and illusions should be fought not only by means of the religious spirit but also by preserving and cherishing knowledge.[69] Moreover, Nadal also insisted that, in addition to this, the Jesuit approach was entirely consistent with the policy of the Catholic Church from the beginning. There was no doubt, he claimed, that it was consecrated with the authority of the Roman Church and immensely important in a period when the church was fighting for the truth against heretics.[70]

If one traces the Jesuit notion of "knowledge as a bridge to salvation" from Favre through Polanco to Nadal, it provides a clue to the difference between their way of constructing the bridge and traditional Thomist approaches to the same question. In the traditional Thomist view, contemplation was conceived of as a bridge between the intellectual search for truth, and the knowledge of God. As previously noted,[71] Thomas insisted that contemplation is achieved only through exercising the faculty of the intellect, the noblest element in man. In the context of Jesuit culture, however, contemplation was endowed with more mystical connotations and became part of a popular, emotional religiosity originating in the tradition of the "spiritual exercises" of Ignatius of Loyola. The architects of the Jesuit educational policy were thus able to draw new boundaries within a broadening concept of religious sensibilities. On the one hand contemplation, prayer, and other religious practices were all placed in the same category.[72] On the other hand was study, held to be a separate path to salvation. This way of drawing the boundary liberated study from the need for legitimation in contemplative, metaphysical, and transcendental terms and in fact sought to create a link between a practical knowledge

69. Ibid., p. 117: "Conserventur igitur summo studio illa arma spiritus, et fiant quasi forma, vita et perfectio scientiarum omnium, ex qua scilicet virtutem scientiae accipiant spiritualem ac divinam; et ne simus seculariter ac ieiune scholastici vel doctores, sed religiose et spiritualiter. Ita fiet, Christo propitio, ut non solum falsitates ac deceptiones mali spiritus, veri spiritus efficacia et luce dissipentur, sed ut purissimo scientiae animo suscipiantur et conserventur . . ."

70. Ibid.: "Persuadeant sibi omnes, ut semper fuit scientia necessaria in Ecclesia Dei, ita nunc esse multo maxime propter insanam quamdam haereticorum scientiae persuasionem, ut orthodoxa veritas ab illorum iniuria vindicetur, atque Ecclesiae catholicae romanae sacrosancta authoritas propugnetur, proprie vero esse nobis necessariam ut finem nobis praescriptum consequi possimus . . ."

71. See my discussion on Thomas, Ch. 4,

72. See above, Ch. 4.

of reality and the higher aim of salvation. This shift away from an abstract intellectualist approach to the problem of the relation between knowledge and salvation thus paralleled the practical justifications for an involvement in teaching and the emphasis upon the usefulness of the concrete knowledge transmitted by the Jesuits.

The Jesuits' contribution to a different conception of the relationship between knowledge and salvation was not limited, however, to the intellectual sphere. The social dimension of their contribution was exemplified in the peculiar position they gradually acquired within the institutional fabric of the Catholic Church. As a religious order, the Jesuits were expected to maintain a certain distance from the mundane world. Their main field of activity, however, was secular society at large, which they attempted to influence, mould, dominate, and control through education. Thus, at the dawn of the modern era the Jesuits became a group aspiring to an intermediate status between the world and the church which would allow them to play the role of the guardians of the frontiers. This aspiration aroused widespread criticism against them. The tactics they developed in the course of coping with the criticism, not only of their own members but also of other subcultures of the church at large, engendered new modes of thought particular to their experience as cultural mediators.

7

Freedom and Authority in
Jesuit Culture

The Jesuits as Cultural Mediators

The special position of the Jesuits as cultural mediators between the
world of politics and the church establishment, between the secular and
the regular clergy, and between non-Catholics and Catholics, is a well-
known fact in the history of the society. It is sufficient to mention just a
few examples of the Jesuits' intermediary status in order to set the stage
for a discussion of their specific strategies as mediators, and of the norms
they adopted as cultural go-betweens.

Historians of the Jesuits have devoted a great deal of space to describing the tactics with which they penetrated into the various European
states and created a network of relationships with the local political
elites.[1] The careers of the group of young people who attended the University of Paris together with Ignatius and became the core group of the society were all marked by missions to various parts of Italy where they were
expected to forge contacts with the local nobility and lay the foundations
of the future political and economic development of the society. Pietro
Favre (1505–1546) was sent to Parma and Piacenza; Alfonso Salmerón
(1515–1576) devoted himself to building up connections in the Kingdom
of Naples; Claudio Jay (1500/4–1552) was sent to Ferrara; Giovanni Alfonso Polanco (1517–1576), himself an aristocrat, worked among the
high society of Tuscany; Girolamo Nadal (1507–1580) had a similar task

---

1. P. Tacchi Venturi, SJ, *Storia della Compagnia di Gesú in Italia*, Rome 1910–
    1951, 2 vols.; M. Scaduto, *L'epoca di Giacomo Lainez. Il governo (1556–
    1565)*, Rome 1964; *L'epoca di Giacomo Laínez. L'azione (1556–1565)*,
    Rome 1974; B. Duhr, SJ, *Geschichte der Jesuiten in den Ländern deutscher
    Zunge*, Freiburg 1907, München-Regensburg 1928, 4 vols.; L. Delplace, SJ,
    L'établissement de la Compagnie de Jésus dans les Pays-Bas, Bruxelles 1887;
    H. Fouqueray, SJ, *Histoire de la Compagnie de Jésus en France*, Paris 1910–
    1925, 5 vols.; A. Astrain, SJ, *Historia della Compañía de Jesús en la Asisten-
    cia de España*, Madrid 1902–1925, 7 vols.

in Messina, and Martino Olave (1507/8–1556), a former courtier of Charles V, was sent to the Council of Trent as the theologian of a cardinal.[2]

A large part of the concessions obtained by the Jesuits for the establishment of educational institutions was provided by secular rulers. Thus, the first Jesuit colleges in Lisbon and Portugal were endowed by the king;[3] the one in Ingolstadt was founded under the auspices of the Duke of Bavaria.[4] The same happened in Padua, Venice, Vienna, and other places. The connections with rulers were reflected in financial and other material assistance to the Jesuits, such as the allocation of buildings. Another form of interaction with secular rulers was the position of personal confessors held by many Jesuits in the courts of secular and religious princes.

The position of the Jesuits as mediators between the secular and regular clergy is demonstrated in their negotiations with bishops over the education of priests in Jesuit colleges. A letter of 1545 written by Claudio Jay from Germany[5] testifies to the opposition of bishops there to voluntary mendicant activities. The letter reflects the conflict of interests, already manifested at Trent, between those who wished to limit the mendicant freedom of action in the sphere of education and the mendicants themselves, who desired to safeguard the privileges which ensured their independence.[6] The bishop was ready to allow the Jesuits to establish a college in order to encourage theological studies, but only on condition that students wishing to join the order would not be accepted and that the college would restrict its activities to servicing the secular priesthood so that the students would return to work in the diocese. The peculiar intermediary position of the Jesuits, however, earned them the resentment and sometimes the hostility of the mendicants. This was particularly true in the case of the Spanish Dominicans, who competed with the Jesuits in the field of education.[7] The secular clergy, for its part, displayed antagonism after the Jesuits were assigned the role of founding and administering the first seminary for the training of priests.[8] Cardinal Cesarini complained

2. The source for all the biographical details of members of the Society is: Tacchi Venturi, *Storia della Compagnia di Gesú* . . . , II.
3. Farrell, *The Jesuit Code*, p. 14.
4. *MP*, I, pp. 405ff., Ignatius de Loyola to Albert Duke of Bavaria, Rome, 22 September 1551.
5. Ibid., p. 361: ". . . quasi a tutti è in odio il nome de fraterie, de compagnie, de mendicità volontaria . . ."
6. Ibid.
7. See historical background to the debate "De auxiliis", ch. 9.
8. Villosolada, *Storia* . . . , pp. 80–83.

that the Jesuits used their contacts in order to obtain commissions at variance with their original goals. Their acquisition of the right to organize the examinations for ordination of priests in the Rome area once more reveals their success in obtaining positions of mediation between the regular and the secular clergy.

The most conspicuous example of the cultural mediation performed by the Jesuits was their missionary activities. They used geographical exploration in China and the Himalayas, and along the Nile, Amazon, and Mississippi rivers, as well as bringing European astronomy to the court of Beijing and medical care to South America and the Philippines, in order to achieve the programmatic objective of conversion.[9]

In their various roles of mediation between secular and ecclesiastical politics, between Europe and the East, the Jesuits were constantly involved in controversies, in the need to compromise between differing views, in the creation of a middle ground between opponents. In order to perform their task, they created norms which allowed the blurring of traditional social, religious, and ideological distinctions as well as of accepted intellectual boundaries. In the following pages I shall focus on one sphere of activity which is particularly instructive for any attempt to characterize Jesuit practices, i.e., controversies with heretics, where the very possibility of mediation was an invention of the society.

The techniques elaborated in the context of these controversies were based upon the need to play down the role of dichotomous concepts such as "true" doctrines versus "false" ones, and to create a conceptual middle ground where traditional oppositions were treated as alternatives, capable of being debated, negotiated, and sometimes even exchanged. What began in the sphere of practice as a way of coping with a challenge to Catholic authority gradually engendered a particular vocabulary. The concepts of "true" and "false" as criteria of judgement were suspended by the notion of the "possible", which was understood as neither true nor false, but usually closer to truth than to falsehood.

This conceptual middle ground, rooted in the actual life experience of the Jesuits, opened up a wide field of possibilities in the intellectual sphere. My attempted reconstruction of the fierce internal debate over the limits of intellectual freedom which took place from the 1570s to the 1590s bears witness to the realisation of some of those possibilities. The conceptual middle ground, constructed by means of the notion of

9. J. Brodrick, SJ, *Origines et Expansion des Jésuites,* trad. française, Paris 1950, 2 vols; P. Lécrivain, SJ, *Les Missions jésuites,* Paris 1991; J. D. Spence, *The Memory Palace of Matteo Ricci,* New York 1984; J. Lacouture, *Jésuites: Une Multibiographie,* Paris 1991, 2 vols.

"the possible", allowed a very broad interpretation of Thomism, the intellectual orientation recommended by the mandates of the Council of
Trent. This relative freedom, however, required the development of some
implicit means of control, in order to ensure a measure of social cohesion
and integration. The Jesuits' experience also provided the practices which
became models for social and intellectual control.

Controversies with Heretics: The True and the Possible

A series of three texts of the 1540s, 1550s, and 1560s testifies to the
emergence of a new style of controversy characteristic of the Jesuits and
justified as referring not to errors of faith as such, but to the individuals
who erred. Favre's *De ratione agendi cum haereticis* is the first among
these texts.[10] Debates with heretics, Favre asserts, should be conducted in
a decorous manner. One should not stress one's differences of opinion,
but lay emphasis on the points common to both parties.[11] In Favre's text,
aggressiveness in argumentation, typical of the medieval literature of controversy,[12] gave way to civility. The uncompromising tone which had been
recommended by the Dominican General Chapters[13] was replaced by attempts at quiet persuasion. The perception of Catholics and heretics as
clear-cut *opposites* between which no middle ground was possible was
played down. Instead, these *opposites* reemerged as *alternatives,* the
choice among which depended upon persuasion. Most significantly, the
possibility of obliterating the opponent, still present in the Dominican
texts of the period, entirely disappeared from Favre's text.

One should not insult heretics, Polanco wrote in 1556;[14] one should
not treat them contemptuously; one should even refrain from opposing
their arguments in a direct and open manner. Instead of direct opposition

10. *MP*, II, Epp. Favre (p. 400); in: Introductio, p. 5.
11. Ibid.: ". . . communicado con ellos familiarmente en cosas quae nobis et
    ipsis sint communes, guardandose de todas disceptationes ubi altera pars
    alteram videtur deprimere; prius enim communicandum in illis quae uniunt
    quam in illis quae diversitatem sensuum ostendere videntur . . ."
12. See for example J. Cohen, *The Friars and the Jews. The Evolution of Medi-
    eval Anti-Judaism,* Ithaca 1982.
13. See Chapter 5.
14. *MP*, I, p. 484, I. A. de Polanco ex comm. sociis ingolstadium adeuntibus,
    Rome, 9 June 1556: ". . . né li sia detta ingiuria alcuna, né si mostri spetie
    di sdegno contra loro, anzi di compassione; né anche apertamente si proceda
    contra loro errori ma che si stabiliscono li dogmi catholici; e di quelli si
    vederà che li contrarii sono falsi.

he suggested persuasion by means of an exemplary adherence to Catholic dogma. Visible behaviour, true action were presented by him as a suitable alternative to direct verbal confrontation.

A further development of the idea of a nonviolent, more indirect confrontation with heretics appeared in a text of Nadal of 1563, where he argued that against heretics one should act with great prudence, especially in places like Germany and France. They should not be subject to insults and called humiliating names; they should not be described as heretics (even if they really are such), but called members of the Augustinian or Protestant creed or other sects such as the Zwinglians, Anabaptists, and so on.[15]

Nadal's text advocated the same tactics as those of Favre and Polanco, namely, the avoidance of any direct confrontation or verbal violence. Here, however, there was not only reversal of the rules of behaviour towards opponents, but also a change in the use of language: the heretics were no longer to be called such. Thus, recognition was implicitly given to the coexistence of alternatives, the decision between which could not be arrived at by means of the law of contradiction (i.e., by calling one "true" and the other "heretical"). In addition to "true" and "false", there now appeared a further possibility – not to be rejected a priori as heretical. For the moment, this represented no more than an effective tactical weapon in the warfare against dissenters, but it soon came to dominate the system of values and the conceptual framework of those who employed it.

At an early stage, the leaders of the Jesuits became aware of the dialectics necessitated by living among heretics and the need of emphasizing similarities as a method of persuasion. The fact that certain Jesuits were always singled out to speak to them, wrote Nadal in his instructions to the college in Cologne,[16] could be beneficial for them and might perhaps persuade some to draw closer to the Jesuit positions. At the same time, it was not without advantage for the Jesuits either, for it enabled them to become acquainted with the heretics' methods of action and to find appropriate responses. Polanco ascribed the remarkable successes of the heretics to their particular activities, which he summarized in a text of 1563.[17] These consisted in the popularization of a mendacious theology and its adaptation to the understanding of the masses, a great deal of preaching, and the dissemination of ideas by means of printed pamphlets

15. *MP*, II, p. 96, H. Nadal, "Distinctio classium etc.", 1563.
16. *MP*, III, p. 82, H. Nadal, "Commissarius instructiones coloniae datae", 1562.
17. Ibid., p. 334 ff., I. A. de Polanco, "Monita qaedam ad religionem catholicam mediante Societate Iesu in Germania et Gallia iuvandam", 1563.

purchased and understood by the masses. By way of a response, Polanco put forward a revolutionary idea: a proposal for a new kind of Catholic theology.[18] Traditional theology, based on philosophy, demanded lengthy efforts to acquire, claimed Polanco. He advocated an abridged form which would be based on essential principles agreed upon by all and adapted to the needs of the people. This kind of practical theology would be taught and transmitted by the method of *loghi communi* (*loci communes*),[19] which would replace Aristotelian logic and would serve as a technique of persuasion rather than a method of proof.

The transition from proving theological doctrines by means of logic to convincing people of theological truths by rhetorical means represented one stage in a process which finally affected Jesuit approaches to various modes of knowledge. A theology based on rhetorical arguments and accepted as true could legitimize "possible" opinions which had not acquired full logical proof. Moreover, the Jesuits, who worked among heretics and attempted to engage them in theological dialogue, came to recognize the "other" as an alternative, not simply as an excluded opposite. Their peculiar life-form engendered a hermeneutic insight into a possible discourse with the "other", whether embodied in a religious opponent or consisting of "other" philosophical opinions. An inclusive mentality was about to emerge which could have implications for the intellectual domain as well as the religious one.

The Debate on "Varietas Opiniorum"

Historical literature has always tended to see the Jesuits as the carriers of scholasticism into modern times. Recent research, moreover, has drawn attention to the close affinities between the methodological and scientific problems which preoccupied the Jesuits, and those that were at the heart of the debates among the founders of modern science.[20] However, the conditions which gave rise to the Jesuits' special mediating position have not been investigated in detail by historians of science. These conditions were formed in the course of a historical process which requires reconstruction in order to reveal the potentialities, but also the limits, of Jesuit

18.   Ibid., p. 335: "Prima adunque, oltra la theologia perfecta che si insegna nelli studi generali . . . si faccia altra theologia sumaria . . ."
19.   Ibid.: ". . . probando a modo di loghi communi con boni testimonii de le scripture et traditioni, concilii et dottori li dogmi che lo ricercano, et refutandoli contrarii . . ."
20.   See Ch. 12, pp. 245 ff.

intellectual activity. The historical episode to be reconstructed is an internal debate on the "freedom of opinion" to be allowed to Jesuit educators. This debate, which continued uninterruptedly for more than twenty years (1570s–1590s), indicated a general crisis of authority within the Society of Jesus. As a result of this crisis, simplistic notions about orthodoxy, uniformity of opinion, and censure were replaced by sophisticated mechanisms for the legitimation of new opinions, accompanied by a system of invisible controls over the limits of the permissible.

The pedagogical texts from the 1550s do not show full commitment to the subject matter suitable for teaching in Jesuit educational institutions. An orderly list of subjects and authors to be studied was indeed drawn up, in addition to the colleges' rules, but the first result, the *Ratio studiorum* of the Roman College which originated in the experience of the College of Messina was devoted to pedagogical and disciplinary matters rather than to any exhaustive discussion of the curriculum. Three texts are exceptional in this respect: Coudret's "De ratione studiorum Collegii Messanensis" of 1551,[21] Nadal's "De studii generalis dispositione et ordine" of 1552,[22] and Olave's "Ordo lectionum et exercitationum in universitatibus" of 1553.[23]

These texts presented a complete academic program, beginning with Latin grammar, which one learned in the first year, and ending with theology and Scripture. They specified the texts suitable for study in each course, placing much emphasis on Latin and Greek authors and on the writings of Aristotle. Thomas Aquinas's *Summa theologiae* and auxiliary studies in casuistry were considered of primary importance in every course of theology.

Despite the richness of the philosophical and humanistic literature read in Jesuit schools in that period, the texts do not suggest any anxiety about the possible effects of the preoccupation with secular knowledge and pagan texts on the orthodoxy of the teachers and students. Insofar as any doubts existed, they were chiefly with regard to literary works such as the comedies of Terence, which displayed problematic moral positions, and the texts of Erasmus and Vives, suspected on account of their harsh attacks against scholasticism and traditional philosophical authorities. These texts were nevertheless included, in spite of the caution with which they had to be read. The naive approach to the problem of orthodoxy is

21. *MP,* I, p. 93 ff., H. du Coudret, "De ratione studiorum Collegii Messanensis", 1551.

22. *MP,* I, p. 133 ff., H. Nadal, "De studii generalis dispositione et ordine", 1552.

23. Ibid., p. 163 ff., M. Olave, "Ordo lectionum et exercitationum in universitatibus", 1553.

well expressed in Ignatius's somewhat naive ideas about controlling negative influences:

> It has always seemed to me, as it seems to me now, that the wisest thing to do would be to take out of these humanistic works the things which are dishonest and new and to insert in their place other things which are more educational: or, indeed, without adding anything, to leave only the good books and to remove the others.[24]

The society's *Constitutions,* issued in 1556,[25] expressed the official educational policy, which assumed the orthodoxy of studies rather than delineating any boundaries to be safeguarded. Here it was stated that the knowledge which best served the needs of the society was that provided by the theological faculty. In theology, Thomas Aquinas was spoken of as a major figure, although not the only one. The *Sentences* of Peter the Lombard[26] were also mentioned, even if this was not finally accepted as the main text, as Thomas's *Summa* was. Generally speaking, the *Constitutions* demanded an adherence to those philosophical doctrines regarded as the soundest and most certain.[27] Of humanistic works, the *Constitutions* said: "One should not teach anything of the humanistic works of the non-Christian peoples which is repugnant to honesty".[28]

Signs of an anxiety caused by a growing sense of deviation from accepted norms were discernible in the 1560s. They appeared in the form of numerous injunctions against a departure from the Catholic consensus in theology, and against the expression of individualistic or novel opinions even in matters unconnected with faith. The following injunction is a good example:

> And, first of all, every effort should be made in order that in all theological matters the interpretation will be accepted which is held by the Roman Catholic Church. Each person should guard himself jealously against being drawn by new opinions, even in matters unconnected with faith, and should simply adhere to the accepted opinion.[29]

The decisions of the regional assemblies in those years reflected concern about the wide variety of opinions to be found both in courses of humanistic studies and in philosophical and theological courses. Thus, the regional assembly of Italy in the year 1568 stated:

---

24. Ibid., p. 390, Ignatius de Loyola to A. Lippomano, Rome, 22 June 1549.
25. Ibid., p. 210 ff., "Capita selecta de studiis in constitutionibus Societatis Iesu, 1547–1556".
26. Ibid., c. XIV, pp. 296–297.
27. Ibid., p. 295.
28. Ibid., c. V, p. 223.
29. *MP,* II, p. 128, H. Nadal, "Instructio brevis quanam scilicet ratione de rebus theologicis his temporibus loquendum sit", 1563.

There should be some directive or general instruction for humanistic as well as philosophical and theological studies in order to prevent the possibility of such a wide variety of opinions and the suffering and damage inherent in so much writing, and the reading of useless books.[30]

At the same assembly a specific call was made to "purify" the humanistic and scholastic literature in the curriculum.[31] Moreover, letters from private members of the society also contained complaints about new and individualistic opinions. The theologian Pérez, of the University of Evora in Spain, wrote to the general, F. Borgia, complaining that both in logical and in philosophical studies there were well-known magisters who held new ideas which differed from the opinion of the Holy Doctors. Especially dangerous in his view were the opinions directed against Thomas Aquinas. Such a variety, he thought, was not to the benefit of the society.[32] The general's reply to this letter testifies to the wide expansion of the phenomenon all over Spain. Similar complaints were reported by P. Canisius, one of the founders of the society in Germany[33] and from E. Mercurian, visitor of the French provinces, who wrote a special report on the wide variety of doctrines and the lack of uniformity of philosophical opinion and addressed it to General Borgia.[34]

Usually, the complaints expressed fear that innovations in logic and physics might lead to a deviation from sound theological doctrine.[35] Ultimately, however, the failure to achieve uniformity was felt to endanger social cohesion. The need to act against disintegration was expressed in a letter of Canisius:

> And I pray that, for the sake of the integrity of our Society, which depends, first of all, on a sound doctrine of faith, there should be proper consultation, because of errors of the faith which we discern, so that those scandals and errors among our members will be avoided in the future, insofar as such a thing is possible, with the help of the Lord Christ.[36]

A look at the writings of some well-known Jesuit deviators reveals what was really at stake. Benedictus Perera (1535–1610), one of the first teachers of the Roman College, was widely accused of individualism,

30. *MP*, III, p. 29, "Ex actis congregationum provincialium Italiae", 1568.
31. Ibid., p. 26.
32. Ibid., p. 396, F. Pérez to F. de Borgia, Ebora, July 1566.
33. Ibid., p. 414 ff., P. Canisius to F. de Borgia, Munich, 26 September 1567.
34. Ibid., pp. 455–456, E. Mercurian to F. de Borgia, June 1569.
35. Ibid., p. 397 n. 2, Epp. Borgia IV, p. 379.
36. Ibid., p. 416, Canisius to Borgia.

nonconformism, and holding dangerous opinions.[37] In 1564 he wrote two short texts in which he expressed his ideas about the way in which philosophy should be taught to students in Jesuit educational institutions.[38] In accordance with the official policy of the order, Perera never tired of stating that the final goal of any philosophical inquiry must be truth, leading up to its ultimate source, namely God. A life of study, he maintained, was in perfect harmony with Christian ideals, in that it preserved one from evil, purified one, and taught one how to withdraw one's attention from worldly things.[39] Also, in accordance with the mainstream of clerical intellectuals of the Counter-Reformation Church, Perera believed that the point of departure for any search for truth is the opinions of others:

> One should not be drawn to new opinions – that is, those which one has discovered – but one should adhere to the old and generally accepted opinions. In one's teaching one should avoid sophistic philosophizing and follow the true and sound doctrine.[40]

Perera's respect for old opinions derived from the rules of traditional philosophical discourse, which assigned a special position to certain ancient authors whose work was constantly commented upon, disputed, argued with, and reinterpreted. Perera never challenged the framework of the interpretive game he was part of, and even provided it with further justification. One should read the authors considered the first in importance who surpass all others in every science, Perera argued, both because of their authority and antiquity and because of their wisdom and intelligence. In the writings of these authors one not only finds the doctrine itself, but it is continued and broadened with authority and evidence, in accordance with the nature of the subject. Thus, one should above all read those ancient writers in whom many things are explained and the doctrine is well-founded and clear. Readers of these books will profit not only in that their minds will become richer and more cultivated through a great deal of knowledge, but also in that they will grow more stabilized, mature, and capable of understanding.[41]

37. Compliants against Perera are to be found in the following letters sent to the General: *MP*, III, p. 415 n. 4, Canisius to Borgia; *MP*, II, p. 476 n. 40: "P. Perera accusatus est a Patribus Ledesma et Gagliardi de averroismo . . ."
38. *MP*, II, pp. 664–669, "De modo legendi cursum philosophiae", 1564; ibid., p. 670 ff., "Brevis ratio studendi", 1564.
39. Perera, "Brevis ratio . . .", p. 671: "Omnia quidem vitia, sed in primis voluptates corporis fugiendae sunt ei qui studet nancisci sapientiam. Voluptas enim obscurat aciem mentis, deprimit animum ad res terrenas, et quasi vinculis quibusdam astrictum retinet . . ."
40. Perera, "De modo . . .", p. 667.
41. Perera, "Brevis ratio . . .", p. 677.

Within the rules of this game, however, Perera aspired to introduce some modifications. In matters which did not concern faith, and where, he believed, one should avoid lightheaded attempts to oppose the most authoritative of all authors, namely Aristotle,[42] he nevertheless believed there existed a still higher authority: truth, arrived at through the application of individual intelligence and human reason.[43] Aristotle himself knew that over and above all authority there was truth itself, and expressed his awareness in the well-known saying, *Amicus Socrates, amicus Plato, sed magis amica veritas*.[44] The search for truth, claimed Perera, permits deviation from the opinions of others.[45] It requires thorough investigation, unperturbed by preconceived notions deriving from hatred, admiration, commitment to a patron, or any other sentiment.[46]

In accordance with the religious-intellectual tradition he belonged to, Perera favoured interpretation as the path to truth. Never did he challenge the position of Aristotle or Thomas Aquinas as the authors to be commented on or disputed, argued or agreed with. What he did question, however, were the tacit assumptions concerning the orthodoxy of some authors and the rejection of others. In this, he challenged the mode of authorization tacitly implied in the official educational program of the Jesuits. In the absence of clear criteria for choice among possible interpretations, Perera was able to opt for the freedom required for any open intellectual exchange. Refusing, for example, to grant an exclusive position to Thomas as the "soundest" theologian and to the Latin commentators of Aristotle as the most philosophically reliable, Perera pointed to a much wider range of interpretive alternatives:

> Although every magister must follow the main authors, such as Alexander, Simplicius, and Tamistius among the Greeks, Averroes among the Arabs, and St. Thomas among the Latins, he must nevertheless

42. Perera, "De modo . . .", p. 667; "Volere reprendere Aristotele nelle cose che non sono contra la fede e che sono communemente approvate da tutti, è segno di leggierezza et ignoranza".

43. Perera, "Brevis ratio . . .", pp. 670–671: "Proximus studiorum finis est cognitio veri, quae quod sit humanae mentis perfectio, ab omnibus naturaliter expetitur. Ultimus autem finis ut omnium actionum et cogitationum nostrarum, sic etiam studiorum est Deus . . ."

44. Ibid., p. 671.

45. Ibid.: "Cuius [veritas] gratia oportet non solum ab aliis dissentire, verum etiam (si veritas id postulat) suas sententias et decreta mutare atque rescindere . . ."

46. Ibid.: "Quocirca oportet eo animo ad capessendas disciplinas quemque accedere, ut nec studio cuiusquam, nec odio nec gratia nec pertinacia, nec alia

avoid being sectarian, especially with regard to the Latin writers who take issue with the Ancients.[47]

In addition, he also expressed his humanistic critical sensibilities by questioning the authority of certain ancient texts which, he claimed, suffered from deterioration of their physical condition and from errors in transcription, translation, and interpretation.[48]

Perera was not unique in questioning procedures for the authorization of opinions. An early letter of F. Suarez (1548–1617), the greatest Jesuit theologian at the turn of the century, also reflected a change in the delicate equilibrium between a demand for doctrinal unity, based on authority, and a spirit of independent enquiry. In response to the accusation that he taught new truths opposed to the opinion of St. Thomas,[49] Suarez contended that while he did not teach new opinions which deviated from Thomism, he admitted deviating from the norm in his manner of teaching, claiming to be more thorough in his expositions.[50]

That the society was facing a crisis of authority in intellectual matters is clear from a letter of Canisius to General Borgia in 1572:

> Our teachers here teach very differently from one another; whereas some, as far as possible, follow St. Thomas, while others, like the pupils of Father Benedictus Perera, say that they follow the Arabs and the Greeks. They are as different from each other as earth is from heaven, even in the most common constitutive principles. For each side claims that his are the common opinions, especially as both opinions are sometimes taught in the society, and because of the fact that the teaching of Father Benedictus, [transmitted] through his pupils, of whom it is said that they have been successful in many places,

---

perturbatione animi adductus aut falsi patrocinium suscipiat aut deserat veritatem . . ."

47. Perera, "De modo . . .", p. 666.
48. Perera, "Brevis ratio . . .", p. 677.
49. *MP,* IV. pp. 809–811, F. Suárez to E. Mercurian, Villoslada, 2 July 1579.
50. Ibid., p. 810: ". . . porque ay costumbre de leer por cartapacios, leyendo las cosas más por tradición de unos a otros que por mirallas hondamente y sacallas de sus fuentes, que son la authoridad sacra y la humana y la razón, cada cosa en su grado. Yo e procurado salir deste camino y mirar las cosas más de rayz. De lo qual naze que ordinariamente pareze que llevan mis cosas algo de novedad, que en la traza, que en el modo de declarallas, que en la razones, que en las saluciones de dificultades, que en levantar algunas dudas que otros no tratan de propósito, que en otras cosas que sienpre se ofrezen; y de aquí pienso que resulta que, aunque las verdades que se leen no sean nuevas, se hagan nuevas por el modo, o porque salen algo de la vereda de los cartapacios".

appears to have become very famous. But others claim that it is to a certain degree dangerous and very unsuited to theological studies.[51] The struggle over interpretive authority forced the hierarchy of the order to reflect upon the old notions of uniformity and orthodoxy, revise the accepted strategies against deviation, and devise new mechanisms for the legitimation of new ideas. One of the first texts to provide evidence of the lessons learnt in the course of twenty years was a document written in 1564–65 by the rector of the Roman College, I. Ledesma. The document grew out of many consultations with the best teachers of the college, who contributed out of their rich store of practical experience in teaching and mutual exchange of ideas.[52] As against the naive assumptions of the previous decade, when the personal orthodoxy and good intentions of teachers and students were conceived of as sufficient barriers against deviation, Ledesma's reflective attitude marks a significant turning point:

> In the teaching of philosophy, one should beware of a two-fold abuse. One is too great a freedom, which is harmful to faith, as is demonstrated by the experience of the academic institutions in Italy. The other is that [people] will be bound only to the doctrine of one author, for in Italy this has led to things which are frightful and contemptible.[53]

In accordance with the official party of the order, Ledesma repeated the normative claim that the aim of studies in the society, and especially philosophical studies, was to support theology.[54] Following the spirit of the *Constitutions* of 1556, he chose to be more explicit in recommending certain interpreters of Aristotle such as Thomas Aquinas and Albertus Magnus and warning against others. Averroes and the Greek commentators, especially Alexander of Aphrodisia, were particularly suspect as liable to lead to heresy, or at least to doubts about religious doctrines.[55] The teachers, he said, should not introduce new opinions in place of those which are accepted without permission from those in charge, or replace those of Aristotle on their own initiative, without the authority of the Ancients. The innovation must be in harmony with the accepted doctrine.[56] For Ledesma, the aim was still defined in terms of *unitas doctrinae* – doctrinal unity. However, for all his traditional rhetoric, Ledesma's document does demonstrate a slight modification of the traditional approach.

51. *MP*, III, p. 415, n. 4, P. Canisius, September 1572, 134 II, f. 338r.
52. *MP*, II, pp. 464–490, I. Ledesma, "Relatio de professorum consultationibus circa Collegii Romani studia", Rome 1564–1565.
53. Ibid., p. 478.
54. Ibid., p. 478.
55. Ibid., 4, 5, 6.
56. Ibid., p. 477, 2.

In demanding that new opinions should be introduced only with permission from those in charge, he already assumed that innovation was unavoidable. Later on, this assumption received more explicit recognition:

> In the case of all other probable opinions which do not concern faith, one should not expect everyone to uphold exactly the same opinion in all matters. For, as has already been said, this led to terrible consequences in Italy and to the contempt of doctrine, and the perfection of doctrine was disrupted. . . . And room should be given to talent, and the external [pupils] should consequently be educated as seems right to the society.[57]

Moreover, in referring to the ideal of doctrinal unity Ledesma saw fit to qualify the statement by saying "doctrinal unity to the degree that it is possible". Together with the recognition that intellectual concerns are bound to give birth to new opinions, Ledesma's text also testifies to an abandonment of direct, visible means of control. Philosophical systems regarded as dangerous were not described as amenable to "correction" – Ignatius's solution with regard to humanistic texts in the 1540s – or as something to be ignored. Also, the option of achieving doctrinal unity through adherence to a single school of thought was rejected. Thus, Ledesma renounced the possibility of demanding absolute adherence to Thomism either in theology or in philosophy.

The rudiments of two mechanisms for the legitimation of new opinions, further elaborated by the Jesuits in the following decades, were also to be found in the document. One was the acceptance of "general opinion" as a means of legitimizing deviation from ancient authorities.[58] The other was the possibility of presenting and discussing opinions considered dangerous or unacceptable in a neutral way, without defending them, but recognizing their existence as possible alternatives to the accepted truths:

> The teachers should be forbidden to interpret the deviations of Averroes, of Simplicius or of anyone else. Their ideas should simply be presented without comment.[59]

Thus, Ledesma left the door open to alternatives which were acknowledged, even if they were not accepted. New ideas could always be placed in the category of the "possible" or "probable" and thus might receive the authorization of the consensus some time in the future.

Ledesma's document, however, did not represent the last word in the debate on freedom of opinion. A counter attempt to prevent interpretive chaos was the decree issued by General Borgia in 1565, "Concerning the

57. Ibid., p. 479, 13.
58. Ibid., p. 478, 10.
59. Ibid., 11.

opinions in philosophy and theology to be held".[60] Its formulation was rigid and hardly left any room for a flexible approach to the possible boundaries of interpretation. Thus, according to the decree, one could not defend or even teach anything which might be construed as slightly opposed to the principles of faith or as tending to diminish its value. One could not defend anything opposed to generally accepted decisions either in the philosophical sphere or in theology. One could not defend ideas opposed to the consensus of the majority. One could not put forward a new idea in philosophy or theology without the agreement of those in charge.[61]

The decree did not mention the distinction between "possible" and accepted opinions, or the permission to discuss in a neutral manner opinions regarded as dangerous. However, even in this problematic document the main criterion for the legitimation of opinions was not the authority of the past, identification with a certain school of thought, or fidelity to a certain interpreter, but the consensus, or "general opinion", referred to by Ledesma.

The debate did not die down as a result of Borgia's decree, but rather increased in intensity.[62] The later 1570s were decisive for the crystallisation of the cultural model particular to the Jesuits. The forms of action adopted and the context in which they emerged were revealed in two texts of P. Hoffaeus, the provincial of Upper Germany, who was nominated by the general of the society to resolve the continuous crisis. The first was a kind of summary of Hoffaeus's recommendations, distributed all over the province after an examination of the claims of the various parties to the debate.[63] The second contained impressions and conclusions on which the recommendations were based, and was sent to General P. Mercurian:

> I know that there is a decree forbidding the acceptance of deviant opinions, but at the same time I know from experience the same decree either does not have the sense that some think, or that it has ever been observed anywhere; nor that it can be observed at all. And therefore, just as one cannot bring it to pass that there will not be any difference of opinion in the entire society, so I think that it will be very difficult to reach in this province full uniformity of opinion. But this is no trivial matter, and it will not be understood by the generations to come if such an opinion, probable for some, is con-

60. MP, III, pp. 382–385, S. F. de Borgia, "Decretum de opinionibus in philosophia et theologia tenendis", Rome 1565.
61. Borgia, "Decretum . . .", p. 383.
62. See MP, IV, pp. 646–648, 703–712, 809–811.
63. Ibid., pp. 703–711, P. Hoffaeus, "Ad omnes provinciae socios", Ingolstadt, November 1577.

demned in such a way as to be invalidated by the canon law, especially when the Church of God itself allows scholars so great a latitude in that area. . . . The same argument, however, does not apply with regard to decisions concerning practical matters and in matters pertaining to speculation, for it is difficult for the intellect, which is further indulging, to be constrained until it remains within the boundaries of faith and virtue. For otherwise, a great opportunity will be missed for the exercise of ingeniousness if such narrow limits are laid down for those who deal with speculation. Indeed, very many professors, and especially the most talented among them, would be frightened if they were not allowed, for good reasons, to publish their new arguments and opinions in order to explain that which they propose, on the account of their arguing for novelty. And thus, for a few years, it will be necessary in various places to publish large, thick volumes of opinions, for it is in the natures of these things and the nature of the best talents that they cannot do otherwise than always to discuss something new. And therefore, the variety of opinions has nothing that offends propriety except in endangering the faith and [causing] scandal. Since no family of theologians in the schools exists, the members of which stick to the same perpetual sense and consensus in all opinions as we see attempted in vain in our society against the universal difference of all opinions . . . [64]

In accordance with the pragmatic ways adopted by the society in all spheres of life – in politics, pedagogy, and even in controversies with heretics – Hoffaeus's text was an attempt to interpret Borgia's decree in the light of the lessons of experience. Experience had shown that Jesuit intellectuals could not be disciplined by explicit laws attempting to impose doctrinal uniformity upon them. Such a policy might jeopardize the whole educational enterprise. There was no alternative, therefore, except to allow a certain liberty within the boundaries of faith and acceptable conduct. Practically, it was very difficult to prohibit everything which was not acceptable to the general opinion.[65] A prohibition which could not be implemented tended to undermine the spirit of obedience more than a lack of doctrinal unity and to produce continual disputes and disagreements.[66]

Hoffaeus's conclusion amounted to an official recognition of the necessity to institutionalize the role of the intellectual as a thinker with a mea-

---

64. Ibid., p. 745, P. Hoffaeus to E. Mercurian, Munich, 2 May 1578.
65. Ibid., p. 746: "Ut autem nihil defendatur contra communem opinionem, hoc facilius erit verbo prohibere, quam facto consequi aut exequi . . ."
66. Ibid.: "Ac ne bona ingenia nimium sine magna necessitate constringantur, neve tandem nullus facile onus in scholis profitendi cum tanta vexatione seu querela suscipere vel diu sustinere velit . . ."

sure of freedom from authority. This conclusion received the sanction of
General Mercurian:

> With regard to what he wrote . . . that he feels that one should not
> excessively limit the boundaries of thought but should allow it room
> to express itself, we also accept that, on condition that a boundary is
> observed. That is to say, unless we have many wide and spacious
> paths on which they can run about and exercise themselves, they are
> liable to lose themselves in dark dead ends, which they will not easily
> find their way out of.[67]

The institutionalization of the role of the intellectual, however, included
the invention of new mechanisms of control, implied in the term "boundary" used by General Mercurian. As a substitution for the atemporal,
abstract concepts of theological truth and philosophical certainty which
justified the demand for doctrinal uniformity, the Jesuits strove to constitute a set of practical rules whose validity depended on the institutional
context in which they were applied. The most visible among the means
of control recommended by Hoffaeus was a system of internal censorship
on various levels, from censorship by the college director of studies, the
rector, and the provincial censorship committee to the censorship of all
the provinces by the general in Rome. The recognition of the need for
internal censorship, which was also expressed in the regional assembly of
Germany in 1576,[68] received a new confirmation here.

The Jesuits' system of internal censorship was one example, among several, of their way of coping with a complex social-intellectual reality. A
code, a system of hidden rules, known and accepted exclusively by members of the society, was invented in order to control the world of multiple
interpretations in which they were forced to live. This code had no universal validity outside the society, and could not be grounded in universal
values, being based as it was on a substitution of "consensus" and the
"possible", for the concepts of "uniformitas doctrinae" and "truth".

Another set of rules for controlling the process of socialization into
the modes of thought accepted by the society was the *Ratio studiorum*,
mentioned in the Italian regional assembly of 1572–73 as an instrument
for achieving unity:

> Because it is known from experience how important it is for the
> achievement of uniformity – to the degree that it is possible to achieve
> in so many provinces – that such uniformity should be preserved in

67. Ibid., IV, p. 747, n. 7.
68. *MP*, IV, p. 282, "Ex actis congregationum provincialium assistentiae Ger-
     maniae", 1576: "Iterum propositum est de tollendis in theologia et philo-
     sophia novis periculosis et minus certis opinionibus; et ad plura suffragia

áll matters pertaining to the teaching of philosophy and especially theology, and to the modes of study in all classes in the colleges of our society, the general assembly has been asked to draw up some common formula or mode [of study] which we shall first consider here in Rome, and then, after hearing the ideas of others, if you so desire, we shall complete and authorise. The declaration of faith will be according to the formulation of the Council of Trent and the apostolic letters of Pius IV, of blessed memory. And with regard to uniformity and the mode of teaching philosophy and theology, the matter has been now thoroughly discussed here in Rome, but the things which are discussed here are to be sent to the provinces first, so that, after their views have been heard, they can respond if they have anything to say since they do not agree with the same everywhere. Only then will the *Ratio studiorum* be more fully acceptable to the Society.[69]

The intellectual and institutional interests which are reflected in the *Ratio* will be dealt with in the following chapters. It is sufficient to mention here that the elaboration of institutional mechanisms for the authorisation and legitimation of a constant stream of new opinions limited the possibilities of the *Ratio* to dictate the contents of studies. The device of a "common formula", mentioned in the Italian assembly document refers, rather, to the rules for the organisation of studies, the boundaries between disciplines and their relative status. In this sense the *Ratio* could control the boundaries of the thinkable, if not the contents of thought itself. Thus, it was comparable with other types of "indirect" control which were designed by the society for coping with the necessity of multiple interpretations.

Authority and Control in the Moral Sphere

The crisis of authority of the 1560s–1570s and the kind of solutions suggested by the Jesuits' experience as cultural mediators were not confined to the intellectual sphere. Signs of crisis in the domain of moral education

---

deputati sunt ex ipsa congregatione tres theologi, scilicet P. T. Canisius, P. J. Torrensis et P. T. Peltanus ut eiusmodi opiniones colligerent et deinde ad nostrum Patrem mitterentur, qui de eis statueret, an ex nostris scholis plane sint explodendae . . ."

69. *MP*, IV, p. 220, "Ex actis congregationum privincialium Italiae", 1572–1573.

were perceived in the 1560s and engendered a model of social control whose application exceeded the boundaries of the Society of Jesus.[70]

Here, too, the period previous to the publication of the final version of the *Ratio studiorum* in 1599 should be divided into two main parts.[71] The first part, the experimental stage, lasted from the opening of the first Jesuit college in the 1540s to the crisis of the mid-1560s, and gave birth to the slogan "studies together with moral formation", expressing the goals of Jesuit education. Accordingly, the moulders of the educational policy, including the first two generals of the society, Ignatius and Laínez, and the architect of Jesuit pedagogy, Ledesma, issued two sets of constitutions for the colleges: one concerned with the curriculum, and the other with moral formation, emphasizing religious practices such as prayers, masses, sacraments, spiritual exercises, etc. During this first stage, the Jesuits succeeded in introducing many pedagogical innovations: organizing the educational material according to levels and dividing the classes according to this arrangement; intensive use of pedagogical exercises; rational division of the day, the week, and the year into periods of study and periods of rest; and the provision of competition, prizes, and punishments as encouragements and deterrents. However, the reports from the colleges (the German College in particular was singled out as an example) indicated that the effectiveness of the teaching was no guarantee for the inculcation of moral values. The opposite may have been true, as a complaint of P. H. Firmani, Rector of the College of Perugia, testifies:

> It appears that the aforesaid magisters are very dry in things of the spirit, and the rectors and others are consequently worried. The reason and cause of this may be the continuous distraction of the books they read, which are. . . . remote from the [religious] spirit, and attract the intellect and affect of the young people, so that at times of prayer and self-examination their brain is full of the concepts and things of Virgil, Cicero, and other similar ones, and they are always preoccupied with them.[72]

It seems that for more than two decades there was no real integration between the two sets of constitutions. The crisis resulting from the recognition of this fact gave rise to a new concept in Jesuit educational thinking. The emphasis changed from pedagogical concerns to a concern with

---

70. G. M. Anselmi, "Per un'archeologia della *Ratio:* dalla 'pedagogia' al 'governo', in: G. P. Brizzi (ed.), *La 'Ratio Studiorum'; Modelli culturali e pratiche educative dei Gesuiti in Italia tra Cinque e Seicento,* Rome 1981, pp. 11–42.

71. Ibid., p. 15.

72. *MP,* III, p. 373, H. Firmani to F. De Borgia, Siena, 17 February 1565.

problems of "control" (*governo*), administration, inspection.[73] A crucial text for the understanding of Jesuit thinking on *governo* was dictated by I. Cortesono,[74] rector of the German College in the years 1564–1569, to Lauretano, his successor. It was here that an awareness of the problem of "control" was first manifested and the problem was isolated from other problems of educational policy. According to Anselmi, Cortesono was among the first to notice that the problem of education in the Jesuit colleges was far more complex than "studies together with moral formation". It required the development of a special corpus of knowledge, *sapere di governo*. "Three things are needed in order to preserve the society: learning, a religious spirit, and 'governo'".[75] The latter was more important than the others, for it determined the manner in which the two other goals – intellectual achievement and the religious spirit – were attained.[76] Cortesono suggested that certain members of the society should be specially trained for the task of administration and control and should assume responsibility for carrying out the rules to be made in this connection.[77]

In explaining the new concept of *governo*, Cortesono indicated three possible areas of application:

(1) In the examination and selection of candidates, which required systems for analyzing their characters, habits, and modes of thought, beginning with general characteristics and touching on the most intimate details of their existence. A system of classification was to be created for determining the place of the individual pupil in the institutional structure of the society. The student thus became an object in the hands of the educators who sought to penetrate to the depths of his being in order to find the points most amenable to influence.

(2) In the development of techniques for the supervision of the individual pupil in order to integrate him into a homogeneous group. The isolation of exceptions and an effective system of punishments and rewards, with flexibility concerning details and firmness with regard to principle, were typical examples of the Jesuits' techniques of supervision.

(3) In the increasing specification of the pyramid of the directional hier-

73. Anselmi, "Per un'archeologia . . .", p. 26 ff.
74. *MP*, II, p. 864 ff., I. Cortesono, "Constitutiones seu monita ad eorum usum qui Collegio Germanico praesunt", 1567–1570.
75. Ibid., p. 869.
76. Ibid.: "Guidico l'ultima più necessaria dell'altre, perché è quello che cava frutto dall'uno e dall'altro, ordinandole al suo fine . . ."
77. Ibid.

archy, each part of which was to concern itself with one aspect of the student's life: studies, prayers, confession, social conduct, etc. At the top of the pyramid were those with a directive capacity, which was something separate either from teaching or from religious guidance. These remained the concern of the teachers and spiritual mentors, whose function became essentially technical. The directional hierarchy had the task of overseeing the delicate points of intersection between intellectual and moral education.

G. M. Anselmi, who first exposed and explained the notion of "governo" and its various applications, claimed the *Ratio studiorum* to be a consequence of the shift of Jesuit preoccupation from pedagogical issues to issues of control. Anselmi, however, tended to isolate the moral and disciplinary sphere from the intellectual one. According to Anselmi, the source of Jesuit dynamism was entirely social, and became culturally significant, although their intellectual programme, crystallised in the first twenty years and accepted by the mid-1560s, was rigidly orthodox.[78] In fact, Anselmi's separation of the intellectual and social dimensions of Jesuit activities is equivalent to saying that they were socially effective *in spite* of their intellectual *irrelevancy*.

I believe that my reconstruction of the debate on the freedom of opinion complements Anselmi's analysis of the social dimension of Jesuit life and justifies further investigation of their intellectual as well as social relevancy. Both the intellectual programme and the mindset of the Jesuits were shaped by their experience as cultural mediators and the practices developed in the course of coping with that position. The underlying structure of their resolution of both the intellectual and moral crisis is worth describing.

The Jesuits' experience as cultural mediators produced an inclusive approach which undermined any transcendental Archimedean point as a basis for drawing the boundaries of the cultural field. The search for new boundaries led to a crisis in the intellectual and moral sphere, and modified the traditional structure of authority. By stretching the limits of interpretation, a typical Catholic strategy, the Jesuits attempted to cope with a complex intellectual and social reality and shifted from a dogmatic to a pragmatic position. Interpretation was applied to widened areas of life and thought. Not only were theological and philosophical texts submitted to interpretation, but also nature – the subject of natural philosophy, and even human nature – the subject of moral education. Very soon the Jesuits found themselves in a world of multiple interpretations, where traditional sources of authority proved ineffective in the face of pluralistic

78. Anselmi, "Per un'archeologia . . .", pp. 30, 36–37.

social and intellectual reality. A mechanism for the legitimation of new interpretations was invented, while traditional presuppositions about truth guaranteeing doctrinal unity and the inculcation of the moral sense guaranteeing discipline were abandoned. The quest for truth was mitigated by the affirmation of the "possible", and doctrinal unity gave way to the "consensus". The treatment of the "possible" as an alternative, and not simply as a negation of truth, was a practice first used by the Jesuits in their role as cultural mediators, especially in the controversies with heretics. This practice affected the epistemological distinction between the "true" and the "probable", which had already been used as a principle for distinguishing the status of different types of knowledge in the time of Thomas. In the context of Jesuit intellectual life, labeling an opinion "probable" did not mean enclosing it within the sphere of "fiction", as was the tendency with some classical and medieval currents of thought.[79] Rather, in this context "probability" was a form of legitimation of new ideas as alternatives, with the possibility of receiving further confirmation from the consensus.

The first twenty years of the society were characterized by a traditional belief in a predominance of a true interpretation, usually the Thomist one, and a tacit demand for a doctrinal unity grounded in truth. The following years engendered practices permitting the legitimation of new interpretations, most notably the constitution of "possible" opinions as candidates for acceptance by the consensus. These practices, however, needed to be counterbalanced by adequate practices of control over the multiplicity of interpretations.

The mechanism of control developed by the Jesuits differed substantially from traditional ones in being indirect, far less explicit, and far more dependent on a strong institutional identity. The Jesuits favoured neither inquisitorial means in the disciplinary domain nor an authoritative philosophical and theological system in the intellectual sphere. Open coercive force was substituted by a hidden system of power relations. Power was exercised through a set of practical rules, meaningful only in the context of the institution, which were to guide interpretations and to serve as a code for choosing between them. In the moral sphere, this code was constituted by the methods of classification and supervision described by Cortesono, and labeled by him *sapere di governo*. Cortesono's text describes the process by which the code was crystallised: "And thus, long experience has taught me everything connected with *governo*."[80] Every rector could take from the constitutions based on experience whatever he

79.   See ch. 2.
80.   *MP*, II, p. 867.

required or add to them in accordance with his evaluation of the needs of his college.[81] For *governo* had to be suited to the time, the place, and the type of people who were to be educated.[82] Experience supported by sound rational arguments and by the authority of senior scholars and educators was *governo*.[83]

The equivalent of *governo* in the intellectual sphere were rules for the classification and hierarchization of areas of knowledge, subjects which will be dealt with separately in the following pages. The main point to emphasize here is that the boundaries between areas of knowledge, rather than being any limitation on the content of opinions, became the most effective mechanism of control. For the rules of the organization of knowledge affected the socialization into modes of thought in the formative stages of the development of the intellect, and were thus much less open to challenge than a particular interpretation in competition with various others.

My description of the crisis of authority in the intellectual sphere converges, finally, with Anselmi's conclusion: When the *Ratio studiorum* – the definitive code of Jesuit education – assumed its final form, it was taken as a model for dealing with problems of control in the broadest sense: not only in the Jesuit colleges, but also in those under the direction of the Tridentine Church as a whole and the secular rulers of various European countries at the dawn of the early modern era.

81. Ibid.: ". . . et acciò il prudente rettore secondo li tempi e occasioni, pigli quello che li parrà farsi più a proposito per utile di suo collegio. et quando ancora li parrà di provar altre cose a simiglianza di queste, con il consenso però di superiori (. . .) et riusciendole alla pruova, le seguiti . . ."
82. Ibid.: ". . . secondo la mutatione delli tempi et degl'huomini, così conviene mutare il modo di governarli . . ."
83. Ibid.: "Oltre che ogni cosa è fondata, oltre dell'esperienze in vivacissime ragioni, et avanti et dopo d'esser approvate, sono state reviste da molti padri esperti nel governare et iuditiosi, et ultimamente dalli assistenti, e da sua Peternità Reverenda . . ."

8

The Thomist Boundaries of
Jesuit Education

In 1622 Ignatius Loyola, the great visionary of the Society of Jesus, and
Francis Xavier, its most daring missionary, were canonized. To celebrate
the occasion Father Guiniggi, Professor of Rhetoric at the Roman College, wrote a play which was put on as a theatrical production at the
college.[1]
    The text incorporated several didactic themes enacted in successive
scenes located in different places. In one act at the heart of this production, Ignatius was depicted torn between the forces of good and evil. The
spectacle told the story of his temptations. It was situated in hell, but hell
and heaven intermingled with the appearance of the Archangel Michael,
accompanied by four celestial princes who furnished Ignatius with arm-
our to fight the demonic, threatening powers. This central act, the third
among five, was a dramatic embodiment of a timeless, universal element
always present in Jesuit theatre, following the traditional form of the medieval "tragedia sacra": a struggle between light and darkness, between
truth and illusion, between faith and heresy, culminating in the victory of
faith and light, the good and the true. Typical, too, was the framing of
the central didactic message with concrete manifestations of power in an
attempt to contextualize the position of the Jesuits in the power structure
of the church and the state. Thus, the first act, taking place in a square in
Monteserrat, portrayed "ecclesia militans" descending from a cloud to
meet the newly elected Pope Hadrian VI, followed by the four continents
which paid homage to the church. The second act, located in a theatre,
showed a tournament between Spanish and French soldiers in honour of
the pope. And the last act was a huge pageant, attesting to the appropriation of elements of popular culture into the highly cultivated environment
of the Jesuits.[2]
    Jesuit theatre serves me as a particularly useful example for the illustra-

1. P. Bjurström, "Baroque Theater and the Jesuits", in R. Wittkower & I. B.
    Jaffe (eds.), *Baroque Art: The Jesuit Contribution*, New York 1972, p. 101.
2. Ibid.

tion of the theme of this chapter: i.e., the challenge to traditional modes of thought and behaviour presented by the rich new elements and variety of practices to which the Jesuit educational program had given rise. This challenge, however, was by no means explicit. The architects of Jesuit education never tired of proclaiming the Thomist spirit of their enterprise. The adoption of the *Summa theologiae* as the main text in the course of scholastic theology[3] and the repeated recommendation in pedagogical documents of the Thomist interpretation of Aristotle[4] has convinced many historians that the intellectuals of the society derived their main inspiration from Thomas Aquinas.[5] This impression is justified in only a very limited sense, however. A short excursion into the varied preoccupations of Jesuit educators, illustrated by a few examples, will reveal the kind of praxis which actually served to undermine the basic premises upon which the Thomist organization of culture rested.

The multiple facets of the Jesuit educational programme will be illustrated through three examples. I shall start by discussing the character of Jesuit theatre as a rhetorical practice and as part of the humanistic course of studies. I shall then turn to the spread of hermetic-emblematic modes of thought which gained support and legitimation among Jesuits long after their decline among other groups of European intellectuals. Some non-Thomist directions in natural philosophy will also be discussed in order to expose the inner tensions which existed in the Jesuit educational system, threatening to undermine it from within.

Jesuit Theatre as a Manifestation of Rhetoric

The representation of Ignatius's life in the production of 1622 exemplifies many of the characteristics of Jesuit theatre in general.[6] Usually, the plays

3. *MP,* I. pp. 15–151, Nadal, de studii . . .
4. *MP,* II, p. 478, Ledesma, Relatio . . .
5. See, for example Anselmi, "Per un'archeologia della Ratio: dalla 'pedagogia' al 'governo'" in: Brizzi (ed.), *La 'Ratio Studiorum'* . . . , p. 37.
6. The sources I have used for my presentation of Jesuit theatrical activities as a rhetorical practice are: E. Boysse, *Le théâtre des Jesuites,* Geneva 1970 (reprint of Paris 1880 edition); Bjurström, "Baroque Theatre and the Jesuits" in Wittkower and Jaffe (eds.), *Baroque Art* . . . , pp. 99–110; A. Beijer, "Visions celestes et infernales dans le Théâtre du Moyen-Age et de la Renaissance" in: *Les Fêtes de la Renaissance, Etudes reunies et presentées par J. Jacquot,* Paris 1973 (1st ed. 1956), T. I., pp. 405–419; M. M. McGowan, *L'Art du ballet de cour en France 1581–1643,* Paris 1978, ch. 12, "La contribution des Pères Jésuites au Ballet", pp. 205–227; M. F. Christout, *Le ballet de cour de Louis*

contained three or five acts. A balance was maintained between the universal moral message on the one hand, and a concrete political and social context on the other, created by the dramatization of well-known historical personalities: Pope Hadrian VI, for example, or Louis XIV. No doubt the representation of the social and political milieu in which the Jesuits operated was a way of finding favour with the Jesuits' great patrons. This was openly admitted in theoretically oriented writing concerning the work of the Jesuit poet and playwright. In a treatise on drama written by the Jesuit Father Lejay, the following instructions to the poet demonstrate that the theatre was partly conceived as a means of catering to the political environment, confirming its "grandeur" and legitimizing it as the protector of the arts:

> Cherchant ensuite un beau tableau d'ensemble, et une occasion d'amener agréablement l'éloge du roi, notre poète imaginera de réunir Apollon, accompagné des Muses, et tous les dieux inventeurs des arts autour de l'effigie royale. Chacun viendra offrir une couronne au monarque, protecteur éclairé et généreux des arts.[7]

In its fully developed form, the spectacles included interludes between the acts, in which singing and dancing were performed. The new art of ballet, a product of the court of France in the seventeenth century, was quickly incorporated into Jesuit theatre, and the Jesuits were the first to invent the appropriate technical terms in Latin by including descriptions of ballet-dancing in their plays.[8]

Not much is known, I believe, of the setting and mise-en-scène of the particular play composed by Father Guiniggi for the canonization of Loyola and Xavier. But, from the testimony of Father Jouvençy, who theorized about the writing of tragedies in his *De ratione discendi et docendi* (1685) it is possible to reconstruct current practices. Our young dramatists believe, he said, "avoir faite une excellente tragédie, quand elle a occasionné de grandes dépenses, quand les décors sont magnifiques, les costumes chargés d'or, et quand la musique est délicieuse".[9] In spite of the irony expressed by Jouvençy, who thought that the sensual taste for decoration should be somewhat controlled, Jesuit theatre was famous for

---

XIV 1643–1672, Paris 1967, ch. 5, "Théories et théoriciens, Claude-François Menestrier (S.J.)", pp. 137–155; F. de Dainville, *L'Education des Jésuites au XVIe et XVIIIe Siècles*, Paris 1978; J. Jouvençy, SJ, *De la manière d'apprendre et d'enseigner* (*De ratione discendi et docendi, 1685*), trans. by H. Ferté, Paris 1892.

7. A paraphrase from *De choreis dramaticis* by Père Lejay, quoted by Boysse, *Théâtre des Jésuites*, p. 39.

8. Ibid., ch. 4, pp. 31–58.

9. Jouvençy, *De ratione . . .* , p. 55.

the intensive use it made of every possible mechanical device to create the illusion of the heavens opening, of the fires of hell burning, of deities flying through the air, and for a developed system of lighting with thousands of weak, adjustable flames, which provided a flood of shadow-free, undirected light—a brilliant ambience where the wooden auditorium simulated gold and precious stones.[10]

The themes of Guiniggi's play were conventional enough. Based on the life story of the two saints, which became part of the history of the church, the themes conformed to the requirements put forward in Jouvençy's *De ratione* for selection of themes: "on fera bien de la tirer du trésor si fécond des Saintes Ecritures, et des annales de l'Eglise, ou l'on trouve une grande abondance de faits utiles et admirables".[11] In fact, however, many Jesuit playwrights drew their themes from the rich variety of classical literature read in the courses of humanistic studies, which they reinterpreted in a Christian spirit. Part of the epic of Odysseus, for example, was interpreted symbolically with Odysseus representing wisdom resisting the temptations of profane love incarnated by Circe.[12]

The thematic choices, no less than the structure of the plays interspersed with singing and dancing, the stage settings and the decorations, all indicate the active participation of the Jesuits in the theatrical culture of the day, adopting its norms and using its conventions. It is the special context in which the plays were performed, however, which is particularly revealing with regard to the intellectual and political significance of this theatre. The *De ratione* considered the writing of tragedies to be an essential part of the vocation of professors of rhetoric in Jesuit colleges.[13] Indeed, the tradition of performing started at the German College in the mid-1560s as a lively method of instruction in ancient languages and literature.[14] It soon gained a broader significance as practical training in the application of rhetorical techniques. The Jesuits did not limit their humanistic education to the reading of classical texts such as Aristophanes' and Terence's comedies, Isocrates' orations, Aesop's fables, Plato's dialogues, Thucydides' histories, Homer's and Virgil's poetry, etc. Nor did they simply teach Quintilian's *Institutiones oratorium*, the pseudo-Ciceronian *Rhetorica*, or the more recent humanistic writers Erasmus, Valla, and Vives.[15] Most significantly, they practiced rhetoric in the form

10.   Bjurström, "Baroque théâtre . . .", p. 107.
11.   Jouvençy, *De ratione* . . . , p. 53.
12.   Bjurström, "Baroque théâtre . . .", p. 102.
13.   Jouvençy, *De ratione* . . . , Ch. II, a. II., pp. 48–58.
14.   Bjurström, "Baroque théâtre . . .", p. 99.
15.   *MP*, I, Nadal, De studii . . . , pp. 138–144; Olave, Ordo lectionum . . . , pp. 168–170.

of playwriting and performing, seeking to move their audiences into actively living a Christian life, engaging them with their sacred message and inculcating them with a moral sense. The Jesuit educators' expectations from the theatre were expressed by Father Jouvençy who insisted that "les exercices littéraires de nos classes servent à apprendre le latin"[16] but also that "il faut que cette action puisse contribuer à former les moeurs . . . quelle que soit la source d'où l'on tire un sujet, il faut qu'il soit toujours sérieux, grave et digne d'un poète chrétien. Une pièce brillante, et dans laquelle les moeurs sont habilement traitées, émeut le spectateur et produit souvent plus d'effet que le discours le plus savant et le plus éloquent".[17]

Jesuit theatre also had a political dimension, of which the portrayal, in Guiniggi's play, of the four continents paying homage to the church was but one example. By definition, "la tragédie a pour but d'instruire les princes et les héros".[18] Hence, most plays included representations of courtly life either of secular rulers or of prelates, which the Jesuits used as a means of rendering service to their patrons. The huge pageants, the reproduction of courtly life including the elegant court of the pope, all served to fix in the imagination of the audience the equivalence between the dignity of the ruling class and that of the high clergy. Among the audience there were usually many grand personalities: "Des princes, des princesses / Des présidentes, des comtesses, / Quantités d'esprits de bon sens / Et des moines plus de deux cents",[19] whose presence was recorded by one feuilletonist of the period. The fashion of the court soon became a fashion among the educators of the ruling class, and justified emulation. Ballet-dancing, in high fashion at the French court, was thus justified by the *De ratione:* "On peut, dans les pièces de théâtre, faire une place à la danse, qui est un plaisir digne d'un homme bien élevé, et un exercice utile à la jeunesse. Ajoutons que la danse dans un drame est une sorte de poésie muette, et qu'elle exprime par les mouvements du corps ce que les acteurs disent dans des vers".[20] Simultaneously, the Jesuits aimed to mould their audiences as a cultural elite identifying with the values of both the ruling class and its educators. Again, it is Father Jouvençy who bears witness to the sophisticated strategic thinking of the architects of Jesuit educational policy: "D'ailleurs, nos théâtres ne doivent pas rechercher toute sorte de plaisirs, mais seulement celui qui est digne d'un spectateur choisi et érudit. Ces merveilles de l'art perdent leur prix quand on les rabaisse au goût

16. Jouvençy, *De ratione* . . . , p. 54.
17. Ibid., pp. 53–54.
18. Ibid., p. 55.
19. Boysse, *Théâtre des Jésuites*, p. 81.
20. Jouvençy, *De ratione* . . . , p. 57.

et au caprice d'une multitude ignorante".[21] While adapting themselves to the utilitarian needs of their clients, the Jesuits simultaneously sought to direct the taste of their audiences to accept the values and hence the authority of the ruling classes who patronized the order.

Jesuit theatre was a place where the theory of rhetoric as a corpus of knowledge, and its practice as a means of action, virtually intersected. The rehabilitation of rhetoric as an autonomous intellectual domain with claims to real knowledge signaled the assimilation by Jesuit culture of the humanistic criticism of scholastic modes of thought, considered too abstract and irrelevant to the solution of concrete problems. In the theatre the Jesuits used rhetoric for the purpose of imparting a moral message, and of bringing the Gospel closer to wide audiences. At the same time they also used rhetoric in the theatre to establish their position as an intellectual elite in the cultural field, to identify themselves with the ruling class in the political field, and to legitimize the political power which patronized them.

It is necessary to remind the reader here that the first stage of Jesuit instruction consisted in a thorough humanistic education which lasted for three or four years, in which grammar, literature, and rhetoric were successively taught.[22] In many ways, the humanistic code of Jesuit education sharply diverged from the traditional scholastic modes of thinking and behaviour. The outstanding contribution of humanist ethics at the dawn of the Renaissance was the promotion of the ideal of the active life (*vita activa*), as against the ideal of the contemplative life (*vita contemplativa*) which had characterized medieval culture and the spiritual universe of Thomism.[23] The Jesuit style of "Christian humanism", representing a fusion of humanist values and Christian thought, put forward the view that the path to the love and knowledge of God was the way of

21.   Ibid., p. 54.
22.   A. Farrell, *The Jesuit Code of Liberal Education . . .* ; Dainville, *L'Education des Jésuites . . .* ; Ganss, *Saint Ignatius's Idea of a Jesuit University.*
23.   H. Baron, *The Crisis of the Early Italian Renaissance*, Princeton 1966, 2nd ed. (1st ed.: 1955); W. J. Bowsma, *The Culture of Renaissance Humanism*, Washington 1973; E. Garin, *Science and Civic Life in the Italian Renaissance* (original title: *Scienza e vita civile nel Rinasciemento italiano*, Bari 1965), trans. by P. Munz, New York 1969; idem, *Italian Humanism, Philosophy and Civic Life in the Renaissance* (original title: *Der italienische Humanismus*, Bern 1947), trans. by Princeton 1966; P. O. Kristeller, *Renaissance Thought*, New York 1961–1965, 2 vols; idem, *Renaissance Concepts of Man*, New York 1972; C. Trinkaus, *In Our Image and Likeness*, London 1970, 2 vols; E. P. Mahoney (ed.), *Philosophy and Humanism: Renaissance Essays in Honor of Paul Oskar Kristeller*, Leiden 1976.

action rather than of contemplation. This claim provided a legitimation for the connection made by Renaissance culture between action and knowledge, and prepared the way for the acceptance of a form of knowledge which was regarded as both practical and real.[24] Such a concept contradicted the classical Platonic and Aristotelian view that practical knowledge was not true knowledge (*scientia*) at all.

The new ideal of living manifested itself in a rejection of the idea that the study of natural philosophy, which served as a preparation for metaphysics and as a training for the contemplative life, was superior to any other kind of knowledge, and that formal logic was the sole instrument for the investigation of truth. By making human reality the centre of interest, humanism created a new order of priorities with regard to the relevancy of different spheres of knowledge. For Renaissance scholars, understanding human experience in rhetorical terms was no less valid than investigating natural experience by means of logical arguments. Although the humanists never attempted to refute the accepted natural philosophy, they tended to neglect universal, theoretical issues in favour of immediate, circumstantial, concrete problems of life. Such a preference led them to pay much more attention to the kind of Aristotelian methods suggested in the *Topics*, and the criteria of reasonableness applied in Aristotle's writings on ethics, poetics, and politics. The quest for certainty and the insistence on the logical necessity of argument, which played such a prominent role in traditional philosophical discourse, was replaced in humanistic writings by an appeal to opinion supported by sound experience – which did not stand up to the traditional criteria of rationality, and which challenged their practical applicability. The coexistence of these two forms of discourse with their competing claims for rationality implicitly questioned the hegemony of the Thomistic system with its preference for the contemplative life, speculative knowledge, and logical rigour.

Hermetic and Emblematic Modes of Thought

Between the years 1596 and 1604 a three-volume commentary on the prophecy of Ezechiel was published in Rome by Juan Baptista Villalpando

24. A. Crescini., *Le origini del metodo analitoco: il Cinquecento,* Udine 1965; N. W. Gilbert, *Renaissance Concepts of Method,* New York 1960; B. Vickers and N. S. Struever, *Rhetoric and the Pursuit of Truth,* Los Angeles 1985; V. Kahn, *Rhetoric, Prudence and Skepticism in the Renaissance,* Ithaca 1985; N. S. Struever, *Theory as Practice: Ethical Inquiry in the Renaissance,* Chicago 1992.

in collaboration with Jeronimo del Prado, both Jesuits. The second volume of this commentary included a reconstruction of the Temple of Solomon, arguing that only by translating Ezechiel's vision into terms of real architecture could the mystical message of the prophet be fully apprehended.[25] Among the many illustrations of this reconstruction which have survived the ravages of time, one in particular exemplifies the kind of hermetic discourse fully legitimized by this commentary. The illustration represents the astrological organization of the temple, and is thus described by Taylor:

> Since the Temple was based on the Tabernacle of Moses, it is shown to have had twelve *castella* or bastions exactly matching the positions of the twelve tribes round the Ark of the Covenant. In addition, there were four bastions corresponding to the Levitical camps. These, being in the center, are to be identified with the sub-lunar world of the four elements, which is the world of man. The remaining twelve bastions correspond to the twelve houses of the Zodiac. Each tribe is allotted its own sign. Thus, Judah, as we would expect is given the sign of Leo, Issachar that of Cancer, and so on. In the same way the seven courts were assigned to the movable planets: the southwestern one to Mars, and so forth. But here again they were not disposed at random. They were placed, as far as it was feasible, in close proximity to the house or houses in which they are enthroned: Saturn in Capricorn and Aquarius, Jupiter in Pisces, Mars in Aries, and so forth. By all this, it may be assumed that the author wanted to convey the idea that the Temple was initiated when the planets were dignified in their own signs. Such an unusual accumulation of beneficent astral influences would have been in keeping with its divine origin.[26]

The astrological organization of the temple illustrated only one facet of the hermetic approach which Villalpando represented. The commentary also incorporated numerology, mystical geometry, Pythagorean musical intervals, mystical properties of colours and precious stones, Cabalic and Pseudo-Dionysian hierarchies. Simultaneously, however, the commentary contained remarks on the center of gravity relevant to practical problems of construction, which probably preoccupied Villalpando no less than the mystical meaning of the structure of the temple.

This intermingling of practical and mystical mathematics was by no means alien to Renaissance intellectuals. From Girolamo Cardano to John Dee, from Copernicus through Kepler to Newton mathematicians of the sixteenth and seventeenth centuries made positive mathematical

25.   R. Taylor, "Hermetism and Mystical Architecture in the Society of Jesus", in Wittkower & Jaffe (eds.), *Baroque Art . . .* , p. 75.
26.   Ibid., p. 79.

discoveries and were simultaneously engaged in preparing horoscopes, making mystical apocalyptic calculations, and contemplating the mathematical structure of the world, without being aware of any contradiction between these two kinds of enquiries. From the point of view of present-day historiography the coexistence of the mystical and the rational, the practical and hermetic in Villalpando's and Del Prado's text comes as no surprise. Today it is possible to speak of a whole historiographical tradition – from L. Thorndike to E. Cassirer, from E. A. Burtt to Frances Yates[27] – which attempts to expose the coexistence of the "rational" and "irrational" uses of mathematics in the Renaissance and the seventeenth century, to explain this encounter in terms of the more or less simultaneous rediscovery of Greek mathematical texts and the complete Platonic corpus, and to argue the value of this conjunction for the scientific revolution. Jesuit examples, however, have hardly been touched by this historiographical tradition, for the Jesuits have been regarded as belonging to the Thomist tradition, known for its suspicion towards magical practices and mystical excesses.

But Villalpando and his collaborator Del Prado were Jesuit priests who took an active part in the educational enterprise of the society. Del Prado, the elder of the two, was a master of arts when he joined the Jesuits, and he then became a bachelor in theology and taught Sacred Scripture in Baeza. He was also a sculptor, and knew a great deal about architecture. Villalpando studied mathematics before entering the order and gained practical architectural training with Juan de Herrera, royal architect to King Philip II. Then he studied arts and theology and became one of a small group of Jesuit architects who created the severe, monumental Jesuit style under the Spanish influence of Herrera. In 1592 Del Prado and Villalpando were probably transferred to the Collegio Romano – the bastion of Jesuit intellectual life – where they stayed until their deaths.[28]

After its appearance, the commentary on Ezechiel provoked a long debate which lasted throughout the seventeenth century and was even echoed in the eighteenth. Villalpando's work was much criticised for be-

27. L. Thorndike, *History of Magic and Experimental Science*, 8 vols., New York 1923–1958; E. Cassirer, "Mathematical Mysticism and Mathematical Science", in E. McMullin (ed.), *Galileo: Man of Science*, New York 1967, pp. 338–351; E. A. Burtt, *The Metaphysical Foundations of Modern Science*, rev. ed., New York 1954; F. Yates, *Giordano Bruno and the Hermetic Tradition*, Chicago 1964; N. H. Clulee, *John Dee's Natural Philosophy: Between Science and Religion*, London 1988; I. Merkel and A. G. Debus (eds.), *Hermeticism and the Renaissance*, Washington 1988.
28. Taylor, "Hermetism and Mystical Architecture . . ."

ing unfaithful to the biblical text in identifying the temple of Ezechiel's vision with Solomon's Temple, for example. It also provoked criticism from rationalists who abhorred the heights of mysticism to be found in the text.[29] At the same time, it elicited much interest and even high praise from the educated public, which was fascinated by the peculiar combination of the practical and the mystical it offered.[30] In fact, the book created a kind of common ground for biblical exegetes, theologians, mathematicians, and architects, a model of exchange in the context of which the old boundaries between mathematical and physical approaches to the universe, between practical and theoretical forms of knowledge, were crossed. That such a dialogue continued and became an integral part of the Jesuit intellectual inheritance is nowhere more manifest than in Athanasius Kircher's works.[31] For Kircher, scientific research meant nothing less than the penetration into the Divine Mind, an enterprise justified by the hermetic teachings. To achieve his goal Kircher deemed it necessary to preoccupy himself with a wide range of branches of knowledge, from theology and metaphysics to physics, mathematics, and mechanics, and from politics and scriptural interpretation to rhetoric and the combinatorial art. His *Ars Magna Sciendi* is nothing less than a living memorial to the attempts at a new synthesis of all knowledge, which some Jesuits aimed at as a result of the opening of boundaries effected by their culture.

Del Prado's and Villalpando's commentary on Ezechiel is a significant example of the actual involvement of some Jesuits, perhaps substantial groups of Jesuit intellectuals, with hermetic practices. How deep this involvement was is still a disputed matter among scholars. Some believe[32] that towards the end of the sixteenth century the Jesuits became wholly committed to an emblematic world-view, which considered nature to be a collection of signs and metaphors carrying hidden meanings and open to any number of interpretations. The opposition to such a world view within the society, however, is well known. It is sufficient to mention the *Disquisitionum Magicarum libri sex* (1599) of Martin del Rio, which vehemently condemned hermeticism, and the stark antipathy towards hermeticism expressed by the bearers of the scholastic tradition, rationalists

29. Ibid., p. 75.
30. W. Herrmann, "Unknown Designs for the 'Temple of Jerusalem' by Claude Perrault" in: *Essays in the History of Architecture Presented to Rudolf Wittkower*, New York 1967, pp. 143–158.
31. J. Godwin, *Athanasius Kircher: A Renaissance Man and the Quest for Lost Knowledge*, London 1979.
32. W. B. Ashworth, Jr., "Catholicism and Early Modern Science", in Lindberg and Numbers, *God & Nature ...*, p. 157.

of the type of Bellarmine and Suarez.[33] It cannot be denied, however, that hermeticism had its imprint upon other Jesuits who clearly identified with this tradition.

The involvement with hermeticism necessarily introduced tensions into an educational program which consciously defined itself in terms of Thomism. The hermetic writings[34] were part of the complete Platonic corpus translated in its entirely for the first time during the Renaissance. Hermetics of different shades and colours all believed that nature was dynamic and full of hidden forces, among which there was a mutual interaction: between higher and lower beings, between the characteristics of various planets and certain human characteristics, between the macrocosm and the microcosm. Not only mutual interaction but analogies between the world as a huge animal and created things, between terrestrial and celestial entities, and between the human body and its natural surroundings were the key to an understanding of both man and the cosmos. In addition, many hermetists believed that numbers and combinations of numbers were symbolical representations of the world and the key to understanding it. The hermetic writings also offered a concept of man as not only created in the image of God, but as similar to God in powers of creation and involvement in the universe. Knowledge of the world consisted in the interpretation of analogies between things and capturing the influences working in the cosmos. This knowledge could be gained through an intimate acquaintance with nature and through technical means, which in the Renaissance nearly always coincided with magical techniques. The "image of knowledge" which associated knowledge of nature with manipulation rather than with the logical analysis of intellectual categories probably had its origins in this tradition. This "image of knowledge" was reinforced by the association of hermeticism with the traditions of artists and technicians, as the cases of Villalpando and Kircher clearly demonstrate.

But there is more to their example. Del Prado and Villalpando on the

33.   Taylor, "Hermetism and Mystical Architecture . . .", p. 66.
34.   For a general view of the literature on hermeticism see, apart from the indispensable work of F. Yates: D. P. Walker, *Spiritual and Demonic Magic from Ficino to Campanella*, London 1958; W. Schumaker, *The Occult Sciences in the Renaissance*, Berkeley 1972; P. Rossi, "Hermeticism, Rationality and the Scientific Revolution", in M. L. Righini Bonelli and W. R. Shea (eds.), *Reason, Experiment and Mysticism*, New York 1975; R. S. Westman and J. E. McGuire, *Hermeticism and the Scientific Revolution*, Los Angeles 1977; B. Vickers (ed.), *Occult and Scientific Mentalities in the Renaissance*, Cambridge 1984.

one hand and Kircher on the other typify the intellectual exchange between theoreticians and practitioners which first became possible through Renaissance culture.[35] The encounter between artists, who developed the ideal of "true description" based on observation, and mathematicians, who provided the knowledge necessary for the theory of mathematical perspective, indicated a new sensitivity to the relevancy of practical knowledge to the theoretician, and conversely, the need of engineers, architects, cartographers, and navigators to legitimize their techniques in terms of mathematical knowledge. In such a culture, *ars* (art) – practical knowledge originating in experience and belonging to the senses – could not easily be separated from *scientia* (science) – speculative knowledge whose source is the intellect. The cultural boundary dividing them, operative on the intellectual and social level alike, now became more precarious and was even endangered. Similarly, the boundary between abstract mathematical entities, the traditional objects of mathematical study, and material substances, the traditional objects of philosophical discourse, had also become more problematic.

The Practice of Natural Philosophy and Mathematics among Jesuits

Natural philosophy constituted the heart of the curriculum of a Thomist educational program. A study based on the examination of thirteen theses[36] in natural philosophy, defended in different colleges and published between 1610–1670, provides further evidence of the potential crossing

35. On the encounter between theoreticians and practitioners in the Renaissance, see: L. Olschki, *Geschichte der neusprachlichen wissenschaftlichen Literatur*, vol. I, Heidelberg 1919; vol. II, Leipzig 1922; vol. 3, Halle 1927 (repr. Vaduz 1965); E. Zilsel, "The Sociological Roots of Science", *American Journal of Sociology*, 47 (1942), pp. 544–562; idem, "Problems of Empiricism", in *International Encyclopedia of Unified Science*, II, 8 (1941), pp. 53–94; R. Hall, "The Scholar and the Craftsman in the Scientific Revolution", in M. Clagett (ed.), *Critical Problems in the History of Science*, Madison 1962, pp. 3–23; P. Rossi, *Philosophy, Technology and the Arts in the Early Modern Era*, New York 1970; G. de Santillana, "The Role of Art in the Scientific Renaissance", in Clagett (ed.), *Critical Problems . . .* , pp. 33–65; A. C. Crombie, "Science and the Arts in the Renaissance: The Search for Certainty and Truth, Old and New", *History of Science*, 18 (1980), pp. 233–246.

36. G. Baroncini, "L'insegnamento della filosofia naturale nei collegi italiani dei Gesuiti (1610–1670): un esempio di nuovo aristotelismo", in G. P. Brizzi

of traditional boundaries inherent in the kind of natural philosophy practiced by the Jesuits.

A few examples of the claims made in the texts will justify my notion of a crossing of boundaries, a notion lurking between the lines of these theses. One example was the claim that celestial matter was fluid and that generation and corruption could be perceived in the superlunary sphere.[37] Such a claim potentially challenged the distinction between astronomy as a mathematical science, dealing with unchangeable mathematical entities, and physics, the science of change and motion. If generation and corruption occurred in the heavens, then astronomical discourse could no longer be confined to geometry. The conditions for the possibility of a "physics" of the heavens were thus created and the demarcation line between astronomy and physics was immediately blurred.

The value of physics as a route up to metaphysics and subsequently to theology was something of a truism in the traditional approach. In many of the Jesuit theses, however, this function of physics was ignored; instead, physics was perceived as useful because it constituted a theoretical basis from which one could "descend" to an understanding of the various practical arts.[38] Such an assertion obviously represented a blurring of the boundary between the theoretical and the practical, the speculative and the concrete. A similar blurring occurred in the concept of nature as a "republic of concrete entities" known only in an empirical manner.[39] Such a concept of nature replaced the Aristotelian notion of a "perfect" world, final and unique, whose a priori unity was perceived as the guarantee of the possibility of its being amenable to "true" explanation.

Many of the theses in natural philosophy revealed an interest in chemical subjects. As a result, the Aristotelian distinction between the investigation of natural operations as the source of speculative knowledge and the investigation of artificial operations as the source of technical knowledge lost its validity. Instead, artificial operations were understood as permit-

---

(ed.), *La 'Ratio Studiorum': Modelli culturali e pratiche educative dei Gesuiti in Italia tra Cinque e Seicento*, Rome 1981, pp. 163–217.

37. Ibid., pp. 176–177.
38. Ibid., pp. 205–206.
39. Ibid., pp. 204–205: "Prioritas naturae, sive causalitas non consistit formaliter in dependentia physica causae ab effectu, neque in necessaria connexione effectus cum causa, neque in aliqua ordinatione constituta a natura tanquam medii ad finem, neque in formali determinatione Dei ad ponendum contingenter effectum sed potius in fundamento istius determinationis per quod scilicet determinatur Deus ad ponendum calorem potius ad praesentiam ignis, quam aquae".

ting an intensification of natural processes.[40] At the same time, the texts also indicated a desire to draw a sharp distinction between magic and technical subjects, and to include the intellectual products of technical operations within the sphere of genuine scientific knowledge.[41]

Most significant was the explicit discussion of the boundary line between physics and mathematics. In a textbook for the teaching of natural philosophy, written in the 1570s by Benedictus Perera, the writer claimed that mathematics could not offer explanations in natural philosophy, because mathematical entities, being abstract, could not be considered causes of natural substances. He therefore concluded that the mathematical disciplines were not to be regarded as true sciences ("Mathematicas disciplinas non esse proprie scientias").[42] This reasserted the Thomist distinction between the mathematical sciences which abstracted from time and matter and the physical sciences wholly concerned with material change. Another well-known Jesuit teacher repeated these assertions and added that all the "mixed" sciences, such as perspective, music, and astronomy, were "more mathematical than natural".[43] However, the mathematician C. Clavius, who wrote at the same period, was of a quite different opinion. Clavius thought that mathematics was more certain than all the other sciences and was essential for an understanding of natural philosophy, and he was not alone among the Jesuits of the period.[44] Some of them showed a belief in the reality of mathematical entities. A rather large number of works focussed on the practical areas of mathematics, traditionally regarded as "mixed sciences" and redefined by their writers as physico-mathematical. This may indicate the existence, among Jesuit educators, of a group of philosopher-mathematicians who worked in the area of practical mathematics, on which they sought to confer a status equal to that of philosophical knowledge.[45] Such a group probably coexisted with another group of philosophers who continued to give priority to Aristotelian physics, both in its Thomist version and in other versions as well, and who constituted the majority among Jesuit philosophers.

One more example throws further light on the problematic nature of the boundary between natural philosophy and mathematics in the Jesuit context. The example concerns one particular attempt to reconcile the Aristotelian philosophical tradition developed in the universities of

40. Ibid., pp. 186–187.
41. Ibid., p. 189.
42. Ibid., p. 192.
43. Ibid., p. 193.
44. Ibid., pp. 193–94.
45. Ibid., p. 195.

Northern Italy[46] with the tradition of practical mathematics revived by Italian humanists and mathematicians in the sixteenth century.[47] This attempt at the reconciliation of different traditions within a Thomistic framework, under the auspices of the Jesuit establishment, stretched to the limit the Thomist principles of the organization of knowledge. While revealing the tensions between practice and theory, between content and form which permeated Jesuit discourse, this attempt also demonstrated its function as a mediatory system between different scientific trends struggling for cultural hegemony.

My discussion is based on the indispensable work of W. A. Wallace who uncovered some crucial texts, written by a Jesuit, containing arguments concerning the nature and value of the mathematical sciences.[48] The texts were written by a disciple of Clavius, Josephus Blancanus, in

46. On the Aristotelianism of Italian universities in the sixteenth century, see: J. H. Randall, "The Development of Scientific Method in the School of Padua", *Journal of the History of Ideas* (1940), pp. 177–206; idem, "Paduan Aristotelianism Reconsidered", in Mahoney (ed.), *Philosophy and Humanism*, pp. 225–282; Gilbert, "Galileo and the School of Padua", *Journal of the History of Philosophy*, 1 (1963), pp. 223–231; idem, "Renaissance Aristotelism and Its Fate: Some Observations and Problems", in J. P. Anton (ed.), *Naturalism and Historical Understanding. Essays on the Philosophy of J. H. Randall, Jr.*, Albany 1967, pp. 42–52; idem, *Renaissance Concepts . . . :* "The Italian Aristotelians"; W. F. Edwards, "Randall on the Development of Scientific Method in the School of Padua – a Continuing Reappraisal", in Anton (ed.), *Naturalism and Historical Understanding . . . ,* pp. 53–68; Crescini, *Le origini . . . :* "Il metodo analitico nella scuola di Padova", pp. 134–189; A. Poppi, *Introduzione all'Aristotelismo padovano,* Padua 1970; C. B. Schmitt, "Towards a Reassessment of Renaissance Aristotelianism", in *Studies in Renaissance Philosophy and Science,* London 1981; L. Olivieri, ed., *Aristotelismo Veneto e Scienza Moderna,* 2 vols., Padua 1983.

47. On the revival of Greek mathematics by Italian humanists and mathematicians, see: P. L. Rose, *The Italian Renaissance of Mathematics: Studies on Humanists and Mathematicians from Petrarch to Galileo,* Geneva 1975; E. W. Strong, *Procedures and Metaphysics: A Study in the Philosophy of Mathematical-Physical Science in the Sixteenth and Seventeenth Century* (repr.), Hildesheim 1966; S. Drake and I. E. Drabkin, *Mechanics in Sixteenth Century Italy,* Madison 1969; P. L. Rose and S. Drake, " 'The Pseudo-Aristotelian Questions of Mechanics' in Renaissance Culture", *Studies in the Renaissance,* 18 (1971), pp. 65–104.

48. W. A. Wallace, *Galileo and His Sources: The Heritage of the Collegio Romano in Galileo's Science,* Princeton 1984. In fact, Wallace exposes the Jesuit context of the debate over the status of mathematics which raged

1615 and 1620 respectively, and entitled: *Dissertation on the Nature of Mathematics (De mathematicarum natura dissertatio)* and *Preparation for Learning and Advancing the Mathematical Disciplines (Apparatus ad mathematicas addiscendas et promovendas)*.[49] Both texts contain traces of the debate over the scientific status of the mathematical disciplines which raged in Italian academic milieus in the sixteenth century. While restating the terms of the debate and the arguments used on both sides, Blancanus's texts also permit a partial reconstruction of a power struggle between groups of professionals – mathematicians and natural philosophers – fighting for recognition and hegemony within the context of Jesuit institutions. The mention of specific names, the reference to the titles of works written by well-known figures affiliated with certain institutions, and the use of extremely polemic language appear to constitute material traces of an actual conflict, which, it seems plausible to assume, concerned prestige and resources no less than an abstract exchange of disembodied ideas.

In attempting to reconcile mathematical and philosophical studies, Blancanus's main strategy was to insist on the materiality, essentiality, and reality of mathematical entities, from which the truthfulness, causality, and certainty of mathematical demonstrations was inferred. The status of a "strict science", which Blancanus claimed for mathematics, followed as a natural conclusion from these arguments.[50]

The subject matter of geometry and arithmetic, said Blancanus, is intelligible *matter*, which means that mathematical entities were not imperfect and false, as his opponents claimed, but *true* beings. His opponents, whose names were specified, were a well-known group of philosophers originating with an Averroist commentator of Aristotle, Alessandro Piccolomini from Siena, whose ideas in the *Commentarium de certitudine mathematicarum disciplinarum* were taken up by Benedictus Perera at the Roman College and spread through his *De communibus*, as well as by

---

among philosophers, mathematicians, and logicians in the sixteenth century. On that debate see also: G. C. Giacobbe, " 'Il Commentarium de certitudine mathematicarum disciplinarum' di Alessandro Piccolomini", *Physis* 14 (1972), pp. 162–193; idem, "Francesco Barozzi e la 'Quaestio de certitudine mathematicarum'", *Physis* 14 (1972), pp. 357–374; idem, "La riflessione metamatematica di Pietro Catena", *Physis* 15 (1973), pp. 178–196; idem, "Epigoni nel Seicento della 'Quaestio de certitudine mathematicarum': Giuseppe Biancani", *Physis* 18 (1976), pp. 5–40; idem, "Un gesuita progressista nella 'Quaestio de certitudine mathematicarum' rinascimentale: Benito Pereyra", *Physis* 19 (1977), pp. 51–86.

49. Wallace, *Galileo and His Sources . . .* , p. 136.
50. Ibid., pp. 141–149.

the Jesuits of Coimbra who transmitted similar opinions through their *Cursus philosophicus*.[51] Of course, Blancanus admitted that intelligible matter abstracted from sensible matter, but he contended that this abstraction did not exclude access to reality, since any object of science presupposes abstraction. Science, he said, abstracts from existence. Thus mathematics, which also abstracts, can nevertheless satisfy the requirements of a true, material cause.

It was possible to arrive at an *essential definition* of mathematical objects, Blancanus stated. Hence mathematics was capable of producing *middle terms* for demonstrations "potissimae", the most valid form of demonstration according to the Aristotelian tradition. This meant that mathematics could furnish formal causes. Being convinced, then, that mathematics can provide *formal and material causality*, he proceeded to show the *causality* involved. He analyzed for this purpose forty-eight demonstrations from the first book of Euclid's *Elements*. Thus, the synthesis between the Aristotelian canons of proof established within the tradition of natural philosophy and the Euclidian requirements of perfect proof was seemingly achieved in any mathematical demonstration.

Arguing of true being, essentiality and causality predominated in Blancanus's attempt to legitimize the scientific nature of mathematics as a "true knowledge through causes" demanded by the Aristotelians. The analysis of Euclidian demonstrations was intended to show the similarities of structure between geometrical and Aristotelian concepts of proof. The synthesis was then invested with authority by enlisting the name of Jacopo Zabarella (1533–1589),[52] a Paduan Aristotelian who summarized in his work the whole Paduan tradition of debates on scientific method.[53] Zabarella's work established the use of "resolutive" and "compositive" to denote proofs from effect to cause and from cause to effect respectively, which, combined in a unified demonstrative moment called the "demonstrative regress", produce a noncircular argument leading to true knowledge through causes.

The problematic nature of Blancanus's attempt to legitimize mathematics in terms of the Aristotelian canons of science emerged as he elucidated

51. Ibid., p. 142.
52. Ibid.
53. Randall, "The development of Scientific Method in the School of Padua"; G. Papuli, "La teoria del 'regressus' comme metodo scientifico negli autori della Scuola di Padova", in Olivieri (ed.), *Aristotelismo Veneto . . .*, vol. 1, pp. 221–277; N. Jardine, "Galileo's Road to Truth and the Demonstrative Regress", *Studies in History and Philosophy of Science*, 7 (1976), pp. 277–318.

the form of a geometrical demonstration in the treatise of 1620. Here he explained the meaning of "resolution" and "composition" in the context of geometry. *Resolution*, he said, is a form of reasoning in which the truth of a theorem is investigated by first supposing that the theorem is true. Subsequently, one deduces consequences from the theorem until a conclusion is arrived at which had already been recognized as false or true. If the conclusion is recognized as true, this is taken as a sign that the supposition was true, on the basis of the logical principle that truth can only come from truth. *Composition* is a similar form of reasoning but in an inverse order: reasoning back from the truth discovered to the conclusion previously sought.[54]

A comparison between Blancanus's words and Wallace's discussion of the treatises of Jesuit logicians concerning scientific knowledge and the ways it is attained[55] cannot fail to demonstrate that Blancanus's description of the process of resolution and composition in the context of geometry is wholly incompatible with the account of resolution and composition in philosophy, suggested by Aristotelian logicians. In fact, resolution and composition in the context of these two entirely separate disciplines (the mathematical and the physical) have little in common, apart from the very general features of an operation in which something composite is separated into its component parts, and vice versa. However, whereas the essential difficulty in reducing the analytic methods of geometry to syllogistic reasoning remains unresolved, Blancanus still insists on resolving it on a declarative level. The function of such an attempted resolution is obvious enough.

The legitimation of the mathematical disciplines in terms of the Aristotelian canons of science opened up the possibility of demonstrating their superiority on the very same grounds as justified the high status of natural philosophy. In other words, once their truthfulness, causality, and essentiality were recognized, it was easy to show that in addition, they were more certain, and therefore satisfied more fully the Aristotelian claims for a knowledge both true and certain. Mathematical demonstrations, said Blancanus, proceed from prior principles, not from effects.[56] The superiority of this procedure was obvious, and required no further emphasis. Such superiority was asserted far more explicitly, however, when mathematics was challenged on account of its being concerned with an accident – quantity – which was regarded less noble than a substance. Blancanus's answer to this charge was bluntly aggressive:

54. Wallace, *Galileo and His Sources . . .* , p. 146.
55. Ibid., p. 126.
56. Ibid., p. 143.

It is much better to arrive at numerous and marvelous truths about such an accident than it is to be concerned with a thousand differences of opinion about material substance, a true knowledge of which will never be attained.[57]

Thus, it became possible to cross the boundaries dividing intelligible and sensible matter, artefacts and natural entities. Blancanus's discussion of the "mixed sciences" suggested the direction in which his discourse was leading. Optics, mechanics, music, astronomy dealt with the *matter* of heavenly bodies, musical sounds, visual rays, causes and forces of machines for the same purpose as the various branches of philosophy dealt with their respective subject matter.[58] Thus, it was possible to study sensible matter by means of geometrical demonstrations, and such an enterprise was equal in status to the enterprises of philosophers. The same theme was repeated in a few remarks on mechanics found in the treatise of 1620: mechanics, he said, was concerned with artefacts in the same way as the natural sciences are concerned with natural entities.[59] Thus, artefacts had become eligible for scientific investigation in the same way as natural entities had always been eligible for philosophical investigation. This claim, whose subversive implications from the point of view of the Thomistic organization of knowledge are obvious, was given respectability through a list of works cited by Blancanus, where the principles of such "science" could be learned: these were the pseudo-Aristotelian *Quaestiones mechanicae,* Archimedes' and Hero's basic works, Jordanus Nemorarius's *De ponderibus,* and the works of Guidobaldo del Monte, Lucas Valerius, Benedetti, and Galileo.[60] Through this list of authors, Blancanus did not simply confer respectability upon mechanics: his list may be read as an attempt to build up a genealogy for a field of knowledge not yet fully recognized as science (*scientia*), in order to convince the reader of its long history. Not only was mechanics a science for Blancanus, but it was indeed a science with a very respectable tradition.

The few isolated examples quoted above should not be understood as offering any generalization about Jesuit natural philosophy. Rather, they are meant to exemplify the variety of approaches to natural philosophy that existed within the Jesuit educational system and their potential to challenge the Thomistic principles of the organization of knowledge on which the system was based. Thus, the boundaries between the physical sciences and the mathematical sciences were challenged by the preoccu-

57. Ibid., p. 207.
58. Ibid.
59. Ibid., p. 208.
60. Ibid.

pation with new content (the assertion of generation and corruption in the heavens, which was no doubt connected to the high level of Jesuit observational astronomy, for example), as well as by the higher status assigned to the "mixed sciences", declared to be not only mathematical but also philosophical, i.e., physical. Likewise, the boundary between the natural – which was said to be grasped theoretically, and the artificial – which was said to be understood practically – was challenged by the investigation of chemical subjects, for example, which until the sixteenth century were considered to have no connection with theory, but only with experience.

The attempt to appeal to large segments of the Catholic secular as well as religious elites by updating the curriculum and catering to the utilitarian needs of their clients, caused the Jesuits to draw upon a variety of traditions. Different intellectual programs which had survived the Renaissance without constituting an alternative to Thomism or without disappearing altogether provided them with a large cultural storehouse which they used and abused for their own purposes. This practice, however, was not without its costs and dangers. The attempt to incorporate into the Thomist framework a whole range of new concepts, as well as new practices, had the effect of constantly calling into question the boundaries which it tacitly presupposed: those between the theoretical and the practical, between the abstract/mathematical and the concrete/physical, and between the necessary/certain and the possible/probable. Consequently, the accepted hierarchy of fields of knowledge and their order of precedence were undermined. In Jesuit culture these boundaries functioned as means of control to ensure the subordination of profane bodies of knowledge to the main purpose of salvation. But the subversive potential of Jesuit education consisted, precisely, in its challenge to these boundaries.

9

Dominicans and Jesuits

A Struggle for Theological Hegemony

In the early 1580s a Jesuit theologian at the university of Salamanca put forward certain theological theses related to Jesus' free will and the forewarning of his death given to him by God. A professor of theology at the university – Dominicus Bañez,[1] a Dominican – opposed the theses of the Jesuit and the matter was referred to the Inquisition at Salamanca. The ideas of the Jesuit were discussed and condemned as contrary to the Augustinian and Thomist doctrine of salvation.

Meanwhile, in the university of Evora another Jesuit theologian, Luis Molina,[2] was also writing on the problem of free will. His book *Concordia liberi arbitrii,*[3] was written in the spirit which had invoked the strong opposition of the Spanish inquisitors, who had objected to the emphasis placed on free will in the interpretation of the doctrine of salvation. Following the publication of Molina's book, a response from Bañez was not long in coming. Eventually, Bañez published a long criticism of the book,[4] and Molina gave his answer to this criticism. The private exchange between these two theologians, Bañez the Dominican and Molina the Jesuit, soon turned into a fierce struggle between the orders. The

1. Dominicus Bañez (1528–1604) studied theology in Salamanca and became a Dominican friar in 1547. For ten years he taught liberal arts and philosophy in that university. In 1581 he acquired the chair of theology there, and became involved in the *De auxiliis* debate.
2. Louis de Molina (1536–1600) was born in Spain to a well-known family. He entered the Jesuit college in Alcalá in 1553, and studied philosophy and theology in Lisbon and Coimbra. The last six months of his life were spent in Madrid, where he acquired the chair of moral theology.
3. L. de Molina, *Concordia liberi arbitrii cum gratiae donis, divina praescientia providentia, praedestinatione et reprobatione*, Lisbon 1588.
4. Bañez's famous response to Molina's innovation was a book written in joint authorship with P. Herrera and D. Alvarez (both Dominicans who later took active part in the debate), and published in 1595: *Apologia fratrum praedicatorum in provincia hispaniae sacrae theologiae professorum, adversus novas quasdam assertiones cuiusdam doctoris Ludovici Molinae nuncupati.*

Dominicans sought jealously to defend the Thomist doctrine of salvation as interpreted by Bañez, while the Jesuits enlisted the support of the entire order for Molina's positions. The matter was brought to the attention of Pope Clement the Eighth (1536–1605) during whose term of office it developed into the well-known controversy *De auxiliis*.[5]

Two stages may be discerned in the development of a major split between the Catholic intellectual elites of the world of the Counter Reformation. Initially the debate was confined to Spain, where the Pope had sent for the opinions of theologians of the two orders, professors in Spanish universities, and bishops, in an attempt to reach some kind of consensus on the question of grace and free will. This consensus, however, was not forthcoming. After the publication of Bañez's *Apologia Fratrum Praedicatorum*, he was invited to Rome, where a committee of theologians was set up to examine the claims of both sides. The committee of 1597 marks the beginning of a second stage, in which the two strongest and most influential orders in the Catholic world engaged in a public struggle for hegemony which lasted for about ten years, until 1607. From a historical perspective, the controversy may be divided into seven periods, beginning with committees of theologians and ending with a direct confrontation in the form of a scholastic disputation between Dominican and Jesuit

5.   My main source for the analysis of the controversy is: *La Storia de Auxiliis del ch. P. Giacinto Sérry dell'Ordine del' Predicatori, tradotta et compendiata da Rambaldo Norimene*, Brescia 1771, hereafter *De auxiliis*. Sérry (1659–1738) was a French Dominican who had studied at the Sorbonne and occupied the chair of theology in Padua from 1697 until his death. He gained his theological fame by becoming the most informed and probably the least impartial historian of the congregation. The first edition of the book appeared in Louvain in 1700 under the pseudonym of Augustin Le Branc. The book became a focus of criticism and polemics for Jesuits who attempted to discredit its historical value. It gives a very detailed account of the controversy in its various stages, including long citations from the complicated arguments of each of the disputants at each of the sessions that was held in Rome. It is not necessary to accept uncritically every statement of Serry in order to acquire the tools necessary for understanding the philosophical, theological, and institutional issues at stake. The revival of Thomism in the last twenty years of the nineteenth century gave birth to a few more apologetic works, among which are two well-known accounts by Jesuits: G. Schneemann, SJ, *Controversiarum de divinae gratiae liberique arbitrii concordia initia et progressus*, Freiburg 1881; and T. de Regnon, SJ, *Bañez et Molina*, Paris 1883. The two others were written by Dominicans: A. M. Dummermuth, OP, *Defensio doctrinae s. Thomae*, Louvain 1895; and Cardinal T. Zigliara, OP, *Summa philosophica*, Paris 1898.

representatives in the presence of the pope. The dozens of congregations which were held in the space of nine years, between 1598–1607, were also interspersed with unsuccessful attempts at a compromise.[6]

Not unlike the trials of Galileo, the *De auxiliis* controversy was a historical event in which intellectual/conceptual issues were intermingled with political/institutional ones. In my analysis of this event, a crucial intersection between knowledge structures and power structures prefiguring the Galileo affair, I shall deliberately separate the intellectual and institutional aspects. The theological claims made by the Dominicans and the Jesuits will be presented in order to expose the epistemological dimension which lay at the heart of the theological discourse. The relationship between the manner in which knowledge was produced, assimilated, and interpreted in the intellectual milieus of both orders on the one hand, and their respective theologies on the other, will thus be clarified. A further analysis focussing on the strategies which conferred on this debate the character of a political event will expose the power structures in which the intellectual practices were embedded.

The Problem of Grace and Free Will:
Its Intellectual and Practical Context

The post-Tridentine history of the problem of grace and free will originated in the decrees concerning justification and original sin,[7] issued by the Council of Trent, and which constituted the basis for the Counter-Reformation doctrine of salvation. The decrees attempted to reconcile two opposite requirements. On the one hand, they had to demonstrate the particularity of Catholic dogma, which sought to avoid the deterministic implications of the concept of predestination proposed by Calvinism. On the other hand, they also needed to exonerate Catholicism from the charge of Pelagianism,[8] according to which free will, representing the moral capacity of man, plays an important role in the attainment of salvation. The work of the Council was rooted in the Thomist teachings, which Molina and Bañez tried to interpret and specify to an even greater extent than did Thomas in his own writings.

6. See *De auxiliis*, pp. 105–118.
7. See Ch. 4.
8. A set of ideas about grace and free will which originated with Pelagius, a monk from Brittany who lived in Rome at the beginning of the fifth century. These ideas were forbidden and their adherents were forced to leave the territories of Western Christendom (A.D. 418) and fled to the East.

From the initial stages of the *De auxiliis* "affair" it was clear that what was at stake was the orthodoxy of Molina's attempt to find a compromise between the principle of free will and the principles of divine grace, divine foreknowledge, and predestination.[9] Essentially, Molina's originality was expressed in an allegedly new concept he introduced into the traditional vocabulary with which the problem of grace and free will had been discussed for hundreds of years. This term, *scientia media* ("middle science") was used by Molina to denote the phenomenon of divine mediation between grace and free will. According to Molina, God, before every act of grace, can discern by means of his "middle science", those individuals who are able to cooperate with him, through the exercise of their free will. It is this divine "science" of man's future actions which finally guides the choice of grace imparted to the elect, and necessarily and inevitably brings them to salvation. Molina's discussion of the kind of "science" which guides God previous to the bestowal of grace – the knowledge characterized by him as a "middle science" – represents the heart of the controversy initiated by his book.

Molina's terminology was a kind of response to the tension, increased by the reformed theology of Protestantism, which existed between divine determinism and anthropocentric humanism. In Molinistic theology, predestination could be regarded as a divine law rooted in the divine will and necessarily acting in accordance with the absolute nature of that will. The deviation from tradition consisted in claiming that God's will is guided by his "middle science" of man's future, contingent actions. Such a guidance of God's will by his "middle science" seemed to recognize the human will as a factor in the process of salvation.

The reaction of traditionalists of the type of Bañez to Molina's innovation was neither slow in coming nor lacking in an originality of its own.[10] As against the concept of "middle science" invented by Molina, Bañez emphasized the Augustinian concept of God's "voluntary decree" (*abso-

9. *De auxiliis,* p. 87: "La intera ragione di conciliare la libertà dell'Arbitrio colla Divina Grazia, Prescienza et Predestinazione".
10. Jesuit theologians of the nineteenth century, among whom G. Schneemann and T. de Regnon raised the charge that Bañez himself was an innovator. Exposing Thomas's ideas by means of such concepts as physical premotion, intrinsically efficacious grace, etc., they claimed, was but one possible interpretation of the Thomist texts. These charges were vehemently rejected by the Dominican theologians (Dummermuth and Zigliara, in particular), who regarded Bañez as standing in the main stream of Thomism. I do not pretend to express any opinion whatsoever on such matters. It seems to me, however, that Bañez brought a largely unkown terminology to his treatment of these issues of fundamental importance to Catholic dogma.

lutum Decretum divinae voluntatis) by means of which predestination is effected. God's will, he claimed, does not depend on his knowledge of man's future actions, but on his absolute will which embodies his absolute – and not "middle" – science of man's actions. Bañez strongly opposed the interdependence, posited by Molina, between the law of divine predestination and man's voluntary acts by means of "middle science". The law of divine predestination, according to Bañez, is an absolute law; its absolute character must be guaranteed by absolute, not just "middle", science, and by God's absolute will. In order to ensure the independence of God's choice from the human will, Bañez went so far as to suggest the concept of the "physical premotion" (*physica praemotione*) of grace, by which free will is led to cooperate with it. In the formal terminology of the Aristotelian theory of motion, the existence of a "physical premotion" could have meant that grace acts on the will in the same way as the mover on the moved. Molina, however, claimed that the motion in question – the cooperation of the will – was the result of two causes: grace and free will. In terms of the Aristotelian theory of motion, he interpreted these to mean two partial causes of the same motion.

Three Key Concepts: "God's Middle Science", His "Voluntary Decree" of Predestination, and the "Physical Premotion" of Grace

Three concepts lay at the center of the controversy between the Dominicans and the Jesuits: God's "middle science", his "voluntary decree" of predestination, and the "physical premotion" of grace. My account of the epistemological dimension of the debate is an attempt to expose the way these concepts functioned in the theological discourses of the Jesuits and the Dominicans.

However, Jesuit and Dominican teaching on grace and free will should not be discussed in terms of disembodied ideas alone. It should also be embedded in the general cultural orientations of both orders in that period. In previous chapters I have described the Dominicans as a traditional intellectual elite, still very attached to the norms and values of medieval culture. The Dominicans favored the ideal of monastic seclusion, stressed speculative knowledge as a means of achieving the contemplative life, and sought to preserve the synthesis arrived at by Thomas through an absolute prohibition of any deviation from his philosophical and theological ideas. The Thomism of the traditional Dominican elite could thus be described as doctrinaire Thomism.

The doctrinaire Thomists were otherworldly in their theological tendencies. Their chief aim was to preserve the idea of the absolute omnipotence of God vis-à-vis his creation through the application of rigorous canons of logic to the theory of salvation. Their intellectual strategies, however, were not dissociated from their traditional role as a small, privileged group which saw itself as guardian of the true faith. They defended old principles of the division and organization of the universe, old boundaries between the celestial and the terrestrial, the intellectual and the practical, in order to support their traditional position as an authoritative elite.

The alternative cultural orientation of the Jesuits, on the other hand, made the active life the centre of their approach to salvation. An involvement in the world through the education of wide segments of society, although justified in practical terms as a training of students for ecclesiastical and secular careers, was also regarded as an apostolic activity which helped in attaining salvation. Accordingly, the curriculum developed in less speculative directions, and the demand for control over the intellectual life was combined with a measure of openness.

The Jesuits' attempt to come to terms with new knowledge was not confined only to scientific and philosophical matters, however. Rooted in the active life, the Jesuits strongly yearned to preserve some role for free will in the path to salvation. In order to guarantee free will without denying God's absolute power the Jesuits had to reconstitute the boundaries between the transcendental and the terrestrial, the intellectual and the practical. Such a reconstitution had far-reaching intellectual consequences, as we shall see in what follows. But it also redefined the position of the Jesuits within the Catholic establishment in institutional terms.

The educational practices, the epistemological and theological premises, and the power status of the Jesuits and Dominicans all played a role in shaping the power–knowledge nexus in seventeenth-century Catholicism. The *De auxiliis* controversy, a major turning point in this process, is thus relevant for any analysis of this nexus.

"Middle Science" (Scientia Media)

The term "middle science" (*scientia media*) was used by Molina to characterize God's foreknowledge of the future acts of man, what Thomas had described as "contingent futures."[11]

For Thomas Aquinas, "contingent" was a grammatical mood denoting

---

11.   Thomas, *Summa theol.*, I, q. 14, a; q. 86, a.; *Summa contra gentiles*, lect. I, c. LXVII; *De veritate*, q. 2, a. 12.

"possible that it will be or that it will not be".[12] A man's actions are "contingent futures" because they depend on a condition. They come into being if God wills them and at the same time knows them through his "foreknowledge" (*praescientia*).[13]

Not unlike Molina, Thomas, in his time, also struggled with the question: with what kind of knowledge does God foreknow the future actions of men? The possibilities for accurately defining God's foreknowledge in terms of the contemporary range of concepts were related to the inexistence of a clear distinction between ontology and epistemology. Thus, the ontological status of the objects of knowledge preconditioned the epistemological validity of the knowledge attained.

In addition, the Thomist solution was limited by two theological premises: (1) For Thomas, God's knowledge and will were one and the same. Both of them existed in the divine essence (*essentia*). (2) The divine essence could only be embodied as *absolute necessity* (what was, is, or will be with an absolute certainty) or as what *hypothetically* could have existed but in fact never will (which is the underlying significance of counterfactual conditional sentences such as: "If the temple had not been destroyed, the Jews would still be living in their country").

In terms of the epistemological distinctions common in Thomas's intellectual milieu divine foreknowledge could be only one of two kinds. It could be knowledge of entities whose ontological status was real, which entailed a certain and infallible epistemological validity. This kind of knowledge was known as *scientia visionis*.[14] Or it could be knowledge of hypothetical objects, of a mere possible existence, which could only be purely hypothetical. Such knowledge of hypothetical entities was called *scientia simplicis intelligentiae*.[15]

Both possibilities were unsatisfactory for Thomas the theologian, seeing that in the Thomist system God's foreknowledge was identical to the divine will. For if one said that the knowledge of man's future acts was *scientia visionis* – certain, infallible knowledge – then it followed that these came into being in the same way as, and simultaneously with, God's knowledge and will in their regard: that is to say, in a necessary and inevi-

---

12. The "contingent" is one of the four categories of modal phrases, the other three being the "necessary", the "possible", the "impossible". See P. Edwards (ed.), *The Encyclopedia of Philosophy*, vol. 2, London 1967, pp. 198–205.

13. The following discussion of contingents in Thomas's texts is based on: A. Stagnitta, *Per una metateoria della logica midioevale*, Palermo 1980.

14. Thomas, *Sum. theol.*, I-I, q. 14, a. 9.

15. Ibid.

table manner. This amounted to a complete determinism, which left no room for the concept of free will. On the other hand, if Thomas said that God knows the future acts of man only by means of hypothetical knowledge – *scientia simplicis intelligentiae* – that detracted from the omniscience of the divine intellect.

Thomas Aquinas considered the unity of God's knowledge and will an uncontested theological truth. Nonetheless, he introduced two logical distinctions which enabled him to avoid either the deterministic or the detractive connotations of God's foreknowledge and will concerning man's future acts. Thomas maintained that the ontological status of the objects of God's foreknowledge – man's future acts – is neither real nor hypothetical but contingent, that is, "possible that it will be or that it will not be". Hence he defined the status of the objects of God's knowledge as "contingent futures". God's foreknowledge of man's future acts – contingent and not hypothetical – could be regarded as absolutely necessary, certain, and infallible, for the objects of this knowledge were not hypothetical objects. However, Thomas also distinguished between two kinds of necessity: *necessitas consequentiae,* or necessity of consequence, and *necessitas consequentis,* or necessity pertaining to the object under discussion.[16] In the case of contingent objects – man's future acts – the latter form of necessity, absolute necessity, does not apply. In accordance with this distinction, God's foreknowledge was absolutely necessary and infallible, but the necessity related only to consequences – that is, to human acts – and did not affect the nature of the object, or, in other words, could not interfere with human free will.

In the seventeenth century the dispute between Molinists and their Dominican adversaries was conducted within the framework of the basic logical distinctions introduced by Thomas. Molinists and Dominicans alike wished to treat the future acts of man as "contingent futures", neither real nor hypothetical, but ontologically conditioned. Both, however,

16. Quoted by Stagnitta, *Per una metateoria,* p. 100–101, from Thomas, *Summa contra gentiles,* 1, 67: "Praetera, si unumquodque a Deo cognoscitur sicut praesentialiter visum, sic necessarium erit esse quod Deus cognoscit, sicut necessarium est Socratem sedere ex hoc quod sedere videtur. Hoc autem non necessarium est absolute, vel, ut a quibusdam dicitur, 'necessitate consequentis; sed sub conditione, vel necessitate consequentiae'. Haec enim conditionalis est necessaria: 'Si videtur sedere, sedet'. Unde et, si conditionalis in categoricam transferatur, ut dicatur, 'Quod videtur sedere necesse est sedere', patet eam 'de dicto' intellectam, et 'compositam' esse veram; 'de re' vero intellectam, et 'divisam' esse falsam. Et sic in his, et in omnibus similibus quae Dei scientiam circa contingentia oppugnantes argumentantur 'secundum compositionem et divisionem falluntur'".

denied that the adversary's theory really justified the status of "contingency" conferred upon the objects of God's knowledge in terms of that theory. The Dominicans contended that in terms of the Jesuits' theory those objects – the future acts of man – were not contingent but hypothetical; the Jesuits claimed that according to the Dominican theory those objects – the future acts of man – were absolutely necessary and not contingent. Both accepted the Thomistic distinction between absolute (*necessitas consequentis*) and conditioned necessity (*necessitas consequentiae*). The Jesuits, however, defended the concept used by Molina – "middle science" – as a definition of the nature of God's foreknowledge (not given a name by Thomas himself). The Molinists argued that if man's future actions were dependent on a foreknowledge which also embodied the will of God, they would no longer be "contingent futures" (i.e., free).[17] Such a dependence would make them real, and hence absolutely necessary, not only with a necessity of consequence (*necessitas consequentiae*) but also with an objective necessity (*necessitas consequentis*). The Molinists thus defied the unity of God's knowledge and his will, posited by Thomas. Rather, they argued, God's "middle science" existed previously to the divine will and independently of it. "Middle science", as Molina had shown, was knowledge of conditional objects ("contingent futures"), whose conditional status stemmed from the fact that they *were not yet willed* by God at the stage of his foreknowledge. In spite of the contingent status of the objects of that knowledge – their lack of ontological reality – God's "middle science" was absolutely necessary, certain, and infallible.[18]

The Jesuits sought to legitimize the Molinist concept of "middle science" by arguing that it did not exceed the boundaries of Thomas's "foreknowledge" (*praescentia*),[19] and that any denial of God's "middle science" amounted to denying his ability to know future contingents.[20] The counterarguments of the Dominicans, however, demonstrated that

17. *De auxiliis,* p. 40: "Servivasi (il Rivio) del principio fondamentale, di cui servonsi ancora i PP. della Compagnia, cioè, che presupposta qualque supposizione antecedente, dalla quale infallibilmente ne segue, che l'Uomo agisca, resta levata la Contingenza delle cose".

18. Ibid. p. 295: "Avanti l'assoluto Decreto della Divina volontà diasi in Dio quella cognizione certa, ed infallibile delle cose contingenti, dependenti da causa libera".

19. Ibid., p. 247: "Preconobbe Dio, che se desse a Pietro gli ajuti efficaci della sua Grazia, Pietro si convertirebbe. Dunque avanti il Decreto della divina volontà deve mettersi in Dio la *prescienza,* per cui certamente conosca questo futuro condizionato".

20. Ibid., p. 298: "Gl' Impugnatori della Scienza Media negavano cumunemente in Dio la cognizione dei futuri contingenti".

"middle science" was, in fact, an innovation, and focussed on its deviations from traditional epistemological norms.

Thomas de Lemos[21] was the Dominican disputant who did more than
anyone else to undermine the concept of God's "middle science". One of
his most successful strategies consisted in an attempt to challenge the status of "contingent futures" assigned by the Jesuits to the objects of God's
foreknowledge. Since the Jesuits refused to admit the identity and simul-
taneity of God's foreknowledge and his will, the ontological status of the
objects of that foreknowledge could only be hypothetical.[22] Ontological
contingency could only be conferred upon the future acts of man *after*
God wills them; before God's act of willing there is not even a possibility
that they may come to pass. Hence God's "middle science" – his knowledge of the future acts of men prior to his voluntary decree and independently of it – was only hypothetical.[23] Any attempt to define such knowledge as absolute and infallible was a categorical error, for there could be
no absolute knowledge of hypothetical objects. The error of the Jesuits,
Lemos contended, lay in their belief that from "middle science", which
was knowledge of hypothetical objects, one could infer an absolute
knowledge of "contingent futures".[24] This was philosophically absurd –
there can be no absolute science of hypothetical entities – and theologically dangerous. For Molina was thus suggesting that the divine will was
only conditional[25] – dependent upon God's knowledge of man's future
acts which were not constituted by that will. The truth was, however, that
God does not wait for our agreement but creates it.[26]

Lemos put forward another argument against the epistemological deviations proposed by the Jesuits. "Middle science", claimed Lemos, by
which God foreknows the future acts of man prior to his voluntary de-

21. See Ch. 2.
22. Ibid., p. 297: "Siccome avanti il suo Decreto non anno per propria natura
    una verità infallibile in ragione di futuro; così da Dio non si conoscono, se
    non dopo il Decreto della sua volontà, con cui vuole ed ordina, che poste le
    tali condizioni se segua il tale effetto; ed in seguito conosconsi da lui colla
    Scienza di Visione, la quale sendo libera in Dio, e supponendo il Decreto
    libero della sua volontà, fa che si conoscano da Dio stesso in ragione di
    futuro."
23. Ibid., p. 298: "Avanti il decreto si conoscono, ma non con una cognizione
    certa ed infallibile".
24. Ibid., p. 296: "E così in seguito fattosi minutamente spiegare, e dimostrare
    ancora, che da questa Scienza Media veniva a dedursi una scienza assoluta
    dei contingenti liberi".
25. Ibid., p. 305: "Vuole, che la volontà Divina sia condizionata".
26. Ibid., p. 306: "Iddio no aspetta il nostro consenso, ma lo fa".

cree, must be wholly speculative, since the objects of that knowledge do not even manifest the possible reality which an act of will would have conferred upon them. How, then, could "middle science" direct God towards an active influence of grace, which by its nature is practical?[27] Clearly the rationale underlying this argument had to do with Aristotelian epistemological canons. The distinction between speculative knowledge (*scientia*), which was certain and true, and practical knowledge leading to action (*ars-techne*), was common among Aristotelians and willingly accepted by Lemos, who sought to maintain and defend it against the erosion of Jesuit innovations. For the Jesuit disputant Arrubal such an objection did not present any real difficulty. "Middle science", he claimed, was neither purely speculative nor purely practical. It was a hybrid which Arrubal described as "guiding" knowledge, both speculative and practical.[28] For a Jesuit, such a hybrid was neither something exceptional nor hard to accept. After all, the blurring of the boundaries between the theoretical and the practical was a prominent feature of Jesuit cultural policies long before the *De auxiliis* controversy manifested itself publicly in Rome.[29] Ironically, perhaps, this violation of the Aristotelian distinctions was more in the spirit of Thomas Aquinas than the Dominicans cared to admit. For it was Thomas himself who, in his *Summa theologiae*, had acknowledged sacred doctrine as both speculative and practical,[30] long before Arrubal defended "middle science" as a combination of the two.

In defending "middle science" against the doctrinaire Thomists the Jesuits violated two epistemological distinctions commonly accepted by contemporary Aristotelians. First, they admitted that there could be an absolute science of hypothetical objects. Although the Jesuits always strove for a recognition of man's future acts as "contingent futures" they were not able to disprove by logical means the Dominican contention that prior to God's voluntary decree these acts remained hypothetical. Nevertheless, they insisted that God's knowledge of them is certain and infallible, and hence absolute, independently of their ontological modality. Thus, the boundary between the true and probable knowledge was blurred in their discourse. Secondly, the Jesuits willingly admitted that

27. Ibid., p. 239: "Questa scienza, essendo speculativa, e non pratica, non poteva influire nell'efficacia della Grazia".
28. Ibid.: "Al che rispose il P. Arrubal, che ciò si verificherebbe, quando si trattasse di una scienza puramente specolativa; ma che trattavasi di una scienza direttiva, dalla quale doveva ricavarsi tale certezza".
29. See Ch. 8.
30. Thomas, *Summa theol.* q. I, art. 4.

God's "middle science" violated the distinction between the theoretical and the practical, while proudly conferring on it the special status of "guiding" knowledge. This violation further destabilized the Thomist principles of the organization of knowledge upon which the whole synthesis of sacred and profane sciences had rested. The documents of the "first trial" of Galileo, which contain traces of Bellarmine's and Lemos's ideas concerning the epistemological status of the "hypothetical",[31] cannot be properly embedded in their context unless the relevance of the De auxiliis controversy is recognized and admitted.

God's Voluntary Decree of Predestination

Both Dominicans and Jesuits agreed that predestination was effected through a voluntary divine decree. The disagreement concerned the time factor in the relationship of God's foreknowledge and his decree. The Jesuits contended that after the decree the future acts of man are wholly determined; after the decree they are no longer free. Lemos sought to prove that their opinion should be understood in terms of the semi-Pelagian error, according to which God's foreknowledge precedes the decree,[32] and predestination is only effected through the mediation of this foreknowledge.[33]

The Dominicans' rejection of the Molinist concept of predestination echoed their need to further defend the necessity of God's law. An absolute decree of God, they maintained, could not be based on a knowledge which, in their opinion was not even certain.[34] A voluntary decree related to God's "middle science" could not be conceived as necessary and was actually only conditional – dependent, to some degree, on man's natural willingness to do good.[35] At the height of the argument, the Dominicans

31. See Ch. 2.
32. De auxiliis, p. 246: "Al P. Arrubal successe tosto il P. Lemos, il quale principiò a mostrare colle lettere dei SS. Prospero, ed Illario a S. Agostino, essere quatro i capi, su cui Semipelagiani fondavano i loro errori circa la Predestinazione. Primo, che la Predestinazione non si fa secondo l'eterno assoluto Divino proposito avanti la prescienza dell'buon usu dell'Arbitrio".
33. Ibid.: "Secondo, essere la predestinazione secondo la prescienza di questo buon uso".
34. Ibid: "Quatro, finalmente farsi la Predestinazione per la prescienza del buon uso dell'Arbitrio futuro; ma però sotto l'ajuto della Grazia; di maniera che la prescienza di questo buon uso sotto la Grazia formi la predestinazione, la cui certezza tutta si fondi sopra questa sola prescienza".
35. Ibid., p. 249: "A ciò rispose il P. Lemos, che Molina connessava benissimo la Grazia efficace, ma che nel tempo stesso asseriva provenire questa effi-

adopted extreme positions, claiming that God's will alone, without reference to his knowledge, was responsible for predestination. Predestination, Lemos maintained, was not related to any foreknowledge whatsoever, whether conditional, absolute, or middle. It derived solely from the efficacity of the divine will.[36] Hence the Dominicans contended that God's decree was, in fact, prior to his foreknowledge and not dependent on it. Only such a concept could ensure the absolute omnipotence and independence of divine will.

The Physical Premotion of Grace

The Dominican emphasis on the voluntaristic aspects of the decree of predestination were closely connected to their ideas about the nature of God's foreknowledge. No certain knowledge of man's future acts could exist before God's voluntary decree, for it was by the willing act of God that those futures were endowed with a measure of reality which would render them objects of infallible knowledge. At the same time the Dominican vindication of the role of God's will affected their ideas about causality in the transcendental world, and by analogy also in the natural one.

The Dominicans held that the movement by which grace moves the will to cooperate with it is a real, physical premotion,[37] a movement which is caused by the intrinsic essence of grace. For them it meant that grace was efficacious of itself[38] and served as a primary cause of the movement.

Against them the Jesuits contended that the efficacy of grace did indeed derive solely from God, but its effect (that is, the activation of the will) also resulted from God's "middle science".[39] They thus maintained that grace and free will operated as two partial causes in producing the effect which is the future acts of man.[40]

---

cacia non dall'influsso della Grazia nella volontà, ma dall'assenso dell'Arbitrio preveduto futuro".

36. Ibid., p. 381: "Aver Iddio avanti qualunque prescienza, o condizionata, o assoluta, o media predeterminato coll'efficace sua volontà ogni buon uso dell'Arbitrio".

37. Ibid., p. 102–103: "Iddio non solo moralmente, ma eziandio come causa prima sovrannaturale col mezzo di questo ajuto predetermina fisicamente, ed efficientemente la volontà".

38. Ibid., p. 98: "L'efficacità intrinseca della Divina Grazia".

39. Ibid., p. 365: "Questa Grazia efficace ha la virtù dal solo Dio, che poi questa Grazia sia per avere il suo infallibile effetto, dipende dalla Scienza Media, e non solamente dalla sua entità prodotta da Dio".

40. Ibid., p. 258: "Il libero Arbitrio, e la Grazia sieno due cause parziali in ordine alla produzione del medesimo effetto".

Finally, the debate focussed on the notion of causality. Molina opposed the idea of the "physical premotion" of grace because it suggested that effects are produced directly through the action of God.[41] Such a doctrine, he argued, implied that grace itself operates as a secondary cause, which he could not accept. For while the divine influence could operate together with secondary causes to produce effects, it could not itself be a secondary cause.[42] Rather than regard an immediate divine influence as a secondary cause, Molina spoke of grace and free will working together as two "partial causes" in the creation of the effects.[43]

Molina's followers, headed by the disputant Bastida, went further and accused the doctrinaire Thomists of clinging to the position that there is a "necessary subordination" of the secondary cause to the primary cause.[44] This conception of the relationships between primary and secondary causes aroused the Jesuits' violent opposition.

The doctrinaire Thomists, for their part, were totally opposed to the idea that secondary causes could operate in a somewhat independent way. They rejected such a concept of causality, in both the supernatural and the natural spheres. Causality, they said, always operates in such a way that the secondary causes do not function on their own if they are not directly moved by the primary cause – that is, by the direct influence of God.[45] The efficacy of grace operates on free will by virtue of the omnipotence of God, in exactly the same way as it operates upon everything else in the universe.[46] The relationship between grace and the will is like

---

41. Ibid., p. 122: "Negavasi che avesse un immediato influsso di Dio nella causa seconda: 'quasi illa prius mota agat, et producat suum effectum'; ma bensì un influsso immediato 'cum Causa in illius actionem et effectum' ".

42. Ibid., p. 88: "Iddio influisce colle cause negli effetti e non già nelle cause stesse".

43. Ibid., p. 122; "Dio, ed il Libero Arbitrio sono come due cause parziali in rapporto all'operazione"; p. 259: " 'partialitate causae: . . . in quanto cioè ciascuna opera per proprio intrinseco principio; di maniera però, che l'una realmente non operebbe, ed influierebbe nell'effetto, quando non concorresse ancora l'altra".

44. Ibid., p. 373: "Primo per la necessaria subordinazione della causa seconda alla prima".

45. Ibid., p. 260: "Se le cause seconde sieno naturali, sieno libere, non sono mosse da Dio, non operano cosa alcuna; mosse poi operano tutte secondo la sua natura: le naturale naturalmente, le libere liberalmente".

46. Ibid., p. 355: "Haec Gratia habet efficaciam ab omnipotentia Dei, et dominio, quod sua Divina Majestas habet in voluntates hominum, sicut in coetera quae sub coelo sunt".

that which exists between a tool and the person who wields it.[47] Grace, therefore, can be regarded as having the status of an efficient cause. There is a complete analogy between the way God applies grace to free will and the way in which he applies natural causes in their totality to the created world.

"Middle science", God's "voluntary decree" of predestination, and the "physical premotion" of grace – these three concepts provided the framework within which discussion of science, law, and causality could take place in Dominican and Jesuit subcultures alike.

In order to glorify God's omnipotence, the Dominicans traced the divine law of predestination to God's voluntary decree which moves the human will, in a truly physical way, to cooperate with it. The intrinsic essence of grace is the sole cause of the cooperation. God foreknows the future actions of the elect who will cooperate with him, but his choice of the elect in no way depends on this knowledge, for his foreknowledge is wholly embodied in his voluntary decree. Prior to the decree God's foreknowledge cannot be considered absolutely certain and infallible, since the objects of that knowledge are then merely hypothetical. However, the decree does not detract from the measure of freedom characterizing the human will, which operates, although under the influence of the motive force of grace, in accordance with its nature created by God – that is, in freedom.

The conceptual constraints within which the notions of science, law, and causality were interpreted in the Dominican milieu can be clearly discerned by now. For the Dominicans, *scientia,* or true knowledge, was knowledge of real objects. A true knowledge of hypothetical objects was excluded as a logical impossibility. This premise was wholly in tune with the Aristotelian-Thomistic canons of logic, according to which the "hypothetical" was identified with the "fictive" rather than with the "true".[48] Hence the limits of the interpretation of the concept of *scientia* – even the *scientia* of God – were constrained by the Aristotelian–Thomist epistemological norms. The Dominican milieu did not give rise to any desire to challenge the commonly accepted canons of knowledge, and they remained basically unchanged during that period.

47. Ibid., p. 357: "Siccome il Bastone si ha in rapporto al movente; di maniera che se non è mosso dalla mano, da sestesso non si muove e niente agisce; così gli uomini si anno in rapporto a Dio".
48. See Ch. 2.

Furthermore, the cognitive interests invested in Dominican theology required that the relationship between man's future acts and God's law which governs them should be one of strict necessity in order that the transcendental moment of God's predestination should be preserved in all its purity. Hence the Dominicans emphasized ontological necessity as the only possible relationship between the law and its manifestations. God's predestination could be granted the status of full necessity only if it was related to his absolute knowledge of nonhypothetical objects. Epistemological necessity was inseparably interlinked with ontological necessity: the one could not be conceived of without the other. Such a connection was also compatible with the Aristotelian-Thomistic canons of logic, and it remained unchanged and was jealously defended by the doctrinaire Thomists.

Finally, the Dominican way of conceiving the causal relationship between God and his creation also encroached upon the legitimate boundaries of natural philosophy. The voluntaristic undertones of the Dominican theology exposed their concept of natural causality to radically contingent implications. If grace operates directly to produce effects by virtue of its intrinsic essence alone, and this action is analogous to the way in which the divine omnipotence operates on all created things, then the possibility of a humanly scientific investigation of natural causes is significantly diminished. There is not much sense in a speculative search for natural causes if those causes are at any given moment subordinate to the direct intervention of God and liable to change, and this is true not only in principle, as Thomas had claimed, but in a truly *de facto* manner.

The Jesuit milieu, in contrast, offered a different theological structure, permitting a deviation from traditional epistemological canons and a broadening of the field of natural philosophical investigation. The Jesuits' main theological objection to the Dominican system concerned the concept of a divine voluntary decree, embodying both God's foreknowledge and his will. The Jesuits separated God's knowledge from his will. Prior to the decree, the Jesuits contended, and separate from it, God possesses "middle science" (*scientia media*) by which he knows with a certain and infallible knowledge man's future acts, although these are not yet predestined by his will. To some extent, God's voluntary decree is guided by this knowledge.

The Jesuits' epistemological deviations concerned both the boundary between the hypothetical and the true and that between the practical and the theoretical. By ascribing certainty and infallibility to God's "middle science", the Jesuits recognized – or at least left the door open for – the existence of a true knowledge of objects devoid of ontological reality, or, in other words, for a science of hypothetical objects. Although such a concept represented a violation of Aristotelian norms, it could constitute a legitimation of a real science of mathematical entities, which according to the same obsolete norms were regarded as abstract/hypothetical and therefore incapable of producing causal explanations in natural philosophy. The need for such a view of science was certainly felt among Jesuit mathematicians who were intensely involved in the practice of the "mixed sciences" (astronomy, optics, mechanics), traditionally classified as "middle sciences",[49] neither purely mathematical, nor purely physical. Disregarding the arguments of traditional philosophers such as Benedictus Perera, for example, who claimed that "mathematicas disciplinas non esse proprie scientias",[50] these mathematicians could assume, on the basis of the new norms created within the sphere of theology, that a real science of hypothetical entities was not a logical impossibility. Likewise, they could assume that true science (*scientia*) did not exclude the combination of the theoretical and the practical, just as God's "middle science" was theoretical and practical at the same time.

The Jesuit separation of epistemological from ontological necessity could also give rise to a different concept of the divine law and, by analogy, of the natural law as well. The Jesuits wished to weaken the notion of the absolute necessity of man's future acts in order to prevent the complete negation of free will. Hence they asserted God's law to be partially conditioned by the hypothetical acts of man, prior to their determination by the divine will. God's law thus became "conditionally" necessary, i.e., necessary "if" such and such conditions were fulfilled. In turn, this concept of "conditional" necessity was not associated with absolute knowledge but with "middle science", the certain and infallible knowledge of hypothetical objects. Analogously, natural law could also be conceived as that law known certainly through the abstract/hypothetical entities of mathematics, whose ontological connection with physical reality is always conditioned, not absolute.

Finally, the possibilities created by Jesuit culture for an autonomous investigation of natural causality should be mentioned. In their arguments against allowing free will to be wholly determined by the operation of grace the Jesuits, like the Dominicans, tended to view the relationship between these two concepts in terms of primary and secondary causes. For the Jesuits, to speak of grace operating directly to produce man's future acts by virtue of its intrinsic essence, was to subjugate secondary causes (free will) to the primary cause (grace) in a hierarchical manner. By refusing this kind of connection between secondary and primary

49. See Ch. 4.
50. See Ch. 8, p. 166; Ch. 11, p. 218.

causes the Jesuits fought for a measure of autonomy of the secondary causes from the primary cause in the operation of nature. The boundaries of natural philosophy were thus widened, and a broader area was created for an independent investigation of nature.

Power Relations between Elites of the Church Establishment

Between 1597 and 1607 the pressure of theological controversies resulted in the opening up of a new conceptual space within which the commonly accepted canons of knowledge were questioned and renegotiated. It was not only theological hairsplitting which was at stake, however, but the relative power and status of the groups involved. The debate brought to its culmination a process by which the Jesuits gradually replaced the Dominicans as the church's educational elite. This development came to a head with the publication of Molina's book, for it was Molinism which finally transformed the position of the Jesuits in the power structure of the ecclesiastical establishment, winning them the status of an elite with an alternative theology crowning its alternative educational program. Thus, the *De auxiliis* debate rendered visible what had previously been implicit. The public consequences of the open clash between the traditionalists and the new militant educators were not unambiguous, however. By exposing the subversive potential of their system the Jesuits were left highly suspect of theological heresy and in deep need for legitimation. The following reconstruction of the institutional aspects of the debate is intended to reveal the limitations, strengths, and points of vulnerability of both Jesuits and Dominicans in the years just prior to the "first trial" (1616) of Galileo. I will try to show that, contrary to common opinion, neither group was omnipotent vis-à-vis the "new science" of Galileo. In coping with the challenge presented by Galileo's Copernicanism, their possible strategies were a priori constrained not only by their intellectual horizons but also by their relative power within the church establishment.

The history of the congregations held in Rome under two Popes – Clement the Eighth and Paul the Fifth – is full of anomalies which betray the power mechanisms governing the actions of the ecclesiastical establishment. Six times, the writings, memoranda, and theses of the Jesuits and the Dominicans in matters concerning grace and free will were submitted for examination to theological consultants. Six times, the Jesuits' theological opinions were almost unanimously condemned. And yet the Jesuits were the true winners of the debate, even if they did not come out uninjured from the contest.

The first examination took place in the years 1597 to 1599, on the basis of memoranda submitted by the Dominican Alvarez and the Jesuit Cardinal Bellarmine.[51] The censors declared that "the Molinist principles clearly contradict the teaching of St. Thomas, St. Augustine, and other holy fathers of the Church, and contain many things contradictory to the Holy Scriptures and the decisions of the councils".[52] Accordingly, they recommended the complete prohibition of Molina's book.[53] A second examination followed, in which Dominicans and Jesuits discussed their disagreements under the direction of a special arbitrator, Cardinal Madruccio, appointed for the purpose by the pope. This ended inconclusively, and the pope decided to appoint two additional arbitrators: Cardinal Bellarmine and Cardinal Bernerio.[54] In the third examination, attempts were made to formulate a compromise. Defenses of the Dominican and Jesuit positions were repeatedly presented for arbitration, but no decision was reached. A year of discussions and exchange of writings followed. Cardinal Madruccio attempted to end the affair by an injunction from the pope, officially confirming the censures of the first examination.[55] His sudden death foiled his plan. The congregations continued to hear the case and censured Molina's doctrine once more in a decision which was finally confirmed on 9 September 1600.[56] The fourth examination began in 1601. In the course of thirty-eight sessions, Dominican and Jesuit representatives debated twenty Molinist propositions before consultants appointed by the pope. These propositions were all censured, with two of the consultants opposing. The censure was finally confirmed in July 1601,[57] but did not have the sanction of a papal decision. In 1602 Clement the Eighth decided that there should be a fifth examination of the Molinist question, this time under his auspices. The examination lasted for about three years, and in that period there were sixty-seven hearings.

51. *De auxiliis*, pp. 80–81.
52. Ibid., p. 91.
53. Ibid.: "Censemus e re catholica esse, ut Liber, qui inscribitur concordia Liberi Arbitrii etc., compositus a Ludovico Molina, et eiusdem Doctrina omnino prohibeatur".
54. Ibid., p. 99.
55. Ibid., p. 106.
56. Ibid., p. 115: "I novanta incirca passi di Molina, notati nella passata più prolissa Censura, furono ridotti, e compendiati in venti capitali proposizioni, le quali furono rispettivamente censurate come eretiche, erronee, temerarie, e pericolose".
57. Ibid., p. 146.

These, too, ended with a censure of all the Molinist propositions, with only two consultants opposing.[58] The sixth examination lasted for two years, from 1605 to 1607, under the auspices of Paul the Fifth. Like all previous examinations, it ended with the censure of the Molinist propositions by an almost complete majority. Paul the Fifth appointed a committee of twelve cardinals to decide whether the interests of the church required an end to the debate in the form of an apostolic decision *ex cathedra*. Ten out of the twelve decided in the affirmative.[59] As a result, the pope appointed advisory cardinals to draw up a papal bull.[60] Whether such a bull was actually drafted has been the subject of a long controversy among Jesuit and Dominican historians,[61] but there is no doubt that such a bull was never published. The debate ended with the convocation of the leaders of both sides by Paul the Fifth, who told them that he would publish the official decision at a time which he considered suitable, and forbade either side to censure the views of the other.[62]

A major contradiction runs through the course of events described above. On the one hand, the popes attempted to bring to a decision, by institutional means, a theological debate of far-reaching philosophical significance. The censure of a papally appointed committee of theologians, enforced by an apostolic decision, was a traditional procedure in the Catholic world. If the debate had ended with an apostolic sanction of the censure, the enunciation, defence, and discussion of the Molinist propositions would all have been prohibited, as had indeed been the intention of the theologians who repeatedly censured them. The Dominicans would then have won a clear victory. In practice, however, things turned out much more ambiguously. Pope Clement the Eighth and Paul the Fifth after him never bestowed the authority of dogma on the decisions of the censors and did not even publish those decisions. By adopting this course of action, which contradicted their initial decision to eradicate theological disagreements in the church, they enabled Molinism to spread, to become known, and to gain a real influence. Accordingly, the Jesuits gained an impressive tactical advantage which they were able to exploit and to turn into a genuine victory.

The ambivalence of the situation was rooted in the different objectives

58. Ibid., p. 65: "La sacra Congregazione giudicò sana, e cattolica la Dottrina dei Domenicani, e non abastanza purgata d'errori quella di Molina".
59. Ibid., p. 391.
60. Ibid., p. 395.
61. Ibid., p. 367.
62. Ibid., p. 408.

the orders set themselves, the tactics they used, the self-image of each one, and their image of the other.

From the very beginning, the Dominicans aimed at obtaining a censure of the Molinist view with the intention of having it suppressed. The memorandum written by the Dominican Alvarez in preparation for the first examination specifically stated this aim. There he condemned Molina's *Concordia* and his commentary on the first part of St. Thomas as works containing an unsound doctrine, and he asked for them to be suppressed.[63] As for the means of attaining their goal, the Dominican tactics were twofold: they first concentrated their efforts on achieving intellectual superiority, and then sought to back up their intellectual superiority with apostolic authority.

The Dominicans were uncompromising in condemning the Molinist ideas as heresy. Accordingly, they were totally opposed to any verdict which might confer upon the Jesuit propositions the status of "probable" opinions. For the discursive rules of the period, common to both Dominicans and Jesuits, dictated that "probable" opinions could be discussed and defended.[64] From the Dominican point of view allowing the discussion and defence of opinions really meant that they were tolerated, and in the Dominican universe heresy was to be eliminated, not tolerated in the form of "probabilities". At every opportunity, the Dominicans stressed the depth of the disagreement between the two sides concerning the very essence of faith,[65] and the harm liable to result from any hesitation in reaching a clear verdict.[66] They saw the main threat to Catholicism chiefly in terms of the danger to the internal situation of the church represented by a failure to achieve uniformity of opinions. At the same time, they defined their apprehensions in doctrinaire terms, insisting on the need to preserve the purity of the faith.

The Dominicans' attitude was that of a traditional intellectual elite whose power was based on its traditional authority. The repeated censure

63. Ibid., p. 80.
64. Ibid., p. 113: "I Domenicani, persistendo nel sentimento, che nella dottrina de' Gesuiti eravi gran pericolo di Pelagianismo, non potenavo acconsentire, che impunemente fosse difesa".
65. Ibid., p. 338: "La controversia procede circa *l'essenza della fede,* di cui trattò S. Agostino coi Pelagiani e S. Prospero coi Semipelagiani". (My italics, R. F.)
66. Ibid., p. 330: "Il non definirsi ne l'una ne l'altra parte sarebbe quasi un approvare gli errori Pelagiani e Calvinisti, lasciando come probabile la congrua ossia morale vocazione".

of the Molinist positions shows that the Dominicans largely succeeded in proving their intellectual superiority, thus maintaining their authority. But it soon became clear that such a victory was insufficient. All the attempts to bring the matter to an apostolic decision did not bear fruit, despite the superiority demonstrated by the Dominicans and the political pressures they exerted on the pope. These were especially expressed through the Spanish ambassador, who communicated to the pope the fears of the king of Spain concerning the dangers both to the church and state represented by the differences of opinions.[67]

The Jesuit point of departure was different from the beginning. Although, when drawn into polemics, they accused the Dominicans of drifting towards the Calvinist heresy, they never attempted to make the Dominican opinions the object of a censure. Their chief aim was a declaration from the pope that the Molinist opinions were recognized as "probable", in order to allow freedom of discussion and defence of their positions.[68] Failing to achieve this, however, the Jesuits concentrated their efforts on an attempt to prevent the publication of the censure and to avert an apostolic decision.

Contrary to the Dominican strategy, intended to create a gulf between the Molinist and the Thomist approach, the Jesuits tried to minimize differences of opinion and hoped to escape the consequences of censure by achieving a compromise. Thus they claimed that the origins of the differences between the two orders were merely semantic[69] and did not affect faith itself. They could not, therefore, be the subject of a dogmatic, unequivocal papal judgement.[70] The Jesuits did not hope to achieve doctrinal unity within the Catholic Church, nor did they attribute the highest importance to such a uniformity. The main danger to Catholicism, they claimed, came from outside, from the heretics. Hence they sought to justify their theological position as the most suitable doctrine for combating

67. Ibid., p. 148.
68. Ibid., p. 107–108: "Pregavano il Papa, che pronunziasse come probabile e sostenibile quella stessa Tomistica Dottrina . . . purchè comme probabile potessero essi pure difendere quella di Molina".
69. Ibid., p. 197: "Si potevano con facilità comporre, e modificare alcuni consettari meno comuni, ed alcune espressioni men rispettuose versi i SS. Agostino e Tommaso".
70. Ibid., p. 337: "Tutta la controversia presente ultimamente consite [*sic*] nel ricercarsi, quale sia stata la sentenza di due uomini circa le proposte questioni, si conchiude, che essendo questa una questione di un fatto umano non può essere il soggetto di una definizione di fede".

heresy.[71] Since they took a broader view of the debate than a merely intellectual and doctrinal one, they were undeterred by the censure to which their opinions were repeatedly subjected. Instead, they looked for effective ways of exerting their full influence in the political as well as the academic spheres in order to attain by tactical means the goal they were unable to reach through intellectual persuasion.

By challenging the legitimacy of the pope's decision in a matter not pertaining to faith, and by ignoring the censure and attempting to bypass it through the exertion of political and academic influence, the Jesuits were in fact casting doubt upon the structure of authority predominant in the church. The Jesuits claimed that the opinion of a handful of theologians could not be decisive in a matter so important in its practical consequences.[72] Just as the opinions of Augustine and Thomas in the intellectual sphere were regarded by them as a "human fact" and not as a matter of faith, so they did not hesitate to cast doubt on the authority of the theological censors who, after all, were only people of flesh and blood whom they accused of bias and even ignorance.[73] But the Jesuits went even further: such an important matter, they said, could not be decided upon without a general council. This demand was first made in 1600,[74] and the Jesuits repeated it in 1601.[75]

In place of the traditional structure of authority, which they did not hesitate to undermine, the Jesuits attempted to create alternative patterns for the authorization of ideas, better suited to their needs. In view of the decisions of the handful of theologians appointed as censors by the pope, they demanded the right to consult with theologians from all sections of the Catholic academic world.[76] This was able to provide them with a large majority because of the large number of academic institutions under their control. While the pope tried to impose secrecy on the Catholic world, they employed every means at their disposal to disseminate their proposi-

---

71. Ibid., p. 107: "La sua dottrina era abbracciata da gravissimi Autori, come la più opposta agli Eretici".
72. Ibid., "Essere troppo pericoloso il voler definirsi una Causa di tanta importanza, sul solo rapporto di otto, o dieci Teologi".
73. Ibid., p. 336: "Che gran parte erano incompetenti per una controversia di questa forta sì per essersi dimostrati troppo parziali, e sì ancora perchè non dotati di tutta la Dottrina necessaria per guidicare sicuramente di cose oscurissme, e di somma difficoltà".
74. Ibid., p. 107.
75. Ibid., p. 157.
76. Ibid., p. 107: "Essere perciò necessario il consultare tutte le Cattoliche Accademie, prima di far l'ultimo passo".

tions, not only in the universities but also from the pulpit and the confessional, not only in the presence of learned theologians but also in front of the general public,[77] and not only in theological writings but also in semipopular works claiming to have been written by divine inspiration.[78]

The challenge to traditional patterns of authority also provided the Jesuits with a much greater possibility of using their political connections to exert influence on the architects of church policies. They organised a powerful lobby in the Molinist affair through the Empress Maria Augusta and the Archduke Alberto, the son of the Spanish ambassador in Rome.[79] In the last stage of the affair, Cardinal De Perron intervened, and that intervention may have been an important factor in the decision of Pope Clement the Eighth not to publish the censure of the Molinist propositions, at least according to Sérry. The Jesuits also attempted to influence the censors[80] and the papal counselors.[81] The diplomatic activity in which they were engaged was vigorous and effective, if not always to the liking of the pope, who saw it as an interference by secular rulers in church affairs.[82]

The Dominicans' failure to perceive the shift of power relations within

77. Ibid., p. 152: "Si fecero a disseminare con tutta la pubblicità, e l'impegno la Dottrina in esse contenuta ed alle Cattedre, e dai Pulpiti, e per fino dai Confessionali; e ciò non solo presso i loro Scolari, ed i Teologi; ma eziandio presso il vulgo più ignorante".

78. Ibid., p. 161: "Sparsero in quest'anno dei libri, gli Autori de' quali spacciavansi per illuminati divinamente a disseminare i Dommi di Molina, quasi fossero sacri oracoli. Di tal farina era un certo Libro stampato a Parigi nel 1601, composto da un certo Prete di Cordova, il quale nel proemio fingendosi uomo idiota, attesta di aver il tutto appreso da luce suprema infusagli da Cristo, cui il Libro stesso vien dedicato".

79. Ibid., p. 93: "Alle Teologiche Dissertazioni aggiungendo ancora le interposizioni, ed i favori dei Principi, impetrarono i Gesuiti Lettere commendatizie dalla Imperatrice Maria Augusta, già divotissima della Compagnia. . . . Altra pure ne scrisse per lo stesso effetto ai Cardinali Aldobrandino, Madruccio, S. Severina, e parimente al Duca di Sessa, Ambasciatore del Re Catholico, raccomandandogli premurosamente gli affari della Compagnia. Lo stesso fece l'Arciduca Alberto suo figlio li 26 Aprile 1599, interessato personalmente in questo affare per la licenza da lui data per la stampa dell'Opera di Molina".

80. Ibid., p. 111: "Si ridussero ad impetrare da alcuni privati alcune Lettere commendatizie presso i Censori Romani, acciò giovassero alla Causa di Molina, e della Compagnia".

81. Ibid., p. 119: "Disperando il Bellarmino di poter piegare l'animo risoluto del Papa, si rivolse a tentar quello dei Consultori".

82. Ibid., p. 161.

the church partly stemmed from their entrapment in the world of traditional images. Having cast themselves in the role of the guardians of faith,
as testified by Alvarez's memorandum to Clement the Eighth, they were
unable to tolerate Molinist concepts alongside the alternative concepts of
Bañez's theology. For them, Molinism was just the innovation of a private
individual. In the world they knew, such opinions were usually isolated,
declared heretical and dangerous, and excluded from the Catholic consensus, suppressed, or driven underground. It was in this spirit that the
Dominican general spoke in the opening speech to the third examination.
The Dominicans were not in disagreement with the whole Society of Jesus, he said, but only with Molina and his supporters, for it could not be
supposed that the entire society would come to the defence of a single
theologian.[83] With the same kind of arguments, the Dominicans tried to
suppress the Jesuits' demand for a general council. The problem before
the pope, they said, was only the prohibition of the doctrine of a certain
theologian. It was not necessary to convene an ecumenical council for
that purpose. After consultation with bishops, theologians, and academic
institutions, it only remained for the pope to deliver his verdict, which
nobody could question as it carried the weight of pontifical authority.[84]
The Dominicans found it difficult to grasp that their authority as a closed
intellectual elite, the sphere of whose influence was restricted to the order
itself, and whose power depended on its traditional status, was being
called in question, if not openly, then at least implicitly, by the tactics of
the Jesuits. They failed to perceive that the powerful influence of the Jesuits, the wide support they enjoyed in academic circles, and their propaganda campaign on the political and popular levels was transforming
Molinism from the opinion of a single man into an alternative theology.
This theology was being spread by a group which regarded itself as an
alternative intellectual elite and used all possible means not only to cap-
ture power bases in order to build up its position, but also to gain recognition and legitimation for that position.

83. Ibid., p. 97: "Il P. Generale de' Domenicani, che fu il primo a parlare, pro-
testò: non aver egli differenza alcuna col corpo della Compagnia, ma solo
col P. Molina, e cò suoi Difensori; non potendo persuadersi, che la Compagnia tutta si fosse abbandonata alla difesa di un Religioso privato".

84. Ibid., p. 109: "Trattandosi di condannare la Dottrina di un Teologo partico-
lare, non era punto necessario un Concilio Ecumenico: che dopo aver con-
sultate ed Accademie, e Vescovi, e Teologi, e dopo uno studiato esame fatto
della materia, poteva sua Santità passar francamente alla pubblicazione
della sentenza, alla quale non averebbe repugnato se non chi avesse posta in
dubbio la Pontificia Autorità".

In face of the blindness of the Dominicans who insisted on regarding Molina as merely an individual theologian, the Jesuits united their ranks and crystallized their institutional and communal interests around the Molinist cause. If, in the 1580s and at the beginning of the 1590s, leading Jesuit theologians had written against Molinism, they soon changed their mind and began to assimilate the Molinist theology into their outlook.[85] In reaction to the pressure of the Dominicans, who sought to eject Molinism from the Catholic consensus by singling it out as an "innovation", the Jesuits strengthened their demand for the recognition of the new elite, which was also the bearer of a new theology.

The Jesuits' insistence that Molina was not an innovator[86] and that his system was more in keeping with the Augustinian and Thomist spirit than Bañez's deterministic theology was the kind of argument the "guardians of the faith" were used to hearing from heretics in every generation, and it was not difficult for them to reply.[87] Together with arguments of this kind, however, the Jesuits used tactics which clearly belied the significance of their claims. Thus, they appeared to abandon the defence of Molina and seek to be accepted as faithful Thomists, but while the Dominicans saw this as a sign of conformity, it soon became apparent that the demand for Molina to be cleared of the accusation of heresy had been exchanged for the much more sweeping demand that Bañez's Dominican theology should be discussed at the same time as, and parallel with, the Molinist doctrines.[88] The meaning of this demand was nothing else than the setting up of the Molinist theology as an alternative to the Dominican theology, with each one regarding the other as a form of heresy.

It was not only by implicitly setting Molina's and Bañez's ideas on a par that the Jesuits revealed their desire for recognition as an alternative intellectual elite. Even more significant was the way in which they used Molinist theology to legitimate their claims as an elite whose chief raison d'être was its education of ever-widening circles of Catholic society. One

85. Among those mentioned by Sérry, the author of *De auxiliis,* were R. Bellarmine, F. Toledo, B. Perera.
86. *De auxiliis,* p. 284: "Molina non cose nuove, ma solo con metodo nuovo aveva insegnato".
87. Ibid.: "Cercavano i suffragi degli Antichi per sostenere una Dottrina, che Molina stesso aveva spacciata per nuova".
88. Ibid., p. 99: "Risposero i Domenicani, che la presente lite verteva solamente contro Molina, ed i suoi Apologisti. Soggiunse allora il Generale de' Gesuiti che ancora nel P. Bannes ritrovavansi delle proposizioni degne di censura, già per tali notate da' suoi Religiosi; esser egli per ciò preparato a preferire la sua sentenza sopra Molina, dopochè i Domenicani avessero esposta la loro sopra di Bannez".

argument of the Jesuit disputant Bastida was particularly revealing in this context. Bastida claimed that when Molina stated that before receiving grace a man must cooperate, he was only referring to the actions of the priesthood.[89] It was through these persons that a Christian could recognise his sins, repent, do penance, and desire God. It was in this sense, he claimed, that St. Augustine had called the representatives of the church "instruments in the hands of God".[90] With these words, Bastida gave vent to a new interpretation of the status of the priesthood which perfectly suited the Jesuits' perception of themselves as the ideal priests. According to Catholic tradition, the members of the priesthood served as channels for the transmission of grace, but they did not participate *actively* in this supernatural function. Molinism, in Bastida's interpretation, conferred a status on the priesthood by virtue of the *active* role it played. Here Bastida gave expression to the Jesuits' self-image of priests fulfilling their apostolic role while the traditional intellectual elite occupied itself with philosophical speculation, which had no connection with the day-to-day works of disseminating and preserving the principles of faith. Another Jesuit disputant named Salas declared that Molina spoke better than any of the other scholastic theologians when he said that grace is given to the man who strives for it with all his might.[91] When Lemos faced him with the theological consequences of this idea, saying that this makes God dependent on the human will,[92] Salas dismissed his reasoning impatiently as a purely intellectual argument. Lemos should not forget, he said, that he was not debating with a learned scholar but with an old soldier.[93] With this statement, in fact, Salas set the intellectual elite of theologians against the active elite of educators, and confronted the one with the other.

89. Ibid., p. 276: "Per quelle cose, le quali necessariamente precedono la Grazia della vocazione, non debbono intendersi gli atti dell'uomo, che fa quanto è in se; ma bensì gli atti dei Ministri della Chiesa, e dell'Arbitrio ancora, con cui si acquistano i fantasmi, e le notizie della fede".

90. Ibid., p. 267: "Stabilì come principio inconcusso essere cosa in se stessa, e secondo S. Agostino ancora certissima, che l'intera eccitazione del peccatore a penitenza dipende massimamente da' Ministri della Chiesa, e dall'Arbitrio, essendo questi come gli stromenti, de' quali Iddio suole servirsi per eccitare i Peccatori".

91. Ibid., p. 257: "Molina parlava meglio di tutti gli altri Scolastici; perchè questi dicono, che Dio dona a tutti gli ajuti sufficienti; e Molina dice che li dona a chi fa quanto è in se colle forze della natura".

92. Ibid.: "Se Iddio fomenta i desideri delle virtù, dunque Iddio vien eccitato dall'uomo".

93. Ibid.: "Che non si pensasse già di parlare con uno Scolare principiante, ma con un vecchio soldato".

The controversy *De auxiliis* exposed the Jesuits as an alternative elite of the church, a modernizing, critical, and even subversive element within the establishment. Guided by the ideal of the *vita activa* the Jesuits were gradually eroding the walls which segregated the traditional intellectual elite from the world. The Jesuit educational programme entailed living with secular students, preparing for secular careers, and opening up the curriculum. New contents were legitimized, new canons of knowledge recognized, new theological possibilities imagined. The sensibilities of the traditional guardians of the faith were thus provoked. Suspicion towards any modification of the power–knowledge relation arose and the boundaries between profane and sacred knowledge were newly problematized.

PART III

Galileo and the Church

10

Traditionalist Interpretations of Copernicanism

From an Unproven to an Unprovable Explanation

For centuries, the "Galileo affair" has played a symbolic role in the history of science and religion. The Tridentine context of the church was only very generally discussed. The story of the Dominicans' and Jesuits' educational systems was not considered relevant for understanding the fall of Galileo, and the tension between their alternative theologies was doomed to oblivion. These, however, constituted the background against which the "Galileo affair" could have been discussed and explained as a historical fact. Against this background, the idea of "conflict" as the central theme of the episode is modified, while the story is enriched with the reconstruction of a dialogue between Galileo and the Jesuits, the intellectual avant-garde of the church.

On the eve of the "first trial" of Galileo the intellectual establishment of the church was institutionally, theologically, and intellectually divided. Differences were brought to a head by the conjunction of two major developments: the *De auxiliis* controversy (1597–1607), and the publication of the *Ratio studiorum* (1599), the educational credo of the Society of Jesus. The resulting theological dilemmas opened up a new conceptual space within which the commonly accepted canons of knowledge were questioned and renegotiated. Moreover, the necessity for the formulation of a *Ratio studiorum* common to all Jesuit institutions revealed the subversive potential of the Jesuits' educational practice and its challenge to the boundaries between disciplines, and between sacred and profane knowledge. The recurring censure of Jesuit theological theses during the *De auxiliis* controversy, and the prohibition of the two first versions of the *Ratio* by the Inquisition were not unrelated to a power-struggle between Dominicans and Jesuits over the supreme intellectual authority in the Catholic world.

Only three years after the culmination of the Dominican–Jesuit dispute, a new intellectual program gained general notice with the publication of *Sidereus nuncius* (1610) by the Paduan mathematician, Galileo Galilei. A questioning of the accepted canons of knowledge, and of the boundaries between physics and mathematics, astronomy and natural philosophy, sacred and profane knowledge not only recurred in all Galileo's works, but were also at the center of intellectual debate in court, in the academies, and in Jesuit colleges as well as in other church institutions. Moreover, the epistemological problems raised by Galileo's scientific program were related to the epistemological problems faced by Dominicans and Jesuits in their theological and educational debates.

In the following section, I will focus on three issues which have not so far been recognized as crucial for the story of science and religion in the Catholic world:

(1) Among the traditional intellectual elite, suspicions against Copernicanism had long before been voiced in terms of the Thomist principles of the organization of knowledge violated by Copernicus's work. A text written by a Dominican – Giovanni Maria Tolosani – in the 1540s echoes the Thomist sensibilities of the traditionalists, who tended to reject Copernicanism as an unproven doctrine. The beginning of the seventeenth century, however, marked a change of mood.

A tendency to epistemological scepticism among the Dominicans following the controversy over grace and free will changed the status of new forms of knowledge like Copernicanism from that of *unproven* doctrines to that of *unprovable* ones. Urban VIII's famous argument against Galileo should be understood against this background. Appearing for the first time in a text by the pope's theologian dealing with God's knowledge of "contingent futures", it reflects the scepticism implied in Thomas de Lemos's arguments against *scientia media*, God's middle science of man's future, contingent, acts.

The transformation in the epistemological attitudes of the traditionalists between the mid-sixteenth century and the beginning of the seventeenth century will be dealt with in the first chapter of Part III. This transformation was reflected in the inner dynamics of church history, and at the same time accounts for the traditionalists' reaction to new forms of knowledge.

(2) The encounter of the alternative intellectual elite with new forms of knowledge was of a quite different character. The Jesuits were well trained in the use of the Thomistic principles for the organization of knowledge as a mechanism of control over the assimilation of new knowledge. These principles were eroded, however, as a result of the Jesuits' openness to the criticism of Thomism by alternative intellectual currents. At the same time the traditional canons of knowledge were also subject to criticism as a result of the alternative theology developed by the Jesuits. Thus, a missed opportunity – a "lost moment" – in the intellectual history of the Jesuits can be reconstructed. At the turn of the century, the Jesuits seemed

about to conceive of a new synthesis of sacred and profane knowledge based on the critique of Thomist epistemology. The failure of the Jesuits to establish an alternative framework in which new forms of knowledge could be assimilated is the focus of the second chapter of Part III.

(3) A detailed reconstruction of the various institutional and intellectual constraints imposed by the church at the time of Galileo constitutes one aspect of the alternative to the "total narrative" of the "warfare of science and religion" in the Catholic world. A complementary aspect is a reconstruction of one scientific debate between Galileo and a Jesuit mathematician over the new telescopic discoveries and their implications for the status of astronomy and for the restructuring of the cultural field shared by Galileo and his adversaries.

The attack of the Dominican friars Niccolo Lorini and Tomasso Caccini in the years 1613–1614 against Galileo is a well-established fact in the history of the Galileo affair. True, the denunciations to the Roman Inquisition in which the friars' campaign culminated figure only very marginally in the history of modern science. But, with regard to the specific Dominican criticisms of Copernicanism, the campaign marks a subtle change of mood which is of considerable significance.

The documents of the Inquisition bear witness to a constant preoccupation with the Copernican doctrine among the Dominicans. Father Ximenes of Santa Maria Novella, for example, while practicing disputation for his theses with another young man, Attavanti, heard from him about the motion of the earth.[1] Father Lorini, overhearing the discussion from the next room hastened to denounce this as heresy,[2] promising to deliver a sermon on the subject. On All Soul's day 1613 he attacked Galileo from the pulpit, denouncing Copernicanism as contrary to the Scriptures.[3]

The use of arguments from Holy Scripture against scientific assertions was not a usual practice of the time. In his biography of Robert Bellarmine, J. Brodrick claims that Lodovico delle Colombe was the first to cite Scripture directly against Galileo.[4] J. J. Langford remarks that Francesco Sizi's *Dianoia astronomica* (1610) applied the same strategy, but only against the claim of the reality of the Medicean stars, not against Copernicanism in general.[5] Both writers, however, assume the historical novelty of this practice. Indeed, within the presuppositions of the traditional Tho-

1. *Opere*, XIX, pp. 316–317. See also Santillana, *The Crime of Galileo*, p. 50.
2. *Opere, XIX*, pp. 299–300.
3. Santillana, *The Crime of Galileo*, p. 27.
4. Brodrick, *Bellarmine*, p. 346.
5. Langford, *Galileo, Science and the Church*, p. 50, no. 1.

mist world view, shared by all Dominicans, it was quite outrageous. A scientific theory could be intrinsically inadequate, or insufficiently proven; it could not, however, by definition, contradict the Bible and still be scientific. For truth was one, whether expressed in the book of God or in the book of nature. Within the Thomist system, each domain of knowledge had its proper subjects as well as its own methods of investigation. The methods of the exegesis of Holy Scriptures could not simply be transposed to natural philosophy, to be used as weapons in the refutation of scientific theses.

A reflection of the contemporary sense of unease caused by the denunciation of Galileo in terms of Scripture can be found in the famous apology of Lorini in a letter to Galileo:

> Please realize that the suspicion that I, on the morning of All Soul's day, entered in a *discussion on philosophical matters* and spoke against anyone, is totally false and without foundation. Not only is it not true, but it is not even *likely, since I have not gone outside my field,* nor have I ever dreamed of wanting to get involved in such matters.[6]

Lorini never tired of stressing that it was not for him, a professor of church history, to enter into a discussion of philosophical matters. Setting aside the psychological interpretation introduced, for example, by Santillana,[7] one may discern in Lorini's words the boundaries of discourse of his time. It was not usual for scholars of his type to go outside their fields. The rules of the game did not really allow for such transgressions.

And yet Lorini was not completely eccentric in his form of attack. On December 20, 1614, another Dominican friar, Father Tommaso Caccini, preached in Santa Maria Novella on the Book of Joshua, beginning with a quote from the Acts of the Apostles (1:11): "Ye men of Galilee, why stand you looking up to heaven?" He declared the theory of the motion of the earth to be close to heresy and condemned mathematics as the art of the devil, calling for the expulsion of mathematicians from Christian states.[8]

The episode of Lorini and Caccini has always been dismissed by historians of the Galileo affair as unimportant. The investigation of the Holy Office, initiated by Lorini's comments on Galileo's *Letter to Castelli* and

6. *Opere,* XI, p. 427, translated by Langford, *Galileo, Science and the Church,* p. 51. (My italics, R. F.)
7. Santillana, *The Crime . . . ,* p. 46: "Lorini had displayed suitable priestly charity. . . . In his heart, however, he felt otherwise; they [the Galileists] were black souls who did not deserve justice, let alone mercy, and nothing should be left undone for their destruction".
8. On Caccini's examination by the Holy Office see *Opere,* XIX, pp. 307–311. See also Langford, *Galileo, Science and the Church,* p. 55.

Caccini's actual examination led to a dead end, and the preacher-general of the Dominicans, Father Maraffi, wrote a letter of apology to Galileo: "Unfortunately", he wrote, "I have to answer for all the idiocies that thirty or forty thousand brothers may and do actually commit".[9] This has been taken as evidence for the contingency of Lorini's and Caccini's reactions, which have not been considered typical of the Dominican position in general. But a different interpretation is not inconceivable. If the assumption is made that Lorini and Caccini did indeed echo something significant about the Dominican milieu, then we may regard this episode as a turning point in one subplot in the history of the Galileo affair; namely, the development of the Dominican attitude towards Copernicanism from the mid-sixteenth century to the beginning of the seventeenth.

A text written by Giovanni Maria Tolosani (1470/71–1549)[10] of the convent of St. Mark in Florence can provide the necessary beginning for such a narrative. A well-known theologian-astronomer in his time, Tolosani was a friend of the magister of the Holy Palace, Bartolomeo Spina – also a Dominican – who had expressed a desire to stamp out the Copernican doctrine. Spina's death in 1546 prevented him from doing so, and Tolosani took the challenge upon himself.

Following the rationalistic tradition of Thomism, Tolosani strove to base his delegitimation of Copernicus's claims on philosophical arguments. Neither the authority of the Scriptures nor that of tradition to which he indeed referred were regarded as sufficient in themselves to cope seriously with an innovative or, at any rate, revived scientific theory. Copernicus's theory was absurd, according to Tolosani, because it was scientifically unfounded and unproven.

According to Tolosani, Copernicus's work was characterized by three errors. The first lay in the procedures of logical argumentation.[11] Copernicus assumed the motion of the earth, and argued that whoever presup-

9. Santillana, *The Crime* . . . , p. 45.

10. The text has been transcribed by E. Garin, "Alle origini della polemica anticopernicana", in *Colloquia Copernicana,* II Studia Copernicana, vol. 6, Wrocław 1973, pp. 31–42. See also R. S. Westman, "The Copernicans and the Churches" in Lindberg and Numbers, *God and Nature,* pp. 76–113; E. Rosen "Was Copernicus's *Revolutions* Approved by the Pope?" *Journal of the History of Ideas* 36 (1975), pp. 531–542.

11. Garin, p. 37: "Praesupponit etiam ille Copernicus quasdam hypotheses quas non probat nisi sub conditione: quae conditio rem in suo esse non ponit, cum dicit libro primo, cap. 8: 'Si – inquit – quispiam volvi terram opinetur, dicit utique ipsius motum esse naturalem, non violentum'. Praesupponit ille quod prius probare debuerat, scilicet volvi terram; quod tamen expresse probatur falsum".

posed such motion had to admit that it must be natural. However, to conclude the *natural* motion of the earth from a hypothesis about its motion in general was obviously not a proof, nor could it even be taken as a correct hypothetical argument. Tolosani referred here to the complete absence of any physical theory from which one could deduce the motion of the earth. However, he chose to formulate his objections in logical terms, not being able to imagine that the new astronomy would eventually require a new physics. Hence, he rejected the Copernican doctrine for being based on unproven hypotheses.

But Copernicus also transgressed against the accepted canons of knowledge. In seeking a knowledge of the universe, one had always to ascertain that products of the intellect were in agreement with things in themselves (*rei*). Only nonhuman invention or contrivance could change this order.[12] This assertion was based on Aristotle's arguments against the Pythagoreans and on the Thomist interpretation of those arguments.[13] The Pythagoreans, according to both Aristotle and Thomas, did not seek causes and principles to explain phenomena, but sought phenomena which would "embellish" the products of their own intellect. The search for causes, according to Thomist Aristotelianism, begins with the evidence of the senses concerning appearances, and it is only through this, and via a process of abstraction, that the intellect can recognize essences. Afterwards, when the intellect recognizes an essence as the cause of a phenomenon, it seeks to infer the effect from the cause deductively. The Pythagoreans, however, changed this necessary order by making the idea, whose source is the intellect, precede the phenomena, and only afterwards looked for support for their opinion in nature.

In Thomist terms, number – the key to all knowledge in Pythagoreanism – was a product of the intellect, an "intelligible being" with no physical reality. Accordingly, numbers could not provide physical causes in the investigation of nature. In attacking Pythagoreanism, Tolosani was in fact at-

---

12. Ibid.: "Est enim veritas adaequatio rei ad intellectum et intellectus ad rem, sicut in facto condita est a Deo, non per humanam confictionem, in qua saepius falsitas reperitur, perverso ordine rei et intellectus".

13. Ibid., p. 36: "De opinione autem Pictagoricorum ibi dicit Aristoteles quod 'non ad apparentia rationes et causas quaerebant, sed ad quasdam rationes et opiniones ipsorum apparentia attrahentes et tentantes adornare'. Ubi divus Thommas hoc exponit, quod Pictagorici 'non quaerebant solummodo rationes et causas, ut applicarent eas ad ea quae sensu apparent, sed in oppositum, ea quae sensu apparent reducere conabantur, et per quandam violentiam attrahere, ad nonnullas rationes et opiniones inextimabiles, quas ipsi praecogitabant".

tacking the possibility of mathematical physics, and his reasoning was based on Thomist principles. Thus, astronomical hypotheses (like epicycles and eccentrics) whose justification was their success in "saving the phenomena" in mathematical terms could not be regarded as physical causes.

Copernicus's greatest offense, however, his transgression of the organization of the sciences into a comprehensive system, was the consequence of both logical and epistemological errors. As Tolosani said:

> An inferior science receives proven principles from a superior science. Thus, all the sciences are interconnected so that an inferior needs a superior and they assist each other.[14]

Copernicus, who used mathematics and astronomy as the basis for his physics and cosmology, transgressed against this principle. Thus, he undermined the system whereby all the sciences were bound up into a single whole.

However, despite his three transgressions, Copernicus was not perceived by Tolosani as a religious dissenter, although he considered his principles to be of some danger to the faith. The source of his errors was rather a defective education. For although he was skilled in the mathematical sciences and in astronomy, he lacked knowledge of physics and logic.[15] No astronomer, in Tolosani's opinion, could be complete in his education if he had not first had a grounding in the physical sciences. A man could not be an astronomer if he was unable, through logic, to distinguish between truth and falsehood and if he was unfamiliar with the various forms of proof used in medicine, philosophy, theology, and the other sciences.[16] Mathematics and astronomy could not serve as a basis for physics and cosmology, for the information they provided about the material world was not essential. This was both because of the abstract nature of mathematical entities and because of the nonessential character of inference from effects to causes.

14. Ibid., p. 36: "Scientia quoque inferior a superiore principia comprobata recipit. Itaque omnes scientiae sibi invicem connectuntur, ita ut inferior superiore indigeat, et se invicem adiuvant".

15. Ibid., p. 35: "Peritus est etiam in scientiis mathematicis et astronomicis, sed plurimum deficit in scientiis physicis ac dialecticis, nec non et sacrarum literarum imperitus apparet, cum nonnullis earum principiis contradicat, non absque infidelitatis periculo et sibi et lectoribus libri sui".

16. Ibid., p. 36: "Non potest enim esse perfectus astronomus, nisi prius didicerit scientias physicas, cum astrologia presupponat naturalia corpora coelestia et naturales eorum motus. Nec homo potest esse perfectus astronomus et philosophus nisi per dialecticam sciat discernere inter verum et falsum in disputationibus, et habeat argumentorum notitiam: quod requiritur in medicinali arte, in philosophia, theologia et ceteris scientiis".

For Tolosani, Copernicanism remained an unproven theory, because it could not meet the criteria of scientific truth set by Thomas and upheld by the inheritors and guardians of his tradition. This is not to say that the Dominicans were defending a purely intellectual concern unrelated to the power structure in which they had vested interest. On the contrary, the insistence on Copernicus's transgression of the principles of the organization of knowledge clearly indicated the nature of their anxieties: they were concerned about the area where knowledge and power intersected and epistemology became inseparable from ideology. The insistence on preserving the boundary between natural philosophy and mathematics and the hierarchical order of which they formed part obviously revealed a desire to retain an intellectual – theological structure which made possible the cultural hegemony of a small, exclusive contemplative elite. Yet, nevertheless, the defense of this structure was intellectually grounded in canons of knowledge which, in principle, were subject to negotiation and change. No transcendental limit was explicitly set to the debate with Copernicanism.

The episode of Lorini and Caccini (1613–1614), however, marked a change in the norms of discussion accepted in the Dominican milieu, as Galileo and his Copernican campaign were perceived in terms of religious heresy. The change from Tolosani's intellectual form of argumentation to Lorini's and Caccini's religious form indicated a new sensibility among the Dominicans, despite the relative unimportance of the personalities involved.

The new Dominican sensibility was not unrelated to the long debate with the Jesuits, and the notion of divine omnipotence to which it gave rise. The emphasis on God's absolute foreknowledge of man's free acts and on its antecedent will allowed for its interpretation as a transcendental limit to human intellectual capacity. Moreover, we have historical evidence of such an interpretation by contemporaries, and, in fact, by none other than the pope himself.

Cardinal Agostino Oregio (1577–1635), a colleague of Maffeo Barberini and his theologian when he became Pope Urban the Eighth was the author of a theological essay, De Deo uno, published in Rome in 1629.[17] In this essay Oregio devoted a chapter to a problem which continued to preoccupy seventeenth-century Catholic theologians after the De auxiliis debate, despite the papal prohibition against further discussion of these matters. The problem was whether God knows the "contingent futures" ("an Deus cognoscat futura contigentia"). Oregio claimed that the divine decree operating with the created will does not

17.   A. Favaro, Oppositori di Galileo: Maffeo Barberini, Venice 1921, p. 26.

interfere with the freedom of the will but assists it. God's infinite knowledge does not prevent created beings from acting freely, in accordance with their nature. All this was as true in the natural sphere as it was in the supernatural.[18]

Oregio's words echoed those of Thomas de Lemos in the *De auxiliis* debate. Lemos had argued there that God operates the created world by means of a continual intervention, both with regard to the human will and with regard to the normal functioning of the universe.[19] On the basis of this concept of God's omnipotence, continued Oregio, Pope Urban the Eighth, during his period as cardinal (1615–1616) put forward an argument to "a very learned man" which was as amazing in its wisdom as it was praiseworthy for its religiosity. The problem which preoccupied the cardinal in his conversations with the scholar (probably Galileo), was that of the agreement between the Scriptures and the theory of the earth's motion. Assuming, for the sake of argument, that all that the scholar had suggested was true, the cardinal asked him if he thought that God could have created a different universe, arranged the stars in another order, and given them different motions from those which had been observed. If he denied this possibility, it was incumbent upon him to prove that any other order implied a contradiction in God's actions, for the omnipotence of God is limited only by the law of contradiction. Because God's knowledge is not inferior to his power, if we acknowledge his power, we are also compelled to recognize his knowledge, and if God knows different ways of ordering the universe, all that scholars could prove was not an "absolute truth" but merely "saving the phenomena", for any other solution would represent a limitation of God's power or knowledge. Oregio ended his account by remarking that on hearing this argument from Barberini, the scholar was silent.[20]

The pope's argument, laced with sceptical undertones untypical of tra-

---

18. A Oregio, *De Deo uno*, Rome 1629. p. 193, quoted in Favaro, *Oppositori*, p. 26: "Cum ergo Deus, quando de facto concorrit ad humanos actus, non auferat libertatem, sed illam adiuvet: neque etiam, quando decrevit cum voluntate creata concurrere, aliquid posuit, quod illi aut adversaretur aut officeret. Nam et infinita sua scientia praecognovit quid requireretur and hoc, ut creata voluntas posset libere operare, tam in actis naturalibus quam supernaturalibus; et iuxta creatae voluntatis naturam et exigentiam, uti scivit ac potuit, sic etiam concurrere decrevit ullo libertatis creatae detrimento."

19. See ch. 9. "Haec gratia habet efficaciam ab omnipotentia Dei et dominio, quod sua Divina Maiestas habet in voluntates hominum, sicut in coetera quae sub coelo sunt."

20. Favaro, *Oppositori,* p. 27.

ditional Thomist theology, makes sense if one takes into consideration
the intellectual climate sanctioned by the Dominican arguments in the *De
auxiliis* debate. From the concept of the divine decree, which simultaneously embodied both the will and the knowledge of God, Maffeo Barberini passed easily to the possibilities of human knowledge implied in
such concept. The divine omnipotence, he claimed, exists both in God's
power of creation and in his knowledge of the infinite possibilities of cosmological orders. He is limited only by the law of contradiction. Accordingly, all human knowledge of nature, because it is limited to the creation
known to us, is merely hypothetical and cannot lead to any ultimate truth.
Any pretension to a final knowledge of reality implies setting a limit to
the divine omnipotence.

Oregio's attempt to elucidate the concept of the divine decree enables
us to situate the position of the pope in the intellectual climate of the
period and especially the Dominican theology. Such a contextualization
also demonstrates the change in the status of Copernicanism after the
theological debate with the Jesuits. In the mid-sixteenth century, the danger represented by Copernicanism had been perceived by the Dominicans
in terms of the particular organization of knowledge which they were
interested in preserving. Within this context Copernicanism was criticized
as an unproven theory. At the beginning of the seventeenth century, however, the argument was put forward that any attempt to know the physical
universe in a definitive manner meant challenging the omnipotence of
God. In such a framework of thought, Copernicanism was necessarily
seen as an unprovable doctrine. Although this form of argumentation was
alien to the spirit of Thomism, it was perfectly in harmony with the theology of the Dominicans and their concept of divine omnipotence as developed in reaction to Molinism.

Maffeo Barberini's argument against Copernicanism, which was linked
by Oregio's *De Deo uno* to the concept of divine omnipotence developed
in the context of the *De auxiliis* debate, was put to Galileo sometime in
1615–1616. This information is suggestive for any attempt to understand
the Dominican interpretation of the official censure of Copernicanism issued in the same year by the congregation of the Index.[21] If the recognition of the divine omnipotence requires that the status of any knowledge
of the cosmological order remain hypothetical, then Copernican cosmology is an unprovable doctrine. Thus, Galileo was not to attempt
to contrive any further proofs of such a theory, although he could go
on using it as an abstract mathematical construction for the purpose
of "saving the phenomena". This position is most in keeping with

21. See ch. 2.

the Dominican sensibilities in the years immediately following the *De auxiliis* debate.

The two types of reaction to Copernicanism exhibit, in my view, the constraints felt by the traditional intellectual elite of the Catholic Church in its attempt to cope with the first expressions of the scientific revolution. The first of these reactions – Tolosani's text – posited the structure of the Dominican organization of knowledge in the mid-sixteenth century as the main barrier against Copernicanism. The other reaction – that of Maffeo Barberini as presented by Oregio – clearly based the opposition to Copernicanism on the religious positions of the Dominicans after the theological debate with the Jesuits. The attitudes of the opponents were thus characterized neither by scriptural fundamentalism nor by sheer authoritarianism.

Copernicanism was regarded as dangerous by the Dominicans because it challenged the Thomist principles of the organization of knowledge on which their own status as a privileged intellectual elite depended. Dominican educational policy had always based its legitimation of intellectual activities on the necessity of training a small, otherworldly oriented elite for a contemplative life. Such a concept of education entailed a grounding in natural philosophy as a path to metaphysics and theology. Any questioning of the supreme status of natural philosophy as the bridge to metaphysics and theology was thus suspicious. Copernicus, with his preference for mathematical argumentation and for inference from effects to causes inverted the hierarchical relationship between mathematics and natural philosophy and thus threatened the conceptual link leading from natural philosophy to metaphysics, and hence to theology. It may be argued, hypothetically perhaps, that the educational program promoted by the Dominicans in the second half of the century was a response to the *kind* of challenge represented by Copernicanism. The Dominicans reacted by stressing their attachment to the Thomist synthesis in its entirety and by demanding an absolute fidelity both to its philosophical principles and to its theological approach. Thus, the dynamics of Dominican intellectual life were restricted not by obscurantism but by the intellectual conservatism of the traditional elite or, in other words, by its vested intellectual interests. The beginning of the seventeenth century, however, witnessed a modification of the Dominican religious outlook which constituted a further constraint on its ability to cope with Copernicanism in its new Galilean form.

Maffeo Barberini's argument demonstrates how the necessity of preserving God's omnipotence, rooted in the Dominican theological position, was in fact interpreted as a transcendental limit to human reason. A brief review of the epistemological framework of Dominican theology may help to reveal more clearly its bearing upon the attitude to natural philosophy and the mathematical sciences.

First and foremost, the Dominicans sought to ensure God's omnipotence by positing his absolute knowledge of the future acts of man. Absolute knowledge was a knowledge of real objects: a true knowledge of hypothetical objects was excluded by them as a logical impossibility. In order to ensure the coherence of their theological structure, the Dominicans had to reject the notion of a true knowledge of hypothetical entities. Thomist doctrine, however, required that the laws of reason and causality which exist in the natural world also apply in the supernatural world. It was not surprising, therefore, that Galileo's claim to equalize the epistemological status of astronomy (which they considered a "middle science" of mathematical, i.e., abstract entities) and natural philosophy (the science of real substances) could find no justification in their eyes.

Moreover, the mode of divine determination envisaged by Dominican theology, which gave the physical universe a highly contingent character, imposed severe restrictions on the human intellect, especially since it was maintained within the rationalistic framework of the Thomistic world view. The contingency of the world may have become an incentive for an empirical investigation of nature, as some interpretations of the fideistic trends of medieval voluntarism have suggested.[22] But there is not much sense in a speculative search for natural causes if those causes are at any given moment subordinate to the direct intervention of God and liable to change. Thus, the decree of the Index of 1620, which attempted to control any further investigation of the motion of the earth, may be seen as a reflection not of scriptural fundamentalism but of the theology of the Inquisitors, with which it was entirely in keeping.

The *De auxiliis* debate alerted the Dominican consciousness to the danger inherent in the intellectual openness of the Jesuits to the criticisms of Thomism, which gave birth to an alternative theology and exposed them to a competition for cultural hegemony. The Dominicans reacted by fiercely adhering to the Thomist system in its entirety, and by upsetting the balance between reason and faith in their theology. Their attitude may account for the peculiar decrees of 1616 and 1620, which made the Galileo affair possible as a historical fact.

22. See, for example F. Oakley, *Omnipotence, Covenant and Order*, Ithaca 1984.

11

Copernicanism and the Jesuits

Cardinal Robert Bellarmine provided history with a unique piece of evidence concerning the Jesuit position vis-à-vis Copernicanism on the eve of the "first trial" of Galileo. It consists of a letter to Foscarini (12 April 1615) written in response to the latter's attempted reconciliation of the heliocentric cosmology and the conflicting scriptural verses. Bellarmine's letter, often quoted in the historical literature on the trials, has often been interpreted as demonstrating two Jesuit characteristics: a fundamentalist approach to Scripture and epistemological instrumentalism. No wonder that historians have tended to relate Galileo's inevitable fall to the theological and scientific attitudes of the Jesuits, as represented by Bellarmine. Galileo, they claim, insisted on a figurative interpretation of the Bible, as well as on epistemological realism. His positions were irreconcilable with those of the Jesuits, a fact which necessarily brought about his tragedy.

If read in the context of Jesuit educational practices, however, Bellarmine's letter expresses neither fundamentalism nor instrumentalism. Rather, the letter delineates the institutional boundaries of scientific discourse, determined in the course of the Jesuits' struggle with the Dominicans over cultural hegemony. An archeology of those boundaries is a necessary condition for a reexamination of Bellarmine's letter.

Archeological investigation of the boundaries of scientific discourse in Jesuit culture is made possible by the survival of two different sources: the final version of the *Ratio studiorum* (1599), which is the official representation of those boundaries; and, secondly, the extensive documentation of the preparatory work for the *Ratio* in addition to an early version of 1586, which has also survived in the Jesuit published literature. On the basis of such an investigation, coupled with a reinterpretation of Bellarmine's letter to Foscarini (1615), the development of the Jesuits' attitude towards Galilean Copernicanism may be reconstructed. In such reconstruction consists the second subplot in the history of the Galileo affair, complementing or rather serving as a counterpoint to the Dominican reaction described in the previous chapter.

1586 – The Perspective of a New Organisation of
Knowledge in Catholic Culture

In dealing with the organisation of knowledge in Jesuit culture, I shall
focus my attention on the problem of the boundary between philosophy
and the mathematical sciences (the "mixed sciences") and the hierarchy
between them.[1] Moreover, I shall limit my discussion to the writings of
Christopher Clavius (1537–1612) as a representative of the trend among
Jesuit mathematicians to confer an official legitimation on the implied
challenge in their work to the Thomist principles of the organisation of
knowledge. By questioning the traditional epistemological distinctions
between the abstract/mathematical and the real/physical, between the
necessary/certain and the possible/probable, and, more generally, between
the theoretical and the practical, Clavius and his fellow mathematicians
sought to reshape the traditional orientation of studies in Jesuit schools.
My contribution, however, does not lie in an analysis of Clavius's position, which has already been done by others.[2] Rather, I would like to

1. In analysing the organisation of knowledge in the Catholic world I have been
    following in the footsteps of R. S. Westman, who was the first to suggest
    the classification of the sciences as the relevant context for the analysis of
    Copernicus's reception. See idem, "The Astronomer's Role in the Sixteenth
    Century", Ch. 2, n. 7; idem, "The Copernicans and the Churches", in Lind-
    berg and Numbers, *God and Nature*, Ch. 2, n. 7; R. Feldhay, "Knowledge
    and Salvation in Jesuit Culture", *Science in Context* 1 (1987), pp. 195–213;
    idem, "Catholicism and the Emergence of Galilean Science. A Conflict be-
    tween Science and Religion?" in: S. N. Eisenstadt and I. Friedrich Silver
    (Eds.), *Cultural Traditions and Worlds of Knowledge: Explorations in the
    Sociology of Knowledge*, Greenwich, Conn. 1988, pp. 139–163. On the so-
    cial status of the mathematician see M. Biagioli, "The Social Status of Italian
    Mathematicians, 1450–1600", *History of Science*, 27 (1989), pp. 69–95.
2. G. Cosentino, "Le matematiche nella 'Ratio Studiorum' della Compagnia di
    Gesù", *Miscellanea Storica Ligure*, II.2 (1970), pp. 171–213; A. Carugo and
    A. C. Crombie, "The Jesuits and Galileo's Ideas of Science and of Nature",
    Annali dell'Istituto e Museo di Storia della Scienza di Firenze, VIII. 2 (1983),
    pp. 3–67; U. Baldini, "Chrisoph Clavius and the Scientific Scene in Rome",
    in G. V. Coyne, SJ, M. A. Hoskin, and O. Pedersen (Eds.), *Gregorian Reform
    of the Calendar. Proceedings of the Vatican Conference to Commemorate Its
    400th Anniversary 1582–1982*, Vatican City, 1983, pp. 137–169; Wallace,
    Galileo and His Sources; Jardine, *The Birth of History and Philosophy of
    Science*, see ch. 2, n. 7. P. Dear, "Jesuit Mathematical Science and the Recon-
    struction of Experience in the Early Seventeenth Century", *Studies in History
    and Philosophy of Science*, 18, no. 2 (1987), pp. 133–175.

expose the affinity between the practical/epistemological stances of Clavius and other Jesuit mathematicians and the positions of Molinist theologians expressed during the *De auxiliis* debate. In so doing I shall put forward the idea that the date of the first publication of the *Ratio studiorum* – 1586 – represented a moment in Catholic intellectual history when the Thomist principles of the organisation of knowledge could have been modified in accordance with new insights affecting philosophy as well as theology. Had such a modification taken place and been recognized institutionally, a new kind of synthesis between science and religion could have become a fact in the Catholic world. The new physical-mathematical science might have been integrated into a new synthesis of science and theology. This synthesis was never realized, however, for reasons which will soon be clarified.

My point of departure is two arguments of Clavius concerning the ontological nature of mathematical entities and the epistemological nature of mathematical demonstrations. Both appear in various editions of Clavius's commentaries on Sacrobosco's *Sphere*[3] and on Euclid's *Elements*.[4] The peculiar ontological status attributed by Clavius to mathematical entities is defined in his contention, made in the preface to the second edition of Euclid:

> Because the mathematical disciplines discuss things which are considered apart from any sensible matter, although they are immersed in material things, it is evident that they hold a place intermediate between metaphysics and natural science, if we consider their subject, as is rightly shown by Proclus, for the subject of metaphysics is separated from all matter, both in the thing and in reason: the subject of physics is in truth conjoined to sensible matter, both in the thing and in reason: whence since the subject of the mathematical disciplines is considered free from all matter, although it [i.e., matter] is found in the thing itself, clearly it is established intermediate between the other two.[5]

This contention is typical of Clavius's practical realism which contained traces of criticism of medieval realism and took its point of departure from fourteenth-century nominalism.

3. C. Clavius, *In Sphaeram Ioannis de Sacrobosco Commentarius*, Rome 1581.
4. C. Clavius, *Euclidis Elementorum libri XV*, Rome 1574.
5. Quoted by Dear in his "Mathematical Science and the Reconstitution of Experience", pp. 139–140 from C. Clavius, *Operum mathematicorum tomus primus* (Mougins 1611), "In disciplinas mathematicas prolegomena", p. 5. See also C. Clavius, *Euclidis Elementorum libri*, 2nd ed. (1589), quoted by Wallace, *Galileo and His Sources*, p. 138; and Carugo and Crombie, "The Jesuits and Galileo's Ideas . . .", p. 34.

In the framework of medieval scientific realism, embodied in traditional Thomism, epistemological certainty depended upon ontological reality: only that which had some ontological reality, it was argued, could
really be known scientifically. Human knowledge of the world, however,
could hardly attain this ideal, since what really existed (i.e., the ontological reality of the objects of knowledge) was not pregiven to man, but
had to be discovered either analytically, a priori, or, else empirically, a
posteriori. The confidence of medieval realism in metaphysical speculation caused its preference for proceeding from self-evident first principles
to particulars and for confirming such knowledge by a movement back
from particulars to first principles. The condition for the possibility of
such a movement was the belief that the philosophical concept constituted
a link between the rational structures of the mind and the real structures
of the world. For the nominalists, such a belief could only be justified
in the realm of metaphysics. Nominalism, which dismissed metaphysics,
strove to disentangle the ontological aspect of the ideal of knowledge
from its epistemological one. That which existed could only be discovered
a posteriori, without any complete certainty. That which could be certainly known was only a product of the mind, whose connection to what
really existed remained unclear. The price to be paid for the penetrating
criticism of nominalism was a complete separation between the rational
structures of the mind and the real structures of the world, resulting in a
lack of confidence in the human ability to attain a certain knowledge of
the world.

Clavius's main effort was devoted to the search for a new way of combining the ontological and epistemological aspects of the ideal of knowledge. Such effort was needed in order to avoid the aporia implied by nominalism, whose legacy probably reached him through the calculators,
namely, the Oxford mathematicians of the fourteenth century who
adopted nominalist positions.[6] Clavius's contention that the mathematical entities are both abstract and immersed in matter, and can yield
true and real knowledge, should be understood against this background. Just as, in medieval realism, the philosophical concept was
perceived as the link between the mind and reality, so mathematical
entities fulfilled a similar function in Clavius's discourse. Clavius affirmed the possibility of true and real knowledge of mathematical entities, whose ontological status was now redefined as mediating between
the mind and reality ("free from all matter, although it [i.e., matter] is
found in the thing itself").

6.   Baroncini, "L'insegnamento della filosofia naturale . . .", p. 198.

Clavius was aware, however, of the limitations of human intellect, which is unable to grasp the real structures of the world in a direct and immediate way that would guarantee the absolute certainty and truth of this knowledge. Knowledge of the world had to be inferred from effects. For example: epicycles and eccentrics had to be inferred from observation of the positions of the planets. Here Clavius's second contention in his commentary on Sacrobosco applies. According to him, astronomy, which reasons from effects to causes by means of mathematical hypotheses, is a true science (*scientia*) in the Aristotelian sense.[7] In other words, reasoning from effects to causes by means of mathematical hypotheses and mathematical demonstrations results in necessary conclusions. This contention raised the problem of the logical relationship between phenomena considered as effects of causes discovered by mathematical hypotheses and the hypotheses themselves (for example, epicycles and eccentrics). Mathematical demonstrations in themselves cannot guarantee the reality and truth of hypotheses in nature. The only guarantee Clavius could offer was the experiences of the astronomers[8] which confirmed the congruity between the mathematical hypotheses and the effects. Nevertheless, Clavius came to the conclusion that astronomy was a real science, that it demonstrated the necessary connection between hypotheses and observed phenomena.

Clavius's contentions had been countered by philosophers who were aware of their subversive implications and chose to remain faithful to the traditional canons of knowledge. In a short treatise, probably written in the first half of the 1580s,[9] Clavius mentioned arguments which he ascribed to the philosophers of the Society of Jesus, against considering the mathematical disciplines true sciences (*scientiae*) in the Aristotelian sense:

> It will also contribute much to this if the teachers of philosophy abstained from those questions which do not help in the understanding of natural things and very much detract from the authority of the mathematical disciplines in the eyes of the students, such as those in

7. Clavius, *In Sphaeram* . . . , quoted by Carugo and Crombie, "The Jesuits . . .", p. 19.
8. See the discussion of N. Jardine in "The Forging of Modern Realism: Clavius and Kepler Against the Sceptics", *Studies in History and Philosophy of Science*, 10 (1979), pp. 141–173, and of Dear, "Mathematical Science and the Reconstruction of Experience", pp. 145, 148.
9. C. Clavius, "Modus quo disciplinae mathematicae in scholis Societatis possent promoveri", *Monumenta Paedagogia Societatis Jesu quae Primam Rationem Studiorum anno 1586 praecessere* (Madrid 1901), pp. 471–474, cited from Cosentino, "Le matematiche nella 'Ratio Studiorum'".

which they teach that mathematical sciences are not sciences, do not have demonstrations, abstract from being and the good, etc.[10]

In the eyes of the philosophers Clavius's arguments, which were no doubt representative of those of other mathematicians as well, challenged traditional epistemological and ontological presuppositions, as they ascribed necessity to astronomical demonstrations based on mathematical hypotheses and attached a peculiar ontological status to mathematical entities, traditionally considered quasi-substances. The implications of such contentions were clear to the philosophers, and they rejected them vehemently. If mathematical hypotheses can be considered physical causes, and proofs from such hypotheses are called real demonstrations, then astronomy – and with it the rest of the mathematical disciplines – acquires a status equivalent to the status of philosophy. Clavius's treatise did not remain silent on this point:

> Physics cannot be understood without mathematics, especially when it is concerned with those parts which deal with the number and motion of the heavenly bodies, the multiplicity of intelligences, the influence of stars dependent on their various convergences, on their positions in relation to one another, and on the distance between them, with the infinite division of continuous quantities, with the ebb and flow of the tide, with winds and comets ... and other meteorological matters, and with the proportion of motions, of qualities, of action, of passion, of reaction and so on.[11]

Some of the problems mentioned by Clavius as "physical" had traditionally been discussed within the framework of the "mixed sciences", namely the mathematical disciplines. By treating them as physical problems Clavius, in fact, blurred the distinction between the mathematical disciplines and philosophy, thus challenging an important boundary of the Thomist organization of knowledge. Other problems which Clavius mentioned – the multiplicity of intelligences, for instance – had always belonged to metaphysics or theology. Here Clavius strove to show the relevance of mathematics to those problems as well.

The significance of Clavius's arguments was not limited to the realm of ideas. The context in which his treatise was written turned his challenge to the cognitive interests of the philosophers into a threat to their institutional interests.

10. Cited from the English translation of A. C. Crombie, in his "Mathematics and Platonism in the Sixteenth-Century Italian Universities and in Jesuit Educational Policy", in Y. Maeyama and W. G. Saltzer (eds.), *Prismata Naturwissenschaftsgeschichtliche Studien,* Wiesbaden 1977, p. 66.

11. My translation of Clavius's text cited by Cosentino in "Le matematiche nella 'Ratio Studiorum ...'", p. 203.

In 1581 the general of the order, Claudio Aquaviva, appointed a committee of twelve to draw up the main lines of a formula for the organization of studies in Jesuit schools (*ad conficiendam formulam studiorum*).[12]
As part of the preparatory work of that committee Clavius wrote two
short treatises, which became the first documents in the history of the
society to treat the problem of mathematics instruction as a problem of
cultural policy.[13] The treatises contained an expanded program of mathematical studies accompanied by a propaganda campaign for the status of
those disciplines and of their professors. The treatises also described the
institutional sanctions which would ensure the implementation of the
new orientation towards the mathematical disciplines. The claim of mathematicians that their discipline should be given the same high status as
was enjoyed by natural philosophy was justified by a conviction of their
equal relevance for an understanding of reality: "It is necessary", wrote
Clavius, "that the pupils should understand that these sciences are necessary and useful for a correct understanding of the rest of philosophy".[14]
The social status of the professors of mathematics within the framework
of the colleges was to be reaffirmed by their participation in all official
occasions such as graduation and public disputations.[15] In addition, passing an examination in mathematics was to become a condition for acquiring a degree not only in philosophy but also in theology.[16]

Clavius's demand for a modification of the boundaries between the
physical and mathematical disciplines and their hierarchical order was
grounded in a gradual change in the epistemological values of traditional
Thomism. Such a change was not limited to the domain of profane
knowledge: it affected theological studies to an equal degree. A comparison between the objections of traditional philosophers to Clavius's ideas
with the objections of the Dominican theologicians to the Molinists of
the Society of Jesus may clarify this analogy. The professors of philosophy
in Jesuit colleges who had remained faithful to traditional epistemological
values denied the possibility of true and real knowledge of abstract (hypothetical) entities applied to the study of nature. Their objections rested
upon the same canons of knowledge as led the Dominican theologians to

12. Ibid., p. 202.
13. The first among these two is *Modus quo disciplinae mathematicae in scholis
    societatis possint promoveri*; the second is *De re mathematica instructio,*
    both published in the *Monumenta Paedagogica* of 1901, see Cosentino, "Le
    matematiche nella 'Ratio Studiorum . . .'", pp. 202, 205.
14. *Modus quo disciplinae mathematicae . . .*, cited by Cosentino, ibid., p. 204.
15. Ibid.
16. Ibid., p. 205.

deny the possibility of an absolute science of the hypothetical future acts of man before God's voluntary decree. It was based on the assumption that there could be no true knowledge of hypothetical entities. Furthermore, just as Clavius's opponents protested that theorizing by means of hypotheses and their testing against facts was no demonstration, so did the Dominican theologians deny the absolute nature of knowledge of objects that are not yet determined by God's will. In both cases the objects of knowledge are not real (or not yet real) and therefore cannot be known with certainty; excluded from the domain of divine absolute knowledge, they are considered merely instrumental in the domain of human knowledge.

The need to sever epistemological certainty from ontological reality played a major role both in Clavius's attempt to account for human scientific knowledge, and in the need of the Jesuit theologians to account for God's knowledge of man's future acts, and hence of predestination.

In striving to provide a basis for astronomy (and the rest of the "mixed sciences") and describe them in terms of the absolute knowledge of causes (*scientia*), Clavius had to take into consideration the limitations of man's intellect. Since man was unable to know the causes a priori in order to infer the effects from them, he had to put forward hypotheses. If known effects could be deduced from a hypothesis, then the certainty of the inference was guaranteed. But what could guarantee the objective truth of the hypothesis? What could guarantee that the hypothesis asserts the real cause, since it is only proved that the effect follows from the supposed cause but not that it is only this cause which would produce this effect? Only empirical confirmation could substantiate the claim, but even this could not prove it. Clavius nevertheless came to the conclusion that there is a necessity between hypotheses and observed effects, since astronomy was a science (*scientia*) and could produce true knowledge.

Obviously, the nature of God's knowledge differed from that of man's scientific knowledge, since God was not limited in his ability to gain direct and infallible knowledge of the future acts of man, and hence did not have to rely on inferences, hypotheses, and deductions to attain a true and real knowledge of them. However, the Jesuits had a clear interest in preserving a meaningful role for man's free will, and it was for this reason that they were ready to place some limitations on predestination by separating God's knowledge from his will. Molinist theologians contended that before the voluntary decree, God knew the future acts of man not with absolute knowledge but with "middle science", the infallible knowledge of future contingents. Similarly to Clavius, who maintained that knowledge of natural phenomena by means of mathematical hypotheses was absolutely certain, these theologians also held that God's "middle science" was absolutely certain and infallible. This meant, however, that the necessity of the law of predestination was guaranteed epistemologically, although no ontological necessity related God's will (the cause) to man's future acts (the effects).

In light of the analogy between the epistemological positions of Jesuit mathematicians and theologians, Clavius's proposals to change the status of the mathematical disciplines and to put them on a par with natural philosophy had much wider implications than this minor modification of the organization of knowledge seems to suggest. By striving for the abolition of the hierarchical order that in classical Thomism underlay the relationship between physics, whose causes could not be defined in mathematical terms, and mathematics, which could not provide real necessary causes, he was paving the way for a physical-mathematical science whose laws were warranted by epistemological necessity. Epistemological necessity could have become an alternative organizational principle, capable of unifying the law of God (conceived in terms of predestination) and the law of nature. This concept, however, imposed limitations on human knowledge as well as on God's predestination.

A major modification of the Thomistic principles of the organisation of knowledge seemed about to take place in 1586, with the publication of the first *Ratio studiorum*,[17] a daring attempt to create a common educational code for all Jesuit schools. Traces of the line of thought represented by Clavius were clearly visible in the pages of this version.

The chapter on mathematics opened with an apology, in the spirit of Clavius, intended to prove the relevance of mathematics for all other spheres of activity in which Jesuits were engaged: salvation through the study of theology considered as the ultimate goal of the society, the teaching of all other sciences to which mathematics is necessary, and the dissemination of practical knowledge useful for civil and religious life.[18]

A revolutionary step followed: in accordance with Clavius's vision, it was asserted that there was a need to prepare cadres of professional mathematicians, especially for teaching purposes. This could not be done

17. Ibid., p. 207, n. 62.
18. Ibid., pp. 207–208, n. 63: "illae . . . poetis . . . historicis . . . analiticis solidarum exempla demonstrationum: politicis artes plane admirabiles rerum bene gerendarum domi militiaeque: physicis coelestium conversionum, lucis, colorum, diaphanorum, sonorum formas et discrimina: metaphysicis sphaerarum atque intelligentiarum numerum: theologis precipuas divini opificii partes . . . ut praetereantur intereas quae ex mathematicorum labore redundant in rem publicam utilitates in morborum curationibus, in navigationibus, in agricolarum studio".

within the framework of a course of studies intended to train theologians. Instead, a special course of professional mathematics would be established at the Roman College, which would teach mathematics for three years to those talented enough to study this subject.[19] The rest would study mathematics for one-and-a-half years, parallel to the first and second year of philosophy. Here, emphasis would be placed on the congruence between the demonstrative principles common to Aristotle's *Posterior Analytics* and mathematics, stressing the fact that Aristotelian logic is hardly comprehensible without examples from mathematics.[20]

There are enough indications in the *Ratio* of 1586 of a deviation, in line with Clavius's suggestions, from the Thomist attitude to mathematics. The relevance of mathematics for both the ascent towards theology and for the descent towards the practical spheres of knowledge provides a justification for making it a central feature of the curriculum. Mathematics is seen as the key to an understanding of physical reality on the one hand and as the model of correct rational procedure on the other. Both functions ensure it a major role in the overall scheme of knowledge leading from physical reality to ultimate (metaphysical) reality and theology. Accordingly, the professional and social status of mathematicians is recognized as equal to that of philosophers.

The *Ratio* of 1586 was printed in only a few hundred copies. It was probably not meant for public dissemination. According to the letter of the general of the order to the provincial superiors, it was intended to be criticised by censors of the various provinces and possibly corrected. The objections, however, were harsh and immediate. A number of memorials were written against it, and a certain Father Enrique Enriquez submitted it to the Spanish Inquisition, contending that it was a "declaration of war on the teaching of Thomas Aquinas".[21]

Enriquez, moreover, was not alone. A number of other memorials criti-

19. Ibid., p. 208: "Professor alter, qui modo P. Clavius esse posset, constituatur, rerum mathematicarum pleniorem doctrinam conferat in triennium, explicetque privatim nostris octo circiter aut decem, qui mediocri saltem sint ingenio, nec e mathematicis alieno, et philosophiam audierint, qui ex variis essent convocandi provinciis, unus ex qualibet, si fieri posset".

20. Ibid., p. 208, n. 64: "[professor mathematicarum] sesquianno quotidianis lectionibus breve curriculum mathematicarum rerum conficiat a nostriis et ab externis audiendum, cuius initium professor auspicabitur post Pascha Ressurrectionis mane primo hora scholarum auditoribus logicae, quia per id tempus fere parant se ad Posteriora Analytica, quae sine mathematicis exemplis vix possunt intelligi".

21. Farrell, *The Jesuit Code of Liberal Education*, p. 231.

cal of the section of theological opinions spurred the Inquisition into action. Whether the *Ratio* of 1586 was found heretical by the Inquisition, as some historians maintain, or whether it was revised as a result of criticisms from within the Jesuit order itself, as others prefer to believe, is by no means irrelevant to the present argument. But even without this piece of historical information, the thesis holds that the *Ratio* of 1586 provided an opportunity for changing the principles of the organisation of studies in a way that might have had long-range consequences for the history of science in Jesuit institutions. A glance at the version of 1599 – the last and definitive one – shows that a major modification along these lines did not take place. The *Ratio* of 1586 remained a historical possibility which was never realized.

1599 – The "Lost Moment" of the Jesuit Educational Program

A long time-lag separated the appearance of the first *Ratio studiorum* from the publication of the last version in 1599: thirteen stormy years of formative experiences with a significant impact on the fortunes of the Jesuit program in the next decades. In the course of those years the conflicts between mathematicians and philosophers at the Roman College grew more intense.[22] At the same time, the theological controversies with the Dominicans, which started as a peripheral issue in the 1580s, crossed the borders of Spain and found their way to the Roman center of the Catholic world. In both cases the boundaries of human and divine knowledge, formatted within the framework of the Thomist worldview, were questioned and even undermined. In both cases, epistemological issues were entangled with institutional interests as well as with the formation of the new professional identities.

Any attempt to characterize the document of 1586 cannot ignore the critical voices of Jesuit intellectuals, transgressing Thomist boundaries both philosophically and theologically. The prolonged process of correction and reformulation in the following years was a sophisticated exercise in which those voices were largely subdued. The final version of 1599[23] reads as the culmination of a tendency towards conformity, reflecting the

22. See U. Baldini, "La nova del 1604 e i matematici e filosofi del Collegio Romano", *Annali dell'Istituto e Museo di Storia della Scienza di Firenze*, 6.2 (1981), pp. 63–98.

23. M. Salomone (ed.), *Ratio studiorum: l'ordinamento scolastico dei collegi dei Gesuiti*, Milan 1979, henceforth *Ratio studiorum*.

vulnerability of innovative Jesuits vis-à-vis traditional theologians and conservative philosophers. In analysing some parts of this final version I shall again focus my attention on the boundaries between natural philosophy and mathematics, and between philosophy and the different kinds of theology. It was through an entrenchment within those boundaries that the Jesuits could use the *Ratio* as both a means of control over dissenters within and as an instrument for gaining legitimation without.

The course of philosophy in a Jesuit university usually lasted for three years, the first of which was mainly devoted to logic.[24] Tactics of exclusion and of implementation of distinctions which insured an assimilation into accepted modes of thought pervade the *Ratio*. The exclusion of a serious discussion of important concepts such as "universals" and "contingents" may serve as a point of departure to the exposition of such tactics. "Universals" were the main target of the nominalist criticism, a concept in which logic and metaphysics intersected and which imposed itself on all spheres of knowledge.[25] The *Ratio* insisted that they were not to be treated in detail, but should be left to the teacher of metaphysics.[26] Similarly, the subject of the "contingents" was to be dealt with only briefly, and one was not to touch the problem of free will pertaining to theology.[27] By excluding a consideration of the concepts of "universals" and "contingents", the questioning of the relationship of logic, metaphysics, and theology, which constituted essential parts of the Thomistic synthesis, could be avoided, at least during the formative years of studies.

The first year was to begin with a preliminary discussion of the nature of Aristotelian science, its division into a speculative part and a practical part, the subdivision of the sciences, the subordination of some sciences to others, and the different procedures of physics and mathematics, abstraction and definition.[28]

24. *Ibid.*, p. 65: "Nel primo anno deve spiegare la logica".
25. See A. Crescini's discussion of the nominalist critique on medieval realism in *Le origini del metodo analitico: il Cinquecento*, Udine 1965, Parte I, cap. I, pp. 21–35; and idem, *Il problema metodologico alle origini dell scienza moderna*, Rome 1972, Introd. cap. I, 2, pp. 5–42.
26. *Ratio studiorum*, pp. 65–66: "Una trattazione completa degli universali la rimandi alla metafisica, accontentandosi di fornire qui un'informazione sommaria."
27. *Ibid.*, p. 66: ". . . trattando però solo molto brevemente quella circa i contingenti e evitando ogni accenno al libero arbitrio".
28. *Ibid.*: "Inoltre per potere riservare tutto il secondo anno alla fisica, organizzi verso la fine del primo anno una dissertazione di una certa ampiezza sulla scienza, in cui concentrare, in massima parte, i prolegomeni della fisica, come la suddivisione delle scienze, le astrazione, la parte speculativa, quella

This concise passage of the *Ratio*, which describes the main themes of an introduction to the heart of the curriculum – physics, the science of reality – contains all the major principles in accordance with which the intellectual program of the Jesuits was organised into a system – a system which was to remain, theoretically at least, an orthodox Thomist one. In the traditional Aristotelian–Thomist world the separation between the theoretical and the practical was based on a distinction between the objects of these two spheres of knowledge: natural substances amenable to theoretical analysis, and artificial products, incapable of substantive changes and therefore given over to mere practical knowledge rooted in experience and the senses but lacking in certainty and truth. Although in the course of its long development, Aristotelian physics had assimilated various spheres of practical knowledge, the latter's status always remained ambiguous and lower than that of pure theoretical knowledge. By maintaining the distinction intact, the *Ratio* actually confirmed the hierarchy of these two areas of study despite the considerable importance which practical knowledge had acquired in the climate of Jesuit culture.[29]

A second major distinction which had become a principle of the traditional organisation of knowledge concerned the boundary between mathematics and physics. Despite an awareness that the two often dealt with the same subjects, Aristotelian thought required a clear distinction between the various discursive practices and the different methods to be used in mathematics and physics. The distinction was based on the ontological difference between mathematical and physical entities. Mathematical entities, like the point or the line, were regarded as quasi-substances, devoid of the characteristics of time and space, in Thomas's terminology intelligible,[30] but not sensible. A point or a line in the physical sense, however, were regarded as natural substances, rooted in time and space. Thus, the mathematical and the physical point were defined as two different objects of investigation. One always remained within the sphere of abstract thought, while the other was deeply rooted in concrete physical reality. Again, despite the attempts of Jesuit mathematicians of the type of Clavius to blur that traditional distinction, the *Ratio* insisted on its implementation during the first year of studies, prior to the acquisition of any philosophical-scientific or mathematical learning. Thus, the assimilation into traditional ways of thinking was taken care of without any need to impose doctrines and specific interpretations of the Aristotelian text.

pratica, la subalternazione, il diverso modo di procedere in fisica e in matematica, di cui tratta Aristotele nel libro 2° della *Physica*".

29. See Ch. 8.
30. Thomas, *Sum theol.*, I. q. 85, a. 1–2.

The second year of philosophical studies, according to the *Ratio*, was to be devoted to the study of the eight books of Aristotle's *Physics, On the Heavens,* and the first book of *On Generation and Corruption.*[31] Here too, exclusion, distinctions between disciplines, and assimilation into accepted modes of thought were common practices. The part of the first book of the *Physics* dealing with the ideas of the "ancients" – the pre-Socratics – was to be summarized, without reading the original text or discussing it. The same also applied to the sixth and seventh books.[32] The sixth book dealt with the relationship of movement, time, and infinity, and consisted in a critique of Zenon. It also treated atomism as a mathematical problem, considering the implications of atomism for questions of continuity and infinity. The seventh book focussed on the relationship between the mover and the moved, the prime mover, and Zenon's arguments. The fact that the *Ratio* recommended a very selective reading of these passages indicates what it tried to avoid, namely, a preoccupation with mathematical or alternately theological problems in the physics course. By this means, it attempted to discard a long medieval tradition of discussion in which mathematical thinking combined with theological thinking within the framework of Aristotelian philosophy.[33] Discussion of the number of intelligences and the freedom and infinity of the prime mover, traditionally related to the eighth book of the *Physics,* was forbidden in this context and transferred to the sphere of metaphysics.[34] Subjects which were relevant to astronomy were also removed from the course of physics. Meteorology, for example, was to be touched on only very briefly.[35]

31. *Ratio studiorum*, p. 66: "Nel corso del secondo anno deve spiegare gli otto libri della *Physica*, i libri de *De coelo* e il primo libro del *De generatione"*.
32. Ibid.: "Per quanto riguarda gli otto libri della Physica, il testo del libro 6° e 7° deve essere riassunto, come, del resto, la parte del primo libro che tratta delle opinioni degli antichi".
33. See J. E. Murdoch, "Mathesis in philosophiam scholasticam introducta. The rise and development of the application of mathematics in fourteenth-century philosophy and theology", in *Arts libéreaux et philosophie au Moyen Age,* Montreal 1969, pp. 215–254; and idem, "From Social into Intellectual Factors: An Aspect of the Unitary Character of Late Mediaeval Learning", in J. E. Murdoch and E. D. Sylla (eds.), *The Cultural Context of Mediaeval Learning,* Dordrecht 1975, pp. 271–348.
34. *Ratio studiorum*, p. 66: "Svolgendo il libro 8°, non si deve dir nulla sul numero delle intelligenze, sulla libertà e sull'infinità del primo motore: questi punti, infatti, devono essere discussi nella metafisica e attenendosi unicamente alla dottrina di Aristotele".
35. Ibid., p. 67: "Il *Meteorologica* deve essere affrontato velocemente nei mesi estivi, nell'ultima ora pomeridiana di scuola".

The third and fourth books of *On the Heavens* were to be taught only in a summary or omitted altogether. The student was to deal with the problems of the elements, the celestial substance, and the celestial influences; the rest was to be left to discussion by the professor of mathematics.[36]

Not only did the *Ratio* instruct avoiding, in the physics course, the reading of all passages of the Aristotelian text pertaining to mathematics; but, in addition, it also sought to remove the cosmological part of natural philosophy from the physics course and transferred it, together with the discussion of astronomical problems, to the domain of mathematicians. Likewise, theological and metaphysical problems were carefully divorced from philosophy, to be treated separately within the confines of their specific disciplinarian framework.

In the second year, together with the study of physics, the students had to take a course in mathematics. The instructions to the professors of mathematics were far less specific than those given to the professors of philosophy. The instructions recommended the reading of Euclid's *Elements* to which some information about geography and the celestial sphere had to be added.[37] The *Ratio* did not even specify from which books these subjects were to be learned.

Not much of Clavius's grand project to improve the status of the mathematical sciences and raise it to equal that of philosophy remained in the last version of the *Ratio*. Instead of involving one-and-a-half years of study, the mathematics course was shortened to one year only.[38] The relevance of mathematics to physical problems was not reinforced. On the contrary, the policy of the *Ratio* was a careful isolation of philosophical problems from mathematical problems and vice versa. The "mixed sciences", namely those specific areas in which physical problems were treated with mathematical methods (astronomy, optics, mechanics), were still subordinated to mathematics, and hence their status as true knowledge pertaining to reality remained ambiguous. Above all, contrary to

---

36. Ibid., pp. 66–7: "I testi del 2°, 3°, 4° libro del *De coelo* devono essere condensati sommariamente, anzi per la maggior parte tralasciati; in questi libri si devono trattare solo alcune questioni sugli elementi. Quanto al cielo, bisogna spiegare soltanto ciò che riguarda la sua sostanza e gli influssi, mentre le altre questioni devono essere lasciate al professore di matematica o riferite in compendio".

37. Ibid., p. 71: "Deve spiegare agli studenti di fisica, per circa tre quarti d'ora, gli elementi di Euclide. Dopo che vi si siano dedicati abbastanza ampiamente per due mesi, deve aggiungere cenni di geografia e sulla sfera celeste".

38. Cosentino, "Le matematiche nella 'Ratio Studiorum . . .'", p. 209.

Clavius's recommendation,[39] no examination in mathematics was required of the students of philosophy and theology. In the absence of any clear external indication of merit, mathematics remained relatively marginal to the curriculum.

The instructions for the third year, devoted to the study of metaphysics, show the same tendency towards control through distinction, exclusion, and a strong emphasis on the Thomist boundaries. The course consisted in reading the second book of *On Generation and Corruption, On the Soul*, and the *Metaphysics*.[40] Again the theories of the pre-Socratics were only to be summarized; problems of anatomy related to medicine were to be avoided.[41] In reading the *Metaphysics*, the discussion of God and the intelligences, considered as theological problems, were not to be touched upon.[42]

The rhetoric of conformity pervaded the *Ratio studiorum* of 1599. Fidelity to the Aristotelian corpus, which was at the heart of the philosophical curriculum, was emphatically declared. Adherence to the subject matter traditionally taught in universities was required of all professors of philosophy: "The professor of philosophy must not stray far from Aristotle".[43] In scholastic theology Thomas was the favoured authority. "The members of our order have to adhere completely to the thought of St. Thomas in scholastic theology, consider him as their master and wholly devote themselves to him, so that the students best dispose their souls towards him".[44] In dealing with scholastic questions it was not enough to present the right opinion, but one had to defend the opinion of Thomas.[45] It was forbidden

39. See above, p. 219.
40. *Ratio studiorum*, p. 67: "Nel terzo anno il professore di filosofia spiegherà il secondo libro del *De generatione*, i libri del *De anima* e della *Metaphysica*".
41. Ibid.: "Nel 1° del *De anima* accenni solo sommariamente alle teorie dei filosofi antichi. Nel 2°, spiegati gli organi della sensibilità, non sconfini nell'anatomia e negli altri argomenti che sono competenza dei medici".
42. Ibid.: "Nella *Metaphysica* devono essere tralasciate le questioni riguardanti Dio e le intelligenze".
43. Ibid., p. 64: ". . . il professore di filosofia non deve allontanarsi da Aristotele . . ."
44. Ibid., p. 57: "I nostri devono attenersi completamente, per quanto riguarda la teologia scolastica, al pensiero di S. Tommaso, considerarlo come loro maestro e applicarvisi interamente, affinché gli studenti dispongano il meglio possibile il loro animo verso di lui".
45. Ibid., pp. 60–61: "Non è sufficiente riportare i pareri dei dottori, tacendo la propria opinione: bisogna difendere l'opinione di S. Tommaso, come già detto, o tralsciare del tutto la questione".

to teach anything which had not been approved by the church and the known traditions in matters not treated by Thomas.[46] In reading the Holy Scriptures, only the version approved by the church must be taught.[47] Again, the interpretation should be that favoured by the church, the one accepted longest and with the widest consensus.[48]

And yet, notwithstanding the conformist language of the *Ratio*, dissenting doctrines were not entirely ignored. The general instructions to the professors of philosophy explicitly ordered them to carefully read those interpreters of Aristotle whose arguments conflicted with the Christian religion, and to combat them.[49] Even Averroes was not completely excluded, although he had to be mentioned without praise.[50] The rule in scholastic theology was to follow Thomas, but one could also deviate from him whenever his opinion was controversial. Reverence for Thomas does not imply a duty to adhere to him to such a degree that in no circumstance does it allow any deviation from him. For even those who declare themselves Thomists sometimes take their distance from him. Therefore, "it is suitable that the members of our order should not adhere to him more than the Thomists themselves".[51] Even in reading the Holy Scriptures one was not completely limited to the version approved by the church, for the *Ratio* stated that the professor must not altogether overlook other versions as well, at least in order to refute them.[52]

Despite their conciliatory tone, the architects of the *Ratio* were not

---

46. Ibid., p. 58: "Non si deve insegnare nulla che non sia in accordo con l'opinione della chiesa e con le tradizioni generalmente riconosciute . . ."

47. Ibid., p. 53: ". . . la piú importante è sostenere la versione approvata dalla chiesa".

48. Ibid., p. 54: "Se invece sono in contrasto fra di loro, egli deve scegliere l'interpretazione preferita dalla chiesa già da molti anni e con grande consenso".

49. Ibid., pp. 64–65: "Deve leggere o citare in scuola con grande prudenza gli interpreti di Aristotele che si sono posti in contrasto con la religione cristiana e stare attento che gli studenti non vi siano attratti".

50. Ibid., p. 65: "Per tale motivo non deve raccogliere in trattazione specifiche i commenti di Averroè, comportandosi analogamente a proposito di altri filosofi del genere. Se da lui si può cavare qualcosa di buono, lo citi senza lodarlo . . ."

51. Ibid., p. 57: "Tuttavia non devono credersi cosí vincolati a S. Tommaso da non potersene affatto allontanare in nessun campo. Quando anche coloro che si dichiarano tomisti di stretta osservanza se ne discostano, è conveniente che i nostri non restino legati a S. Tomasso piú dei tomisti stessi".

52. Ibid., p. 53: "Deve prendere in esame per confutarli soltanto gli errori piú importanti o all'apparenza piú probabili, contenuti nelle altre versioni

oblivious of the stormy debates over freedom of opinion which split the ranks of the Society of Jesus during the 1570s and 1580s.[53] At the same time as confirming the traditional orientation of the curriculum and crystalising it around a well-recognized, Aristotelian–Thomistic canon, the *Ratio* also made room for some deviating voices. It legitimized at least a neutral presentation of alternative interpretations of both Aristotle and Thomas, sometimes even recognizing the utility of nonorthodox commentators like Averroës, and the necessity of relating to different versions of the Holy Scriptures. Thus, the *Ratio* succeeded in maintaining an atmosphere of relative openness, essential for vital exchange within a community of intellectuals such as the Jesuits.

The mechanisms of control envisaged in the *Ratio* were neither visible nor direct. They did not involve explicit limitations on contents, opinions, or interpretation. Rather, they consisted of a discriminate use of exclusion, and a series of rigorous distinctions, which allowed for the transmittance of traditional epistemological values to which the society had tacitly committed itself.

Exclusion was used only in those cases where the interrelationship between two or more disciplines was threatened in a way which could lead to the undermining of the system as a whole. Such was the case with the nominalist critique, which could instil scepticism concerning the possibility of ascending from physical to metaphysical reality, or with the "contingents" with regard to the theological problem of free will and predestination. By isolating metaphysical problems from logical problems, or logical problems from theological ones, controversies could be contained within the boundaries of the different disciplines, without harming the basic presuppositions which sustained the structure. Formally, at least, the system kept the general Thomist framework.

The deployment of a series of distinctions between disciplines – the main strategy of control and socialization in the Jesuit intellectual system – was derived from the old Aristotelian rule prohibiting "the transference of methods from one discipline to another."[54] The distinction between the different procedures to be used in mathematics and in physics, which, according to the *Ratio* was to be taught as part of the first-year course in logic, reflected a policy applied again and again in other disciplines

---

latine più recenti, nella caldea, nella siriaca, di Teodozione, di Aquila, di Simmaco".

53. See Ch. 7.

54. A. Funkenstein, *Theology and the Scientific Imagination from the Middle Ages to the Seventeenth Century*, Princeton 1986, pp. 36–37; 296–297; 303–307.

as well. Thus, the *Ratio* insisted that scholastic methods should not be used in questions pertaining to the Scriptures.[55] Likewise, one must not use the historical method in scholastic theology.[56] And no one should treat in detail casuistic questions[57] within the framework of scholastic theology.

Like most documents involved in the construction of cultural realities, the *Ratio* is a text which betrays tensions and contradictions, despite the coherent structure that conceals them. The areas of exclusion indicate points of sensitivity, places where the Jesuits were susceptible to harsh criticism from without. Such areas were the exclusion of physical problems pertaining to mathematics from the course in natural philosophy, and the exclusion of "contingents", pertaining to theology, from the course of logic. The first related to the conflicts between mathematicians and natural philosophers over status and professional identities. The second indicated the vulnerability of Molinist theology to the criticism of orthodox Thomists. In both cases the *Ratio* confirmed the official positions of the church establishment, seeking to suppress a specifically Jesuit critical voice in the face of orthodoxy. The language of conformity was adapted to the necessity for legitimation when confronted with criticism from without.

However, the strong emphasis on the boundaries between disciplines and the conscious use of distinctions were mainly directed against subversion from within. For, in fact, many Jesuits were involved in cultural and intellectual activities which tended to erode these boundaries, thus signaling the possibility of transcending the Thomist world view and entering modernity. Such were the Jesuit activities in the theatre, in architecture, in the occult sciences, which undermined the distinction between the theoretical and the practical. Such also were the innovations of Jesuit mathematicians, which transgressed the boundaries of their discipline.[58] The mechanisms of control activated by the *Ratio* were mainly intended to check such dangerous tendencies.

But the *Ratio* did not construct a reality in which they could be alto-

---

55. *Ratio studiorum*, p. 55: "Non deve affrontare con il metodo scolastico le questioni proprie della sacra scrittura".
56. Ibid., p. 59: "Nelle loro trattazioni ogni qualvolta queste capitino nelle varie parti della *Summa* di S. Tommaso, i professori devono attenersi al metodo scolastico piuttosto che a quello storico . . ."
57. Ibid., p. 60: "Al quarto gruppo si ricollegano i casi di coscienza. A questo riguardo, essi devono evitare una spiegazione troppo accurata e minuziosa dei casi".
58. See Ch. 8.

gether eradicated. Mathematics is an example which demonstrates this clearly. The removal of the sixth and seventh books of the *Physics*, the cosmological part of *On the Heavens*, and the *Meteorology* from the domain of the philosophers and their appropriation by the mathematicians meant the delineation of a new discursive field within which physical problems were analyzed mathematically, and mathematical problems gained physical significance. This field did not become the center of the curriculum, as Clavius had hoped, and it failed to achieve the status reserved for natural philosophy in the Thomist system. By clearly distinguishing it from philosophy, and confining it within the boundaries of the mathematical disciplines, the *Ratio* attempted to subdue its claims to truth and reality. But within its own confines, the value and significance of mathematical–physical discourse had been legitimized through the assimilation of new problems and new areas of discussion.

Nonetheless, from the point of view of the history of Western science the year 1599 – the date of publication of the last *Ratio studiorum* – was the lost moment of the Jesuit intellectual program. The language of conformity prevails and the subtle means of control reflect the necessity for legitimation from without and some unity within. The text betrays the historical possibility of a breakthrough to modernity, but this possibility remained largely unactualized.

The Letter to Foscarini: Science Institutionally Constrained

Bellarmine's letter to Foscarini[59] should be read as public exposition of the Jesuit attitude towards Copernicanism. It confirms both the associative fields relevant to the discussion of Copernicanism and the criteria of validity for its judgement within each field. Thus, it permits a glance at the institutional constraints on Jesuit scientific discourse, constraints made visible through the reconstruction of the boundaries between theology and natural philosophy, and between natural philosophy and astronomy. These constraints have traditionally been represented by historians as deriving from fundamentalism in the interpretation of Scriptures and from epistemological instrumentalism. Can a historization of Bellarmine's positions change the judgements of historians?

59. *Opere*, XII, pp. 171–172.

The Holy Scriptures and Copernicanism

In his attempt to establish a limit to possible readings of the Scriptures Bellarmine invoked the decree of the Council of Trent concerning the Holy Scriptures,[60] and went on to interpret this decree in the light of the reality that had evolved in the course of some seventy years since its publication:

> Second, I say that, as you know the council prohibits interpreting Scripture against the common consensus of the Holy Fathers; and if Your Paternity wants to read not only the Holy Fathers, but also the modern commentaries on Genesis, the Psalms, Ecclesiastes, and Joshua, you will find all agreeing in the literal interpretation that the sun is in heaven and turns around the earth with great speed, and that the earth is very far from heaven and sits motionless at the centre of the world. Consider now, with your sense of prudence, whether the church can tolerate giving Scripture a meaning contrary to the Holy Fathers and to all the Greek and Latin commentators. Nor can one answer that this is not a matter of faith, since if it is not a matter of faith "as regards the topic," it is matter of faith "as regards the speaker"; and so it would be heretical to say that Abraham did not have two children and Jacob twelve, as well as to say that Christ was not born of a virgin, because both are said by the Holy Spirit through the mouth of the prophets and the apostles.[61]

The severe judgement of Morpurgo-Tagliabue on the principles enunciated by Bellarmine is worth quoting, as it is fairly representative of the broad consensus among historians, who have condemned such strategy as both irrelevant to the real needs of Catholicism and artificially revived in order to suppress the new science of Galileo. "Whereas for the Reformers", says Tagliabue, "who were enthusiastic readers of the psalms, the Book of Job, and Ecclesiastes, it is clear that the literal rendering of the Scriptures could have a religious significance, it was not so for the Catholics, who were not permitted to read these books".[62] And he goes on: "A number of circumstances brought it about that the attitude to science was determined, in the final analysis, by an obscure decree of the Council of Trent which dealt with the use of the Scriptures: a restrictive decree, which was now interpreted in an even more restrictive way than in the decree itself".[63] The decree, according to Tagliabue, must have been a

60. See Ch. 2.
61. *Opere*, XII, p. 172; trans. by Finocchiaro, pp. 67–68.
62. Morpurgo-Tagliabue, *I processi di Galileo*, p. 51.
63. Ibid., p. 53.

dead letter, for it was probably produced for the purposes of polemics with the Reformers, not for internal Catholic consumption. Its invocation by Bellarmine was a short-sighted political tactic, necessitated by the immediate circumstances of the argument with Galileo. In the long run, it actually restricted the ability of Catholic culture to assimilate new scientific theories, thus condemning it to obscurantism and cultural backwardness for centuries to come: "With this weapon, Bellarmine responded to Galileo's argument . . . and not only that, but he also precluded any possible counter response in the future. . . . Bellarmine already at that stage adopted an interpretation of the decree which made any answer whatsoever unacceptable".[64]

This is a sweeping judgement not only on Bellarmine's letter, but on the cultural requirements of Catholicism in the seventeenth century, the interests of the intellectual elite Bellarmine was part of, and more specifically, the entire spiritual and intellectual universe of the cardinal. A reassessment of this judgement in the light of a contextualization of the milieu in which Bellarmine was acting seems appropriate.

Before condemning Bellarmine's invocation of the decree of the Council of Trent, some remarks must be made concerning the circumstances of its composition and its cultural role. The crisis of sensibility in the Catholic world, whose roots may be traced back to the fifteenth century, and which culminated in the monumental work of the council, cannot easily be overlooked. The renewed interest in the Holy Scriptures was not simply a reaction to the Protestant challenge. Rather, it betrayed a fear of alienation from the sacred text common among Catholics long before the Reformation. The hope of finding in the Scriptures a moral source for religious revival was partly an expression of Christian humanism, partly a reaction against an institutional emphasis on "good works" which had come to be conceived as over-automatic and hence as inauthentic. Thus the Holy Scriptures were among the four main doctrinal elements which constituted the body of legislation of the Council of Trent. The doctrine reformulated at Trent, however, was accompanied by a number of disciplinarian decrees which were meant to provide institutional tools for the implementation of the doctrine. These decrees ordered the establishment of seminaries for the training of the priesthood next to all churches and cathedrals, and the establishment of chairs of Scripture next to every church, monastery, and school for the public.[65] As a result, new facts evolved, which were reflected, for example, in modifications of the curriculum of educational institutions.

64. Ibid., pp. 53–54.
65. See Ch. 4.

The *Ratio studiorum* prescribed a daily lesson in the Scriptures during the first two years of study in the faculty of theology.[66] From the second half of the sixteenth century, the General Chapter of the Dominicans also gives expression to a new emphasis on the study of the Scriptures. The General Chapter of 1564 emphasized the necessity of teaching the Scriptures regularly.[67] In 1615, the General Chapter decided on the setting up in Italy of a central seminary for members of the order which would be concerned solely with the study of the Scriptures.[68] This last decision seemed to reflect the need to train priests who would be concerned with the reading and interpretation of the Scriptures from the pulpit before the people.

The immediate circumstances in which the accusations against Galileo were made provide living evidence of the effects of the new orientation towards the Scriptures after the Council of Trent. T. Caccini, who accused all the "mathematicians" of heresy, did so in the course of a sermon on Joshua, chapter 10, as part of his task as the occupant of the chair of Scripture next to the Cathedral of Santa Maria Novella in Florence.[69] N. Lorini, who denounced Galileo's letter to Castelli to the Roman Inquisition, justified his action on the grounds that the letter was passing from hand to hand and causing a controversy among the preachers of the monastery of San Marco. He demanded the Inquisition to adopt a clear position on the question of the motion of the earth in order to prevent controversies among preachers on the interpretation of the Scriptures.[70]

Bellarmine's invocation of the decree on the interpretation of the Scriptures was not, therefore, a return to a forgotten and irrelevant principle. In fact, Bellarmine was referring to a law closely connected with the needs created by the cultural activity of the Tridentine church: namely, transmitting the word of God while preserving the monopoly of the church on interpretation, and securing doctrinal unity. To reinforce that monopoly and unity, however, Bellarmine added his own comment to the decree. With regard to the Scriptures, he said, it was impossible to distinguish between matters of faith (*materia di fede*) and other matters. The scriptural text constituted the theological boundary which was permitted in interpretation. Bellarmine here restated the preference of Jesuits for literal

66. *Ratio studiorum,* p. 55
67. *Acta,* 1564, pp. 52, 64.
68. *Acta,* 1615, p. 247.
69. *Opere,* XIX, p. 307.
70. Ibid., pp. 297–298.

interpretation, also manifested in the *Ratio studiorum*.[71] Can Bellarmine's demand for literal interpretation be labeled "fundamentalist", with all the obscurantist connotations assumed by the term in modern times? In an article written by U. Baldini on Bellarmine's astronomy,[72] the writer puts forward an interesting thesis concerning the function played by the demand for literal interpretation in the cardinal's system of intellectual considerations. Bellarmine was well grounded in the astronomy of his age and deeply involved in the scientific activities of the Roman College. Quite early, he gave expression to deviations from traditional Thomist astronomy by maintaining, among other things, the thesis of *liquiditas coelorum*, acknowledging the motions of the stars in noncircular paths, recognizing physical phenomena in the stellar regions analogous to known phenomena in terrestrial physics, and so on. Though his writings do not contain a coherent set of hypotheses about the real structure of the cosmos, they do contain many new tendencies which signal his willingness to make a serious revision of traditional Thomism. Oddly enough however, the justification for these deviations from Thomism were looked for in the cosmology of the Book of Genesis.

Like Perera, who practiced interpretation of the Scriptures and was far from being a strict Thomist Aristotelian,[73] Bellarmine believed in the impossibility of a contradiction between the scriptural text and scientific theories, but he wished to fix a boundary to the permitted interpretations, which he found in the literal meaning of the text. The significance of Bellarmine's "fundamentalism", Baldini contends, was not that it constituted a major hindrance to the development of scientific theories. His insistence on literal interpretation was a means of controlling philosophical ideas, but by implication it also sanctioned deviations from the philosophical ideas of Thomism. Moreover, in a period when the collapse of the Aristotelian cosmos was already visible on the horizon, Bellarmine sought to preserve one definite principle of unquestionable truth which he found in the unequivocal expressions of the Bible. Thus, this principle, despite its limitations, served as a barrier against the definite collapse of

71. *Ratio studiorum*, p. 53: "Sappia che il suo maggior compito è spiegare, con spirito religioso, impegno e dottrina, le sacre scritture, secondo l'interpretazione autentica e letterale".
72. U. Baldini, "L'astronomia del cardinale Bellarmino", in P. Galluzzi (ed.), *Novita' celesti e crisi del sapere*, Florence 1984, pp. 293–305. See also G. V. Coyne and U. Baldini, "The Young Bellarmine's Thoughts on World Systems", in: G. V. Coyne, S. J. M. Heller, and J. Zycinski (eds.), *The Galileo Affair: A Meeting of Faith and Science*, Vatican 1985, pp. 103–109.
73. See Ch. 7.

the belief in the capacity of the human intellect to understand the universe. It represented a possibility for the survival of the rationalistic tradition in Catholic thought.

That Bellarmine was not essentially an obscurantist is demonstrated by an additional argument, with which he modified his demand for literal interpretation:

> Third, I say that if there were a true demonstration that the sun is at the center of the world and the earth in the third heaven, and that the sun does not circle the earth but the earth circles the sun, then one would have to proceed with great care in explaining the Scriptures that appear contrary, and say rather that we do not understand them than that what is demonstrated is false.[74]

It is obvious that Bellarmine did not make the Scriptures the sole criterion for judging the truth of a scientific theory. For him, too, rational truth was autonomous. And if the Copernican theory could be unequivocally proved, it would become a physical truth, and would even require a reevaluation of the significance of the passages in the Scriptures in the light of it.

Copernicanism Unproved

Bellarmine did not choose to leave the discussion of Copernicanism on the theological level, however. Not only did he attempt to draw the boundaries between theological and scientific discourse, but he also wished to restate the conditions of mutual dialogue and exchange across those boundaries. Literal interpretation of the Scriptures was a limit to possible readings of Copernicanism until such thesis was truly proved. Copernicanism had not yet acquired the status of a proved theory, however. For:

> Nor is it the same to demonstrate that by supposing the sun to be at the center and the earth in heaven one can save the appearances, and to demonstrate that in truth the sun is at the center and the earth in heaven; for I believe the first demonstration may be available, but I have very great doubts about the second.[75]

Was Bellarmine an instrumentalist? Was he an epistemological sceptic who maintained that there was an immanent limit to human capacity to know? By making the distinction between two kinds of scientific proofs, Bellarmine was obviously referring to the distinction between demonstration "ex suppositione" and true demonstration. A few lines above, in the

74. *Opere*, XII, p. 172.
75. Ibid.

same letter, he had associated the first with the methods of mathematicians, the second with scholastic philosophers and theologians. It seems
that Bellarmine here restates the traditional boundaries between astronomy – confined within the mathematical disciplines – and natural philosophy and theology, in terms of the different methods used in each of these
branches of knowledge. We have seen that – in terms of the Thomistic
organisation of knowledge – confining astronomy to the mathematical
disciplines did not mean epistemological scepticism, but rather a low
status in the hierarchy of the sciences, due to the nature of the object
(hypothetical or real), and to the kind of demonstrations usually used.[76]
Astronomical knowledge, in that framework of mind, was considered
"probable", not yet strictly proven, nor fictive however; neither derivable
deductively from a general principle, nor unprovable.

By the standards of his day Bellarmine, therefore, was no instrumentalist. His letter demonstrates his insistence on a firm barrier between astronomy as a mathematical discipline on the one hand, and natural philosophy on the other. The discussion of Copernicanism, according to him,
should be restricted to astronomers, and should not have claims to a degree of truth and reality similar to the claims of philosophers. Bellarmine's
letter seems to have set serious limits to the scientific status of Copernicanism. By confining the discussion to astronomers, however, Bellarmine
had to ignore the claims of Jesuit astronomers to a completely different
set of criteria for establishing their science. Jesuit astronomers of the type
of Clavius and Blancanus would never have accepted the argument that
astronomy, being based on demonstrations "ex suppositione", fell short
of true science. My discussion of their works[77] in fact attempts to trace
their struggle to free astronomy from the confines of the Thomistic organisation of knowledge by claiming for their proofs the status of scientific
demonstrations. In doing so they could have, and probably did, make use
of methodological discussions among Jesuit logicians of the type of Valla
"who holds that this type of demonstration . . . is not as perfect as the
type made from immediates. . . . Yet the argument it produces is truly
scientific, and not merely probable, and so is capable of producing a conclusion that could not be otherwise".[78]

Bellarmine's letter gives vent to the institutional constraints imposed on
Jesuit scientific discourse by the official policy of the order, but not in

76. See Ch. 4.
77. See Ch. 8 on Blancanus, and the discussion of Clavius earlier in this chapter.
78. Wallace, *Galileo and His Sources*, p. 114.

terms of either fundamentalism or epistemological scepticism. My reading of it joins with my reading of the last version of the *Ratio studiorum,* and attests to the unwillingness or inability of the Jesuits to manifestly change the Thomistic organisation of knowledge, with its hierarchy between the "mixed mathematical sciences" and natural philosophy and the boundaries between them. By asking Galileo and the Copernicans to distinguish between a demonstration producing "cognitio certa per causas",[79] and a demonstration "ex suppositione", whose conclusion falls short of the Aristotelian ideal of knowledge, Bellarmine sanctioned the traditional Thomist position according to which the status of astronomical truths reached by mathematical methods was only "possible", not strictly proven although not fictive either.[80] This position could serve to justify the boundary between mathematics and the "mixed mathematical sciences" subalternated to it and natural philosophy with its higher status in the hierarchy of knowledge, a knowledge pertaining to truth, not to mere "probability". This policy, however, was not congruent with the practice of Jesuit mathematicians and astronomers who constantly endangered those boundaries, and claimed a higher status for their discipline than the Thomistic system allowed for. Moreover, Jesuit culture has always tended to grant legitimation to truths which had not been fully proven according to the traditional canons of knowledge, and thus to blur the distinction between absolute and probable truth, and to modify the accepted standards of proof. The Molinist claim that God's *scientia media* produced infallibly certain knowledge of man's future acts not yet determined by God, is one prominent example.[81] The decision to allow consideration of philosophical ideas which deviated from Thomism by classifying them as "possible" or "probable" is another.[82] Bellarmine's words thus remained ambiguous, and allowed for the continuation of the dialogue between Jesuit mathematicians and the new science of Galileo, in spite of the admonition of 1616.

79. Ibid., p. 99.
80. See Ch. 4; also see Thomas, *Sum. theol.,* Ia, xxxii, I, ad 2.
81. See Ch. 9.
82. See Ch. 7.

12

The Cultural Field of Galileo and the Jesuits

Josephus Blancanus, the Jesuit author of the *Dissertation on the Nature of Mathematics* was,[1] on his own admission, an admirer of Galileo. In a letter of 14 June 1611 to Christopher Grienberger, Clavius's successor at the Roman College, he wrote: "I love and admire Galileo, not only for his rare doctrine and inventions, but also on account of the old friendship I contracted with him in Padua, to the courtesy and affection of which I have remained attached".[2] The language used by Blancanus was that of Italian court culture and its system of patronage. It presupposed loyalties and an exchange of favours and connections through which personal status was established.[3] A friendship had been "contracted",[4] which implied a long-term loyalty one was supposed to be "attached" to.[5] Blancanus's language indicates that he was familiar with court circles. Indeed, the rest of the letter records specific occasions in which he took an active part in scientific discussions in court – which became, at that period, part of social routine:[6]

> I do not believe there has been anyone who has confirmed and defended his [Galileo's] inventions more than I have, in public and in private, both in this Court of Parma and in that of Mantua, by showing with the telescope the moon, the Medicceans and others [stars], to these same princes of Mantua; and to Cardinal Gonzaga

1. See Ch. 8.
2. Biancani to Grienberger, *Opere*, XI, p. 126.
3. R. S. Westfall, "Patronage and the Publication of the Dialogue", in *Essays on the Trial of Galileo*, Vatican City 1989, pp. 58–84; M. Biagioli, "Galileo's System of Patronage", *History of Science*, 28 (1990), 1–62; idem, "Galileo the Emblem Maker", *Isis*, 81 (1990), pp. 230–258.
4. Biancani to Grienberger, *Opere*, XI, p. 126: "... per l'antica amicizia che già *contrassi* con lui". (My italics, R. F.)
5. Ibid: "... dalla *cortesia* et amorevolezza del quale *restai legato*". (My italics, R. F.)
6. See Biagioli, "Galileo's System of Patronage".

have I confirmed many such inventions with the highest praise to Galileo.[7]

And yet Blancanus's letter was apologetic in tone. He was attempting to justify to Grienberger his part in a critique of Galileo written by another Jesuit, the author of the "Problema",[8] whom Blancanus assisted in revising and editing his text, but with whom he did not identify completely. There is no doubt that the difference between the two was a difference of style, not unrelated to a generational gap. Blancanus clearly objected to the offensive tone of the author of the "Problema" with regard to Galileo.[9] From other sources we also know that Blancanus saw himself as the spokesman for a new generation of Jesuit astronomers and the elaborator of Clavius's program.[10] But the author of the "Problema" had the advantage of seniority,[11] which apparently gave him a certain authority. Blancanus, however, mentioned two points of disagreement with Galileo one of which he shared with the author of the "Problema". The first, which according to him was not significant, was a criticism of the structure of Galileo's mathematical demonstration concerning the measurement of the height of the moon's mountains. The second was more serious. It concerned Galileo's claim that the mountains continued all the way to the circumference of the moon. As against this, Blancanus and his colleague maintained that the mountains did not touch the outer circle, which was "entirely lucid, without any shadow or sign of inequality".[12] This, according to Blancanus, was demonstrated by observation.

Although this exchange between Galileo and Blancanus was conducted through Christopher Grienberger, Blancanus mentioned a letter he had written to Galileo, but which had not been preserved among the Galilean

7. Biancani to Grienberger, *Opere*, XI, p. 126: "Nè credo sia stato alcuno che habbia più publicato, confirmato et difeso le sue invenzioni di me, in publico et in privato, tanto in questa Corte di Parma quanto in quella di Mantova; col far vedere con canocchiale la luna, le Medicee et l'altre, sino anco alli stessi principi di Mantova; et al Card.¹ Gonzaga confirmai molto tali invenzioni, per tutto con somma lode del Galilei".

8. The text: "De lunarium montioum altitudine: problema mathematicum", whose author is not known, was published by Favaro in *Opere*, III, pp. 301–307.

9. Biancani to Grienberger, *Opere*, XI, p. 127: "L'avvisai di nuovo che avvertisse di cancellare quell'insulto contro al Galileo".

10. See Wallace, *Galileo and His Sources*, Chapter 3, c. pp. 141–148.

11. Biancani to Grienberger, *Opere*, XI, p. 127: "Io non poteva far altro, perchè egli e Padre, *et aetatem habet*". (My italics, R. F.)

12. Ibid.

manuscripts.[13] This mention attests to the direct contacts which obviously existed between the two mathematicians.

Galileo, for his part did not leave Blancanus's criticisms unanswered. In a letter to Grienberger dating from 1 September of the same year[14] he provided additional evidence of the kind of intellectual and social exchange which took place between Blancanus and himself. Galileo began his communication to Grienberger by mentioning Blancanus's letter, a copy of which he seemed to have acquired, and the pleasure he derived both from Blancanus's expression of affection towards him and from the objections expressed by the Jesuit to the malignant tone of some of Galileo's critics, including the author of the "Problema".[15] If Blancanus's letter hinted at a possible generational gap between different Jesuit astronomers, Galileo emphasized this cleavage by dividing astronomers into two categories: those with whom he had a dialogue and those with whom he fought. He also implied that he regarded himself as belonging in the same group as Grienberger and Blancanus: "I have preferred," he wrote "to respond to those friends, from whose kind familiarity it seemed that I could obtain the greatest security".[16]

Turning to the Jesuits' criticisms, Galileo chose to invert their order of presentation and to answer the more serious contention first. Blancanus had spoken about the "mountainousness" ("montuosità") of the moon as being limited to its central area and not existing at its margins. Also, he had justified his claim by an argument from observation which, in this particular case, he regarded as a demonstration. Being as consistent in his attempts to invalidate the reasoning process of his critics as in disproving the arguments themselves, Galileo challenged Blancanus's claim to demonstrate his case by direct sensory evidence. It was not simply by direct observation, he said, that one can ascertain the "mountainousness" of the moon. This kind of knowledge can be established only within the context of a certain kind of *discourse,* combining observations, sensory appearances, and arguments with a coherent whole:

> How, then, do we know that the moon is mountainous? We know it not simply by the senses, but by copying and combining discourse with observations and sensory appearances.[17]

Only after introducing the notion of "discourse" combining arguments with observations and sense experience did Galileo express his definite

13. See Favaro's note 2 to Blancanus's letter, *Opere,* XI, p. 126.
14. Galileo to Grienberger, *Opere,* XI, pp. 178–203.
15. Ibid., p. 179.
16. Ibid.
17. Ibid., p. 183.

conclusion concerning the mountainous surface of the moon. Then he again stressed the dependence of such a conclusion on a variety of strategies which establish it as a fact:

> These are the appearances and phenomena, which posited as the suppositions and hypotheses of discourse, necessarily convince some to hold without any doubt that the surface of the moon facing the earth is mountainous and unequal.[18]

Having established the "mountainousness" of the moon as a physical fact, Galileo turned to the first objection. This had to do with the structure of his mathematical demonstration concerning the measurement of the height of the lunar mountains. The author of the "Problema" argued that Galileo had posited the diameter of the moon together with the height of the mountains, while pretending to calculate the same height in one demonstration. Had he not posited the diameter – which cannot be accepted as a proper procedure – he would not have a demonstration.[19] Galileo vehemently denied such an accusation. In fact, he argued, the demonstration which the author of the "Problema" claimed as his own was the very same demonstration he had made and developed in his *Sidereus nuncius*.[20] Galileo admitted, however, to a slight difference between his demonstration and the one suggested by the author of the "Problema":

> Against the custom of the geometricians, [he] marks on the figure three squares, for no purpose whatsoever, but only in order to render it more visibly pregnant; while I, assuming I am speaking with people of intelligence, do not use any words which are unnecessary, especially since the demonstration is very simple and short in itself.[21]

In this attempt at self-justification, sometimes bordering on the offensive, Galileo drew attention to the differences between himself and his Jesuit opponent: the way facts about physical reality were established, and the procedure of mathematical demonstration. Substantially, there was no difference in the structure of their mathematical arguments; but the Jesuit, Galileo suggested, attempted to appear more oriented towards math-

18. Ibid., p. 184: "Queste sono le apparenze e fenomeni, li quali fatti, suppositioni et ipotesi del discorso, necessariissimamente convincono altrui a tenere senza niuna dubitatione che la superficie lunare, che risguarda verso la terra, sia montuosa et ineguale".

19. Biancani to Grienberger, *Opere*, XI, p. 127: "... si pigli il diametro lunare, corre la dimostrazione".

20. Galileo to Grienberger, *Opere*, XI, p. 199: "... la dimostratione, posta dall'autor del *Problema* per suo trovato, esser a capello la medesima che io pongo nel Nuntio Sidereo".

21. Ibid., p. 200.

ematics than himself, unnecessarily amalgamating geometrical figures in order to gain more authority, perhaps, certainly not for the sake of a better mathematical understanding. As for the way facts about physical reality were established, Galileo indicated the limits of observation and its dependence upon discourse. However, he was aware of the difficulties involved in deciding between the two possibilities: that the whole surface of the moon was mountainous, or that the mountains were in the center, but not near the circumference. It was due to those difficulties that he felt in need of gaining the support of other Jesuit mathematicians, whose unique authoritative position in the world of science and letters he recognized and sought to recruit on his side:

> Because the testimony of one of the Brethren of a certain community of the highest distinction in letters and perfection of doctrine, which has already gained an absolute authority in persuasion and arbitration in matters relating to all the sciences, must be of no small esteem . . . It therefore seems to me that I will need no less defence than that of somebody of the same Brethren, who is Father Blancanus, Your Reverence, and some other professor from your most famous Collegio.[22]

In 1611, the first year after the publication of the *Sidereus nuncius,* a somewhat intense exchange between Galileo and some top Jesuit mathematicians testified to their mutual investment in a common field of intellectual interests. However, the objects of their respective discourses, their limits, and their relation to existing disciplines and traditional institutional structures were fluid and as yet unclear. The letters of Blancanus and Galileo to which I have just referred illustrate the dialogue between Galileo and the Jesuits, the type of problems they commonly dealt with, and the strategies they both used in order to find acceptable solutions. This exchange signaled their involvement in the transition from one scientific paradigm to another.

Explication of the background common to Galileo and the Jesuits is a necessary precondition for an understanding of the rivalry which was yet to come. The rivalry grew out of a split within a new field which appeared in the gaps between traditional disciplines, such as physics and mathematics, astronomy and cosmology, metaphysics and theology; a split that cre-

22. Ibid., pp. 179–180: "Perchè l'attestatione di uno de i Frattelli di una Congregatione, per somma scieltezza di lettere et perfettione di dottrina già fatta di assoluta autorità nel persuadere et arbitra nel determinare circa i particolari di tutte le scienze, deve essere stimata non poco; . . . onde pare che di non minor difesa mi fosse necessario che di quella di alcuno de i medesimi Fratelli, quale è il Padre Biancano, la R.V. et qualche altro professore del vostro famosissimo Collegio".

ated both the new science of Galileo and the new science of the Jesuits in the image of an ego and an alter-ego, engendering the peculiarly bitter enmity typical of such love–hate relationships.

A Historiographical Digression: A Reconstruction of the Critical Dialogue between Galileo and the Jesuits

In traditional histories of the "Galileo Affair" the Jesuits have always played the role of the villain. Whether defending the interests of scholastic philosophy, or pulling the strings of political power, the Jesuits have hardly ever escaped their central role as the "other" of modern science. Yet the evidence accumulated in recent years all indicates that at the turn of the seventeenth century the Jesuit educational system displayed a considerable potentiality for reconstituting itself along the path taken by Galileo (i.e., the mathematization of natural phenomena, and the systematic observation of phenomena). Galileo, it has been shown, was not so much an enemy as a disciple of the Jesuits, even though he had never attended any of their schools.[23] According to William Wallace, the leader of this new trend in Galileo studies, the key to a proper interpretation of Galileo's science lies in his use of the term *scientia*. "Galileo employed the term scientia and scineza repeatedly. Never once did he depart from this ideal of certain and irrevisable knowledge as the goal of his investigations".[24] *Scientia* as an ideal of science, however, was not something new, and was definitely not invented by Galileo. "Indeed it was the commonly accepted doctrine of the schools".[25] Yet, there were different interpretations of this ideal. A comparison of the treatises, commentaries, textbooks, and "reportationes" (lecture notes) written by a number of Jesuit logicians, natural philosophers, astronomers, and theologians with an analysis of Galileo's early logical and physical questions,[26] shows that

23. It seems to me that this has been clearly shown in Wallace's *Galileo and His Sources.*
24. Ibid., p. 99.
25. Ibid.
26. These texts are hardly represented in the National Edition of Favaro, as he believed them to be student notebooks of Galileo dating from his period of studies in Pisa. They have been recovered, partly transcribed, reconstructed and summarized by W. A. Wallace in his *Galileo's Early Notebooks: The Physical Questions. A Translation from the Latin, with Historical and Paleographical Commentary,* Notre Dame 1977; in the first part of *Galileo and His Sources;* and in *Galileo's Logical Treatises. A Translation with Notes and Commentary of His Appropriated Latin Questions on Aristotle's "Posterior Analytics",* Dordrecht 1992.

about 75 percent of Galileo's text was borrowed directly from the Jesuit professors of the Roman College. A perusal of Galileo's later work and its interpretation in the light of the earlier writings, based on Jesuit sources, led Wallace to believe that Galileo never departed from the methodology of the schools as understood and developed by the Jesuits of the Collegio.

Further impressive evidence of the continuity between the mixed mathematical sciences of the Jesuits and the new experimental philosophy is provided by the work of Peter Dear. In his "Jesuit Mathematical Science and the Reconstruction of Experience in Early Seventeenth Century",[27] Dear has argued that the mathematization of nature led to the isolation of natural phenomena and their establishment as "experience", the basis and source of first principles from which it was then possible to deduce further results. Although for the Jesuits the model of scientific thinking had remained Aristotelian, and the quest for a valid scientific demonstration – namely deduction from general principles – did not change, the method of establishing first principles was gradually altered within the tradition of the mixed sciences. Traditionally, first principles were based upon generalizations derived from common sense. The validity of those generalizations sprang from their being evident to all, in the manner of geometrical axioms such as: the whole is greater than its parts, the law of "tertium non datur", etc. First principles anchored in common sense could not provide a basis for a mathematical science of nature, however. In the new science, "experience" had to be "constructed" out of the particular observations of experts whose creditable evidence could be universally accepted. In fact, Dear contends, Jesuit texts written by mathematicians, astronomers, and opticians testify to a modification of the meaning of "experience" understood as a generalization from common sense. These mathematicians endowed the concept "experience" with a new significance, it being now understood as the product of particular manipulation accepted on the evidence of experts.[28] Among the Jesuits quoted by Dear, Christopher Clavius, Josephus Blancanus, Christopher Scheiner, and Oratio Grassi played a particularly prominent role. Not only were they all interlocutors of Galileo, but it was in the course of a continuous dialogue, periodically interrupted by crises, bitter arguments, and personal rivalries that both Galileo and his interlocutors established new patterns for the constitution of scientific objects, drew up new boundaries between disciplines, asserted their authority, and laid the groundwork for

27. See Ch. 11, n. 2.
28. See Dear, "Jesuit Mathematical Science . . .", pp. 148 ff.

an organised transmission and reproduction of their knowledge in the future.

The work of Wallace, Dear, and others[29] has drawn our attention to the amorphous, relatively undefined discourse of Jesuit mathematicians, forming part of the tradition of the schools and yet exceeding its boundaries. This discourse, precisely because of its fluidity and uncertain status, was open to dialogue with speakers from other intellectual and institutional environments.

Galileo's new position in the cultural field introduced a new kind of challenge to the Jesuit mathematicians' quest for identity. Galileo was a special case, since he was grounded in the teachings of the schools, which he left in order to acquire a new professional identity[30] through the patronage of the Medici and the Accademia dei Lincei. At the university of Padua Galileo's status as a professor of mathematics had been secure and self-evident, but his work developed an uneasy dichotomy. On the one hand he stayed within traditionally accepted boundaries – the *De motu*, *Le mechaniche,* and the *Trattato della Sfera,* despite their innovative tone still remaining within the framework of the mixed mathematical sciences. On the other hand he started to develop new mechanical concepts which he communicated to close friends only, in long letters.

Galileo's new role of court mathematician and philosopher entailed new forms of scientific writing, as well as different scientific practices. In a series of papers Mario Biagioli has recently shown how Galileo's new position in the cultural field encouraged but also constrained his scientific work. Biagioli skillfully argues for the framing of scientific beliefs by social conditions.[31] The patronage of the Medici, however, was also a constraint on Galileo's science, in spite of Biagioli's tendency to emphasize the congruence of their interests:

> Because Medici patronage rewarded marvels that would fit the discourse of the court but not scientific theories or research programs, Galileo tended to present the satellites of Jupiter not as astronomical

29. Cosentino, "Le matematiche nella 'Ratio Studiorum' "; Dainville, *L'Education des Jesuites;* Baroncini, "L'Insegnamento della filosofia naturale . . ."; J. L. Heilbron, *Electricity in the 17th and 18th Century: A Study in Early Modern Physics,* Berkeley 1979, pp. 101–114, 180–189; Feldhay, "Knowledge and Salvation in Jesuit Culture"; Harris, "Transposing the Merton Thesis . . .".

30. My use of the concept of "professional identity" follows Biagioli's "The Social Status of Italian Mathematicians".

31. Biagioli, "Galileo's System of Patronage"; idem, "Galileo the Emblem Maker".

discoveries supporting a new cosmology but as dynastic emblems, and himself not as a discoverer, but only as the mediator of an encounter.[32]

It seems, then, that Galileo's intellectual interests were perhaps connected with, but not solely shaped by, the needs of the Medici. Galileo had abundant scientific plans which he confessed to many, but particularly to Vinta, the secretary of the Medici:

> Two books on the system and constitution of the universe – an immense conception full of philosophy, astronomy and geometry. Three books on local motion – an entirely new science in which no one else, ancient or modern, has discovered any of the most remarkable laws which I demonstrate to exist in both natural and violent movement; hence I may call this a new science and one discovered by me from its very foundations. Three books on mechanics, two relating to demonstrations of its principles, and one concerning its problems; and though other men have written on this subject, what has been done is not one-quarter of what I write, either in quantity or otherwise. I have also lesser works on physical topics.[33]

The Medici were not interested in such plans, which did not directly relate to their desire for power. In order to pursue his own interests, Galileo needed an exchange with other speakers in a common cultural field where he could benefit from the prestige of his position as court mathematician and philosopher. He also needed to find a system of patronage less desirous of spectacles and marvels and more ready to invest in the products of philosophy and mathematics. Jesuit mathematicians of a philosophical orientation thus became his natural interlocutors. The academy of the Lincei, already in existence for some nine years but still in search of a focus and prominent intellectuals offered additional opportunities for patronage which the Medici were unable to supply.

Galileo's dialogue with the church, as well as his fall, took place within such a context. It was not the tale of a hero who grew out of a tradition, rebelled against it, and was then silenced by authority. It was rather the story of the restructuring of a cultural field.[34] The path taken by the Jesuit mathematicians and their discourse, which after all remained a "mixed

32. Biagioli, "Galileo the Emblem Maker", p. 253.

33. Galileo to Vinta, *Opere*, X, pp. 351–352, trans. by S. Drake, *Discoveries and Opinions of Galileo*, New York 1957, p. 63.

34. My use of the concept of "cultural field" is borrowed from the writings of P. Bourdieu, especially "The Field of Cultural Production", in *Distinction: A Social Critique of the Judgement of Taste*, trans. by R. Nice, Cambridge Mass., 1984; idem, *The Field of Cultural Production: Essays on Art and Literature*, ed. and intro. by R. Johnson, New York 1993; and idem, "Social Space and the Genesis of the Group", *Theory and Society* (1985).

science", as well as that of Galileo and the type of "new science" he was engaged in were both determined in the course of this restructuring. Unfortunately, this context also determined the conditions in which some twenty years later Galileo's trial could become a historical fact.

Galileo's Telescopic Discoveries and the Reconstruction of the Cultural Field

On 24 April 1611 a group of four mathematicians from the Roman College[35] responded to Bellarmine's request for their opinion concerning the new telescopic discoveries of Galileo. Their letter[36] confirmed Galileo's discovery of new stars previously unseen, four heavenly bodies that moved around Jupiter, which could not possibly be fixed stars, but were nonetheless stars, seemingly planets. Also, the letter emphatically affirmed the truth of Galileo's observations of the phases of Venus, and the oval form of Saturn, which could have indicated the existence of two small stars on either side of the large one. As for the "inequality of the moon" – the mathematicians admitted the existence of certain appearances that seemed to point to some kind of apparent inequality, whose nature, they declared, was as yet unknown. Whatever disagreements may have existed between the Jesuits and Galileo about the interpretation of these phenomena, one thing is clear: the letter indicates the incorporation of new objects discovered by Galileo into the discourse of Jesuit mathematicians.

This well-known letter of the mathematicians to Bellarmine reveals the emergence, in the areas between the disciplines, of a common cultural field, in which speakers of different status, from different intellectual and institutional environments, attempted to take up a position.[37] Obviously, the evidence of the process is not confined to a single letter. Traces of it can be found throughout the pages of Galileo's correspondence of 1610–1611.

The first letter written by Galileo from his new location in Florence was addressed to Christopher Clavius. Galileo excused himself for his long silence lasting several years, but assured Clavius of his continuous

35. The four were: Christopher Clavius, probably the most famous Jesuit mathematician of the period; Christopher Grienberger, the heir of Clavius's chair of mathematics; Odo van Maelcote; and G. P. Lembo.
36. I Matematici del Collegio Romano a Roberto Bellarmino, *Opere*, XI, pp. 92–93.
37. See Bourdieu, "The Field of Cultural Production", and "Social Space . . ."

devotion. Through Antonio Santini[38] Galileo had become aware of Clavius's telescopic observations, which had failed to discover the Medicean planets. Galileo assumed this failure to have been due to technical difficulties in operating the telescope. Clavius's intellectual cooperation, however, and an exchange of other favours were taken for granted by Galileo: "It remains for me", wrote Galileo, "not wishing to weary you any longer, to beg you to keep me in your favours, which through your courtesy and compliance in learning has been granted to me a long time ago, and I assure you that there is nothing within my power that you cannot absolutely rely upon".[39]

Galileo's rhetoric made no attempt to separate questions of truth from questions of power, intellectual exchange from types of patronage relationships. Biagioli has pointed out this nonseparation to be a structural feature of a search for professional identity under the specific conditions of patronage systems.[40] I would like to add further that such strategy is indicative of a situation where a cultural field is in process of being restructured. Copernicus, in his time, could easily argue in terms of "mathematics for the mathematicians". A few years later Galileo would attempt to assert authority by delineating the boundaries of what he called "philosophical astronomy", and by excluding "mathematical astronomers", theologians, and philosophers. Galileo, however, failed to create an autonomous field for his science – a dream which only began to come true with the efforts of the Royal Society – and the attempt cost him a great deal of humiliation and much of his freedom. Between the two phases of relative autonomy, however, the cultural field allowed nonprofessional patrons or their representatives to make their voice heard and to take up a position among the professionals. Thus, two of Galileo's most important "scientific" letters of 1611 were addressed to Cardinal Dini,[41] and to Gallanzone Galanzoni,[42] a member of Cardinal Francesco di Joyeuse's court,

38. Antonio Santini (1577–1662), a Venetian gentleman interested in mathematics, was for many years involved in business. In the process he made the acquaintance of Galileo and some of his disciples. He then took holy orders, moved to Rome, and in 1644 was elected to the chair of mathematics at the Sapienza. See *Opere*, XX, Indice Biografico.

39. Galileo to Clavius *Opere*, X, 432: "Restami, per non tediarla più lungamente, il supplicarla a ripormi in quel luogo della sua grazia, il quale dalla sua cortesia et dalla conformità degli studii mi fu conceduto gran tempo fa, assicurandosi, niuna cosa essere in poter mio, della quale ella no possa con assoluta potestà disporre".

40. Biagioli, "Galileo's System of Patronage".

41. Galileo to Dini, *Opere*, XI, pp. 105–116.

42. Galileo to Gallanzone Gallanzoni, ibid., pp. 141–155.

in response to their request for scientific clarifications and evidence concerning the Medici planets and the mountains of the moon.

As well as representatives of princes, there was also room in the field for university professors of mathematics, Antonio Magini of Bologna being perhaps the most famous among them. These too, however, were actively involved in patronage networks transcending the limits of localities and covering the whole of the European "republic of letters". Magini's letter to Galileo dated 28 September 1610,[43] for example, opened with a report of the positions of the Medici's planets on the 20th of that same month. The largest part of the letter, however, was devoted to Magini's report on his services to various patrons, and first and foremost to Rudolph II, to whom he had dedicated two books, in addition to a gift of a concave/convex mirror. In return he had acquired "recognition" in terms of money – three thousand thalers, and an additional thousand later on. Magini went on to tell Galileo about his negotiations with the prince of Mantua, whom he wished to present with a similar mirror, but from whom he had to accept a much smaller reward partly in cash, partly in diamonds, and partly in the form of some vague promises of favours in the future.

Magini's case further testifies to the nonexclusiveness of a cultural field in which university professors routinely used objects of knowledge in order to practice rituals of exchange, while courtiers pursued scientific questions in order to promote or sustain their position in the world of power politics.

In the cultural field under investigation, Jesuit mathematicians generally occupied the most prominent and most authoritative positions, as Galileo frankly admitted in his letter to Grienberger.[44] From September 1610, the date of the letter to Clavius, Galileo sought to keep open regular channels of communication with them. Sometimes he addressed them directly, as we see in his letters to Clavius, to Grienberger, and perhaps even to Blancanus. More often, however, the exchange took place with Cigoli[45] in Rome, Santini in Venice, Gualdo[46] in Padua, and later Welser in Augsburg acting as intermediaries.

The complex position occupied by Jesuit mathematicians in the cultural field is sensed in these letters without ever being explicitly stated.

43. Magini to Galileo, *Opere*, X, 437–438.
44. Galileo to Grienberger, see n. 22.
45. Ludovico Cardi da Cigoli (1559–1619), a Florentine painter and architect who lived many years in Rome and held a regular correspondence with Galileo.
46. Paolo Gualdo (1553–1621), a jurist and writer who took holy orders in 1579, was secretary to Urban VIII, and lived most of his life in Padua.

Significant in this respect was an early letter by Cigoli in which he mentioned a group of "Clavisti" – disciples of Clavius – who made fun of Galileo's new discoveries.[47] The very act of naming the Jesuit mathematicians "Clavisti" was significant as an act of differentiation, the sign of an emerging identity, a position in the making. Later on, Galileo reported to Gualdo in Padua that he had managed to convince some Jesuit fathers, disciples of Clavius in Florence, but not the philosophers of the same order.[48] The special position of the Jesuit mathematicians was thus further distinguished from that of the philosophers. Gualdo in turn, wrote to Galileo saying that the more he (Galileo) relied upon the Jesuits' testimony – probably referring to the mathematicians' letter to Bellarmine – the more the philosophers remained entrenched in their old obstinacy.[49] Welser, who represented a point of view outside the profession, but was definitely one of the most active people in the field, represented the acceptance of Galileo's ideas as being in the professional interest of mathematicians.[50] Indirectly, then, the letters contain traces which confirm a theme much emphasized in previous chapters – namely a continuous tension between the mathematicians and the philosophers within the Society of Jesus. The appearance of new objects such as lunar mountains and sunspots as well as the emergence of a court mathematician claiming the title of philosopher challenged the stability of the cultural field created by the Jesuits in the latter part of the sixteenth century and invited reactions.

Clavius, Grienberger, Blancanus, and later Grassi of the Roman College, and Scheiner from Ingolstadt – to mention but a few representative names – were all engaged in a struggle to redefine the boundaries of their discourse within rigid, and sometimes even contradictory, institutional constraints. The competition of the Society of Jesus for the status of the leading intellectual elite of the church had provided an opportunity for a relatively free intellectual inquiry, permitting a crossing of the traditional boundary lines between disciplines and a transgression of their hierarchical order.[51] Wallace has shown that even within natural philosophy a combination of intellectual traditions – the Aristotelian, the pseudo-Aristotelian, and Archimedean – was not uncommon, and permitted the combination of a philosophical treatment of motion and a mathematical approach to the subject. Examples of similar crossings of boundaries have

47.   Cigoli to Galileo, *Opere*, X, p. 442.
48.   Galileo to Gualdo, *Opere*, X, p. 484.
49.   Gualdo to Galileo, *Opere*, XI, p. 56.
50.   Welser to Galileo, ibid., p. 52.
51.   See above, Chs. 8, 11.

been pointed out in the work of other historians.[52] Recent studies have increasingly drawn attention to the beginnings of a program of physical–mathematical science which was gradually taking shape within Jesuit institutions.

A rather severe institutional limitation hampered such a development, however. I believe that the theological debate between the Dominicans and the Jesuits, which raged in the years 1597–1607 exposed the subversive potential of the Jesuit educational program and thus forced its architects to seek legitimation by making a full commitment to the Thomist organization of knowledge.[53] This commitment was institutionalized in the *Ratio studiorum* of 1599, and it resulted in official sanctions against transgression of the traditional boundaries between disciplines. Most probably, this official policy was unable, in itself, to prevent a continuation of the previous intellectual tendencies towards the blurring of boundaries. However, peripatetic philosophers could and did exploit it in their struggle to maintain their identity and traditional position of preeminence vis-à-vis the mathematicians, who demanded promotion in the institutional hierarchy.

Jesuit mathematicians thus found themselves in need of a dual legitimation: as Jesuits confronting the traditional intellectual elite of the church and as mathematicians versus the philosophers. With the discovery of the telescope their discourse entered a state of fluidity. The status of the traditional objects of astronomy (epicycles and eccentrics, crystalline spheres, and the like) became more questionable, the boundaries of astronomy more uncertain, the justificatory strategies more unsatisfactory as the astronomers stressed their somewhat novel claim to truth. As the field developed a stage of great fluidity, the need of Jesuit mathematicians to communicate with other practitioners of their profession became acute. Exclusion of interlocutors from the field could not be regarded as an option in this situation. Inclusion, however, meant greater exposure to criticism, especially on the part of philosophers.

After 1610, one set of constraints in the cultural field was closely connected with the status of Jesuit mathematicians within the educational establishment of the society: the tensions between mathematicians and philosophers, the subtle interdependence between the Jesuit mathematicians and the other members of the "republic of letters", the necessary boundaries of their discourse, and the strategies required for its legitimation.

52. See Dear, "Jesuit Mathematical Science . . ."
53. See above, Ch. 11.

Another set of constraints was related to the peculiar position of Galileo. Biagioli has argued that, as a client of the Medici, Galileo was obliged to produce spectacular, highly visible discoveries as well as to adopt aggressive strategies of argumentation over major issues.[54] Prince Cesi and his entourage, while less dependent on power fetishes were no less insistent on publicity and the rituals of contest. Thus Cesi, in an early letter to Galileo, urged him to write quickly, and not to delay presentation of his discoveries:

> You have not yet written anything about the horned Venus or the triple Saturn. Please do it as soon as possible so that your sons may not find an impudent father that dares to adopt them.[55]

Galileo's need to gain legitimacy as a mathematical philosopher outside the traditional university system, and within a system of patronage that demanded visibility and aggressivity, further limited his possibilities of dialogue with the Jesuits. Galileo needed the Jesuit mathematicians as intellectual interlocutors no less than they needed him for the legitimation of the new status they claimed for their profession. The Jesuits, however, had to compromise this need with the commitment of their order to the Thomistic organization of knowledge. Any support they received from the outside had to be treated with circumspection and acknowledged discreetly. In contrast Galileo was obliged to stress and dramatize the dialogue for the purpose of gaining maximum visibility and publicity. Two different discourses, then, emerged within the same cultural field, evolving through a dialogue into an "ego" and an "alter ego", determining their difference within a common matrix, and preparing a structure within which the trial of 1633 could become a fact.

The dispute on sunspots, the first among a series of encounters between Galileo and Jesuit mathematicians, provides a mini case study of the emergence, within the same field, of two discourses displaying similarities of structure but very different constraints.

The circumstances of the dispute (1611–1613) are well known and do not need to be described at length. It was initiated by three letters of Christopher Scheiner,[56] who had observed certain spots on or near

54. Biagioli, "Galileo's System of Patronage . . ."
55. Cesi to Galileo, Opere, XI, p. 175: "Ella non ha ancor scritto cosa alcuna della cornuta Venere e del tripplice Saturno. Faccialo, per gratia, quanto prima, acciò i suoi figli non trovino qualche sfacciato padre che ardisca adottarseli".
56. Christopher Scheiner (1573–1650) studied philosophy and mathematics in Ingolstadt, and taught mathematics and Hebrew there from 1610–1616. In 1624 he became Rector of the college of Neisse in Rome. He was called in 1633 by the emperor Ferdinand II to return to Germany, and then stayed six years in Vienna. In 1639 he returned to Rome where he stayed until his death. Opere, XX, p. 534.

the sun, and attempted to interpret their location, material essence, and movement. Scheiner sent the letters – written under the pseudonym "Apelles" – to Mark Welser, a widely educated nobleman of Augusta, who held several political positions there, and also owned a banking house and financed both Rudolph II and the Jesuits. In addition, Welser was a patron of the arts and member of the Accademia dei Lincei and the Accademia della Crusca. Welser sent the letters to Galileo, urging him to reply. Rather slowly, Galileo reacted with a letter of his own addressed to Welser, which in turn gave rise to a more detailed response from "Apelles" in the form of three additional letters which he named *A More Accurate Disquisition*. Galileo's remaining letters to Welser were written in response to the challenge of Appelles both in his letters and in the *Disquisition*.

It was during this first confrontation between Galileo and a Jesuit astronomer that characteristics of Galileo's "new science" and of the revised mathematical science of the Jesuits first emerged, with the constraints peculiar to each. Patterns of constitution of objects within well-defined boundaries of a discourse, strategies of inclusion and exclusion, of authorization and reproduction were established and in many respects determined future possibilities.

13

The Dispute on Sunspots

The Constitution of Objects in Scientific Discourse

Galileo's third letter to Welser[1] contains a methodological principle, later perceived as foundational for modern empirical epistemology:

> In our speculating we either seek to penetrate the true and internal essence of natural substances, or content ourselves with a knowledge of some of their properties. The former I hold to be as impossible an undertaking with regard to the closest elemental substances as with more remote celestial things. The substances composing the earth and the moon seem to me to be equally unknown, as do those of our elemental clouds and of sunspots. . . . But if what we wish to fix in our minds is the apprehension of some properties of things, then it seems to me that we need not despair of our ability to acquire this respecting distant bodies just as well as those close at hand – and perhaps in some cases even more precisely in the former than in the latter. . . . Hence I should infer that although it may be vain to seek to determine the true substance of the sunspots, still it does not follow that we cannot know some properties of them, such as their location, motion, shape, size, opacity, mutability, generation, and dissolution.[2]

The passage is indeed foundational, but far from supporting an empiricistic epistemology, if read in the context of Galileo's practices during the dispute on sunspots. Against Blancanus, Galileo contended that knowledge of the heavens was not gained simply by the senses, but by combining discourse with observation.[3]

Galileo refused to discuss objects in terms of "substances" or their "essence". This, however, should not be regarded as an advocacy of unmedi-

---

1. Unless otherwise stated, all my citations from Galilei's *Istoria e Dimostrazioni intorno alle Macchie Solari e loro accidenti*, *Opere*, V, pp. 71–260 are taken from *Letters on Sunspots*, trans. by Drake in his *Discoveries and Opinions of Galileo*, pp. 87–144. Henceforth *Letters on Sunspots*.
2. Ibid., pp. 123–124.
3. See above, Galileo's letter to Grienberger, Ch. 12.

ated sense experience – Galileo was no empiricist – but as part of an attempt to reconstitute the objects of scientific inquiry in terms of a new set of categories, i.e., to place them within a different discourse. Four of these categories – location, motion, shape, and size – were traditional enough where astronomy was concerned, and represented a rationalization of experience through mathematical or geometrical models. The other three, however – mutability, generation, and dissolution – were not generally the concern of astronomers, being as qualitative as any physical "accident" in the peripatetic discourse of the Aristotelians. Galileo included these "qualitative" categories, hinting that his reconstitution might involve more than a modification of the conditions of enunciation, from qualitative to quantitative discourse, or from "linguistic" to mathematical analysis. In fact, it might involve a different view of what was "observable" and what was "manipulable", requiring a change in the conditions of "observability" and perhaps even of "manipulativity".

Christopher Scheiner,[4] disguised as Apelles, was concerned no less than Galileo with the need to reconstitute the objects of astronomical discourse, a need that became more acute after the telescopic discoveries. In his work on Jesuit mathematical science Peter Dear pointed out Blancanus's and Scheiner's awareness of the "constructed" nature of the objects of astronomy and optics. Blancanus was the first to distinguish between "appearances" or "phenomena" which:

> may be perceived by all, and "observations" which are certain items of knowledge [cognitiones] provided from experiences which do not become known by everyone as appearances do, but only by those who, skilfully labouring hard at it with diligent work, and instruments, apply themselves seriously to the science of the stars (stellarum).[5]

Similarly, Scheiner spoke of things "which either don't occur or don't become evident [non patescunt] without the industry of special empirics [peculiari Empirici]."[6]

The incorporation of sunspots into science – the discourse of the real and the true – was the aim of both Scheiner and Galileo. According to both, sunspots were no illusions. They were real, material objects, indicating regular motions in heaven. In his *Three Letters on Solar Spots* written before Galileo's response to the newly discovered phenomenon, Scheiner

---

4. All citations from Scheiner's work are translated by me from: Apellis latentis post tabulam, *Tres epistolae de maculis solaribus,* and idem, *De maculis solaribus et stellis crica Iovem errantibus accuratior disquisitio,* in *Opere,* V, pp. 20–70. Henceforth *Letters on Solar Spots* and *Disquisition.*
5. Dear, "Jesuit Mathematical Science", p. 149.
6. Ibid., p. 156.

concluded the reality of sunspots by eliminating the possibility that they were the product of a "defect either of the eyes, or of the tube, or of the air".[7] Galileo recognized this basic agreement between Scheiner and himself at the opening of his first letter:

> They are real objects and not mere appearances or illusions of the eye or of the lenses of the telescope, as Your Excellency's friend well establishes in his first letter. . . . It is also true that the spots do not remain stationary upon the body of the sun, but appear to move in relation to it with regular motions, as your author has noted in that same letter.[8]

Likewise, the practices of establishing the existence of such phenomena were similar for both Galileo and the Jesuit. Being aware of the exclusiveness of telescopic observations, they both insisted on the continuous and repetitive nature of theirs, and recruited witnesses to confirm their statements: "And thus" Scheiner remarked, "we have consulted the eyes of a great variety of people, who saw the same appearances, without exception, in the same position, order and number".[9] In his second letter, describing a conjunction of Venus and the sun, Scheiner informed the readers that: "I scrutinized the sun carefully, and not only I, but together with me many other people, celebrated throughout that whole day the conjunction of the sun with Lucifer".[10] Galileo wrote in much the same spirit: "I have observed them for about eighteen months having shown them to various friends of mine, and at this time last year I had many prelates and other gentlemen at Rome observe them there."[11]

And yet, both Scheiner and Galileo knew that simply observing and identifying phenomena did not suffice to secure their status as objects of science. Both devoted much space to detailed descriptions of their mode of observation, the new means available for the purpose (specifying, for example, the type of telescope used, the lenses chosen, etc.), and the correct way of using the necessary instruments. Moreover, both were very concerned with specifying the particular techniques of representing the results of observations. The whole last part of Scheiner's first letter was devoted to such details, and in the last part of his second letter Galileo recommended a method, invented by Castelli, of representing sunspots on paper. What Scheiner and Galileo shared was the consciousness of a

---

7. Scheiner, *Letters on Solar Spots, Opere*, V, p. 25: "Ne forte id latente quodam vel oculorum vel tubi vel aëris vitio accideret".
8. Galilei, *Letters on Sunspots*, p. 91.
9. Scheiner, *Letters on Solar Spots, Opere*, V, p. 25.
10. Ibid., p. 28: "Sedulo inspexi, non ego solus, sed et alii mecum quamplurimi, Solisque cum Lucifero coniunctionem toto die celebravimus".
11. Galilei, *Letters on Sunspots*, p. 91.

need to make their method known to wider audiences. Neither could assume a nonproblematic acceptance of the new objects by their readers. Both needed to transform observation from a private into a public event, in an attempt to gain it legitimation through a wide measure of consent.

The status of scientific objects was not assured, however, even after the mechanism of observation was explained. Much depended upon the ability to integrate the new objects into a network of terms which constituted a structure, a meaningful whole. Thus both Scheiner and Galileo proceeded to construct their respective networks attempting to give meaning to the newly discovered phenomena.

In Scheiner's first letter, the motions of sunspots were reconstructed in the context of the motions of other stars, Venus and Mercury in particular. This context remained the focus of Scheiner's discussion throughout the three letters and the *Disquisition*. Scheiner's strategy in determining the context opted for maximum rigour. Well aware that the context of reconstruction would become the context of interpretation, Scheiner proceeded by eliminating all other possibilities in order to gain acceptance for his own. Thus, one would have thought that spots observed to travel across the sun's disk in approximately fifteen days (an observation confirmed by Galileo as well) should return "in the same order and with the same position amongst them and with regard to the sun. But, on the contrary", observed Scheiner, "until now they never came back".[12] From this argument Scheiner deduced that the spots could not be reconstructed in the context of the body of the sun itself.

The other possibility – that the spots were either in the atmosphere of the earth, or in the heavens of the moon, Mercury, or Venus – was also rejected by Scheiner. Had the spots been in the air, or in either of those heavens, a parallax would have been detected. In the absence of a parallax the spots, he said, should be located in a region close to the sun, though not in the sun itself:

> Therefore, I judge that there are no real spots, but that there are parts eclipsing the sun for us, and that these parts are consequently stars that are under or about the sun.[13]

The spots, then, were interpreted by Scheiner as shadows of real stars projected upon the sun's surface while eclipsing it and moving, like all celestial bodies "by their own movements". Such a hypothesis had the advantage of being useful for checking another astronomical problem of

12. Scheiner, *Letters on Solar Spots, Opere*, V. p. 26: "Eodem ordine et situ inter se et ad Solem; at nunquam adhuc redierunt".
13. Ibid., p. 26: "Quin, nec veras maculas esse existimaverim, sed partes Solem nobis eclipsantes, et consequenter stellas, vel infra Solem vel circa".

the utmost significance: namely the center of the motions of Mercury and Venus.

The Copernican theory, presupposing the sun to be the center of all planetary motions, had predicted – without empirical evidence – the phases of Venus as a necessary corollary of the location of its orbit in relation to the sun. For Galileo, a Copernican long before his telescopic observations, the phases of Venus counted as empirical evidence of the truth of the Copernican system. But Copernicanism expected a great deal to be granted without sufficient proof, an idea too radical for Jesuit astronomers. The difficulties of reconciling Copernicanism with the principles of the Thomistic organisation of knowledge and with biblical exegesis resulted in the suspension of "global" Copernicanism. In these circumstances Scheiner was naturally enthusiastic about the possibility of checking the location of the orbit of Mercury and Venus without having to presuppose the Copernican theory, and independently of Galileo's observations of the phases of Venus.

If the spots were shadows of stars projected upon the surface of the sun by tiny invisible stars, then visible stars like Mercury or Venus must likewise have projected their shadows during conjunction with the sun. Astronomical data about a diametrical conjunction of Venus and the sun, predicted by the best astronomers of the age (especially Magini) on the basis of the most recent astronomical calculations, should have provided an opportunity of checking the location of Venus's orbit. Scheiner reasoned that although its shadow did not appear on the sun's surface in the shape of a spot, a clear indication that Venus was hidden behind the sun proved it to be revolving around the sun, and not simply passing under it while revolving around the earth. The proof was thus presented by Scheiner in his second letter to Welser as follows:

> We did not see Venus under the sun, although, according to the calculation, Venus was under the sun. It clearly became red, and hurried on, so that we did not observe its change of form. What had happened? I say nothing. I myself feel my way. Although we lack all other arguments, the point has been carried that the sun was orbited by Venus. I do not doubt that the same can be said unambiguously about Mercury, and I shall not omit to investigate this in the same way as soon as an opportune conjunction has taken place.[14]

14. Ibid., p. 28: "Venerem sub Sole, quae tamen secundum calculum erat sub Sole, nequaquam vidimus. Erubuit scilicet, et proripuit sese, ne suas intueremur nuptias. Quid hinc sequatur, non dico; ipsemet palpas: etsi careremus omnibus aliis argumentis, hoc uno evinceretur, Solem a Venere ambiri: quod item a Mercurio fieri nullus ambigo, neque id simili modo investigare omittam, quam primum opportuna se obtulerit coniunctio".

Scheiner's attempt to incorporate sunspots into his astronomical dis-course led him into an implicit analogy between hitherto unknown stars with their newly observed projected shadows and some well-known stars (Mercury and Venus) with their shadows which were not yet observed but nevertheless expected to appear. The failure of the spots to appear at the right moment (the diametrical conjunction of Venus and the sun) provided the proof for the motion of Venus around the sun, and not merely under it.

After eliminating in the first letter the possibility that the spots be-longed either to the heaven of the sun or to that of the earth, moon, Mer-cury, or Venus, and after developing in the second letter the analogy be-tween the spots and other stars by using it as a means of checking a hypothesis concerning the orbits of Venus and Mercury, Scheiner put for-ward three other arguments by which he further suggested that the spots moved around the sun like the other planets:

1. Spots observed around the edge of the sun become smaller, he claimed. He then proceeded to support this assertion by a geometric proof, dem-onstrating that the brightness of the sun makes less of the shaded part visible. This could only happen if the spots are not in the sun but around it.
2. Spots, he said, are seen to unite and divide. He claimed that this phe-nomenon showed that the spots moved around the sun.
3. Spots, he said, moved faster in the middle of the sun than on the perim-eter. He considered this too an indication that the spots moved around the sun.[15]

For Scheiner, the significance of the discovery of sunspots lay in the innovative or even revolutionary thesis, in terms of medieval cosmology, that Venus and Mercury revolved around the sun – a conclusion arrived at without having to presuppose the Copernican system. The suggested analogy between the motion of sunspots and the motions of Venus and Mercury helped to support Scheiner's hypothesis that sunspots were shad-ows of stars unseen by the naked eye, moving close to the sun but not contiguous with it. Thus, the accepted cosmological picture was indeed modified – the sun now being considered the center of the motions of

15. Ibid., p. 29: "Primum, omnis macula seorsim spectatae, circa Solis limbum, sive in ingressu sive in exitu, gracilescit: phaenomenon hoc defendi nequit, nisi per motum maculae circa Solem: ergo. Secundum, duae vel tres aut plures maculae circa limbum Solis videntur coire in unam magnam, in medio sese diducunt in plures: hoc defendi nequit, nisi per motum earum circa Solem: ergo. Tertium, medio celerius moventur quam circa perimetrum Solis: hoc defendi nequit, nisi per motum circa Solem: ergo".

other stars – but not wholly unvalidated, as no generation or dissolution occurred in the heavenly region. But, to consolidate his hypothesis Scheiner needed to eliminate another possible analogy, namely that sunspots were like clouds, and hence introduced change in heaven:

> But what are they finally? Not clouds. Because who would posit clouds there? And if they were clouds, how many would there be? Why would they always move in the same way and with the same movement? And how could they produce so many shadows?[16]

The blackness of the spots, Scheiner concluded ("these spots are blacker than the ones seen on the moon ever are, with the exception of a single small one"[17]) indicated a much more solid, opaque, and dense substance than clouds, namely that of stars.

Scheiner's discursive strategies were even more clearly exposed in his *Accuratior Disquisitio*,[18] written in response to Galileo's criticisms. In his first letter Galileo accused Scheiner of believing that the spots were outside the sun because they were black, whereas the sun is known to be bright: "For it proves nothing to say, as this author does in his first argument, that it is unbelievable that dark spots exist in the sun simply because the sun is a most lucid body."[19] And he went on to ask his readers to reject any attempt to accommodate the essence of things to names, instead of inquiring into the essence of things and then naming them: "For names and attributes must be accommodated to the essence of things, and not the essence to the names, since things come first and names afterwards".[20] Galileo also complained of confusion and inconsistency in Scheiner's argument, stemming from his inability "to detach himself entirely from those fancies previously impressed upon him which his intellect returns to and assents to out of long use and habit".[21] The Jesuit astronomer could obviously not leave such a challenge unanswered. The very title of his new letters, promising a "more accurate disquisition", was a subtle recognition of the need to substantiate claims made by legitimate argumentation.

The text opens with a lemma, a long explanation of an elementary geometrical proposition concerning the sum of angles on the one side of

---

16. Ibid., p. 30: "Sed quid eae tandem sunt? Non nubes: nam quis illic poneret nubes? et si essent, quantae essent? quare eodem modo et motu semper agerentur? quomodo tantas umbras efficerent?"
17. Ibid., p. 26: "Easque nigriores multo quam sint in Luna unquam visae (praeter unicam parvulam)".
18. Henceforth *Disquisition*.
19. Galilei, *Letters on Sunspots*, p. 92.
20. Ibid.
21. Ibid., p. 96.

a line produced through the intersections of the extended sides of a triangle, and parallel to the opposite side. There follows an investigation of Venus's motion under the solar disk on 11 December 1611 (a morning conjuction with the sun), deduced by calculations and geometrical demonstrations, from which Scheiner concluded the true location of the planet's orbit. Scheiner then produced a series of arguments, including the phases of Venus, in order to strengthen his thesis, and proceeded to explain the changing form of the spots through an optical argument, proving geometrically that modifications of their form occur when illumination reduces the shaded area of the spot.

As Galileo correctly pointed out, "the observations of Venus", in the *Disquisition*, "(are) explained more at length than in the first letters".[22] Far more important, however, was the rationale – unrecognized by Galileo – for such a detailed account, which lay in Scheiner's deliberate attempt to emphasize the mathematical basis of his arguments. The lemma, whose unnecessary length was deplored by Galileo, signified Scheiner's attempt to secure the status of truth for his claims within the framework of the mixed mathematical disciplines. As Wallace has shown, in the Jesuit tradition of those disciplines the writer's authority depended to a major extent upon his ability to prove the propositions he borrowed from geometry, and not simply to accept them, in which case his conclusions were considered "mere opinion" and thus not really true.[23] Scheiner's strategy should be read in the context of the sixteenth-century debates over the claims to truth of opticians and astronomers. By inserting the lemma he both harmonized himself with the Jesuit tradition and reassured Galileo of his professional identity as a mathematician, and hence as someone not to be included among those who were accused of merely accommodating things to names.

The geometrical proof with which the *Disquisition* begins, the detailed investigation into the motion of Venus by deductions from calculations and geometrical proofs, as well as the optical argument used for explaining the changing shape of the spots were all strategies by which sunspots were constructed as legitimate objects of astronomical investigation. By using such strategies Scheiner and his followers could claim sunspots as objects suitable for mathematical analysis and relevant for solving more general astronomical problems, such as the order of celestial orbits.

Full legitimation of the spots, however, also depended on the contention that their "heavenly" qualities were unquestionable. Unfortu-

22. Ibid., p. 192.
23. Wallace, *Galileo and his Sources*, p. 133.

nately, this was by no means the case. First of all because sunspots had
never been perceived in the heavens before the discovery of the telescope,
or at least have never been recognized by astronomers. This objection,
however, was relatively easy to answer: the Medicean planets of Jupiter
were, after all, confirmed by the Jesuits only a short time before the discovery of sunspots.[24] More problematic were the discernible changes sunspots clearly passed through while crossing the sun's disk, a peculiar phenomenon which they did not share with the moons of Jupiter. The trouble
with sunspots was that they changed shape all the time, appearing and
disappearing in whatever part of the heaven they happened to be located,
with very little order or system. Scheiner therefore made a particular effort to develop an analogy between spots and stars, an analogy already
suggested by him in his three previous letters, and which had also been
implied in the method he had suggested in order to prove that Venus ro-
tated around the sun.

Scheiner's strategy in the *Disquisition,* however, unlike the one adopted
in the letters, was to refrain from putting forward a clear hypothesis about
the spots' location and to suspend judgement concerning their substance.
At the same time, he strove to dispel any doubts about their heavenly
nature by stressing the similarities between them and other stars. Thus he
could now admit – contrary to his previous contention in the letters – that
the spots were not absolutely regular in their shapes, or even that they
changed their shape when close to the sun, providing he could find analogies among the stars for such phenomena. Accordingly he chose to com-
pare the spots to the newly discovered moons of Jupiter, the shape of
which, he suggested, was not, perhaps, so regular either, and whose periods it was impossible as yet to calculate.

Scheiner's categories for discussing sunspots were not very different
from those suggested by Galileo: location, motion, shape, and size were
primary and required mathematical language as a basic mode of enunciation. As for the substance of the spots, Scheiner adhered to the traditional
strategy of the astronomers, who preferred to have recourse to "invisibles" whenever observations led them to deal with substances unrecognized by Aristotelian physics and cosmology. Much as the irregularities
of the planetary motions were explained by means of epicycles and eccentrics – "invisibles" the substance of which remained outside the boundaries of astronomical discourse – so sunspots were explained by invisible
stars whose "solar" substance was deduced by analogy, but could not be
determined solely by astronomers.

Galileo began his interpretation of sunspots by rejecting the context of

24.   Letter of Jesuit mathematicians to Bellarmine, see Ch. 12, p. 249.

the discussion constructed by Scheiner. First, he suggested that this context was in fact trivial, for, in contrast to what Scheiner thought (that the direction of the motion was from east to west) "the spots describe lines on the face of the sun similar to those along which Venus and Mercury proceed when those planets come between the sun and our eyes. Hence they move with respect to the sun as do Venus and Mercury and the other planets, which motion is from west to east".[25] Then, he further protested that the context suggested by Scheiner was also superfluous:

> Next Apelles suggests that sunspot observations afford a method by which he can determine whether Venus and Mercury revolve about the sun or between the earth and the sun. I am astonished that nothing has reached his ears – or if anything has, that he has not capitalized upon it – of a very elegant, palpable, and convenient method of determining this, discovered by me about two years ago and communicated to so many people that by now it has become notorious. This is the fact that Venus changes shape precisely as does the moon.[26]

For Galileo, his discovery of the phases of Venus constituted a necessary demonstration that it revolved about the sun:

> With absolute necessity we shall conclude, in agreement with the theories of the Pythagoreans and of Copernicus, that Venus revolves about the sun just as do all the other planets. Hence it is not necessary to wait for transits and occultations of Venus to make certain of so obvious a conclusion.[27]

This conclusion of Galileo did not stem simply from observations indicating a connection between the planet and the sun, but above all from the discovery of a compatibility between observation and theory. The Pythagorean and Copernican theory which he had adopted before observing the phases of Venus was empirically verified by his observations. Scheiner, however, was not really interested in an alternative theory, and preferred to limit the discussion to the connection of the spots, the sun, and other stars without embarking upon a cosmological debate. Galileo accepted these constraints on the explicit level. Implicitly, however, he was looking for a different context in order to suggest an alternative framework of interpretation.

Fixing the sun as the frame of reference within which the spots should be discussed was Galileo's first target. First he located them on a network of coordinates, which he drew on the surface of a sphere, thus making possible the measurement of their breadth and length. Observation showed, he argued, that whereas the spots tended to have the same length

25. Galilei, *Letters on Sunspots*, p. 91.
26. Ibid., p. 93.
27. Ibid., p. 94.

wherever they appeared on the sun, their breadth changed as they moved from the edges to the central parts of the sun's disk, being very little at the moment of appearance and disappearance and growing towards the center. Such behaviour could best be explained by the geometrical rules of perspective, which described the fixed length and expanding breadth of an object drawn on a spherical surface as "foreshortening". The point of maximum foreshortening, he went on to explain, was also the point of maximum thinning, the third dimension of the spots, a fact which indicated that the spots must be located on the sun, or at an imperceptible distance from it, otherwise maximum foreshortening would occur outside the face of the sun.[28] Two additional mathematical arguments were advanced, to show that the spots must be on the sun, and not at a distance from it: one concerned the decreasing distances they travelled in the same time as they approached the edge of the sun, and the other exposed the mode of variation in the distances between one spot and another, a variation which could be accounted for only by circular motion on the sun itself.[29]

The successful location of the spots on the sun immediately raised two problems: the first concerned the source of their motion. Galileo referred this motion to the rotating motion of a spherical sun, or its ambient. Galileo, however, lacked any physical principle from which such celestial motion could be deduced. Not surprisingly, perhaps, his argument for the rotational motion of the sun or its ambient – indicated by the spots – came from his experience with the motion of physical bodies upon the earth. The sun's motion was described in terms of "inertia", of which he had learned by watching the motion of physical bodies on earth, and the principle of which he here formulated for the first time:

> For I seem to have observed that physical bodies have physical inclination to some motion (as heavy bodies downward), which motion is exercised by them through an intrinsic property and without need of a particular external mover, whenever they are not impeded by some obstacle. And to any other motion they have a repugnance (as the same heavy bodies to motion upward), and therefore they never move in that manner unless thrown violently by an external mover. Finally, to some movements they are indifferent, as are these same heavy bodies to horizontal motion, to which they have neither inclination (since it is not toward the center of the earth) nor repugnance (since it does not carry them away from that center). And therefore, all external impediments removed, a heavy body on a spherical surface concen-

28. Ibid., pp. 107–108.
29. Ibid., pp. 108–109.

tric with the earth will be indifferent to rest and to movements toward any part of the horizon. And it will maintain itself in that state in which it has once been placed; that is, if placed in a state of rest, it will conserve that; and if placed in movement toward the west (for example), it will maintain itself in that movement.[30]

The sun, having neither a disinclination nor a propensity for rotational motion, could participate in such a motion of the ambient, or continue forever in such a motion of its own, just as a ship would continue to move around the earth forever once it were set in motion and not arrested by external impediments.

Galileo's explanation of sunspots' motions exposes the large gap that opened between his ability to deduce the location of the spots on the sun mathematically, and his inability to use such mathematical deduction as a physical account of the motion. This gap was erased by the analogy between the "indifferent" motion of heavy bodies on earth and the rotational motion of the sun inferred from the phenomena of sunspots.

Scheiner, who considered the visible motion of the spots in terms of the motion of invisible stars, argued that those stars had their proper motion like all other celestial bodies. Galileo, who argued from the inner position of a Copernican ran the risk of exacerbating the weakness of his theory in accounting for the invisible motion of the earth, by having to assume another invisible motion, that of the rotating sun. The analogy between the inertial motion of heavy bodies on earth and the inertial motion of the sun or its ambience thus served to evade both difficulties by using the visible (sunspots' motion) to justify the invisible motions implied by his theory.

The second problem raised by the proof of the spots' contiguity with the sun concerned the spots' failure to return after fifteen days, although they were clearly seen to cross the sun's disk in approximately that period. Scheiner considered this fact as the ultimate proof that the spots were *not* on the sun. Galileo deemed it an argument for their generation and dissolution on the sun, and used an even more audacious analogy to substantiate his claim. He contended that the spots resembled clouds more than anything else:

> If proceeding on the basis of analogy with materials known and familiar to us, one may suggest something that they may be from their appearance, my view would be exactly opposite to that of Apelles . . . I find in them nothing at all which does not resemble our own clouds.[31]

30. Ibid., pp. 113–114.
31. Ibid., p. 98.

And he went on to describe exactly how the analogy worked in terms of the periods of generation and decay, condensation, expansion, shape, colour, size, opacity, number, and the like.

At first, the analogy seemed innocent enough. It was very tentatively suggested, and was accompanied by repetitious remarks about our inability to know, strangely untypical of Galileo's polemical style.[32] Its radical significance was soon revealed, however, when Galileo further suggested that the sun and its spots resemble the earth and its clouds, if the earth is imagined as having light of its own:

> There is no doubt that if the earth shone with its own light and not by that of the sun, then to anyone who looked at it from afar it would exhibit congruent appearances. For as now this country and now that was covered by clouds, it would appear to be strewn with dark spots that would impede the terrestrial splendor more or less according to the greater or less density of their parts. These spots would be seen darker here and less dark there, now more numerous and again less so, now spread out and now restricted; and if the earth revolved upon an axis, they would follow its motion . . . In a word, no phenomena would be perceived that are not likewise seen in sunspots.[33]

Apparently, a rotating earth and its clouds were presented as a model for understanding the sun and its spots. Sunspots generate and dissolve like clouds and are carried by the motion of a body rotating on its axis. In fact, however, since the rotation of the earth on its axis could not be deduced from any physical principle, the observation of the movement of sunspots and their interpretation by means of such a model was a way of arguing for the unobservable motion of the earth.

Galileo's explanation of the spots' motion, as well as his account of their substance, led him to invoke "invisibles", which were somehow justified by observed facts. Such invisible entities were not unknown to traditional astronomical discourse. Astronomers have always attempted to account for the observable (e.g., retrograde motions of the planets, or their unequal velocities) in terms of the unobservable motion of epicycles and eccentrics. These unobserved entities, however, remained qualitatively different from observed ones. Their status as legitimate objects of science did not depend upon their visibility. Galileo's strategies differed radically from traditional astronomers', as he systematically strove to create a continuity between the observable and the unobservable through analogical arguments, invoked to erase the gap left by his mathematical demonstrations.

32. See, for example, his remarks on p. 98: "I see nothing discreditable to any philosopher in confessing that he does not know, and cannot know, what the material of the solar spots may be".

33. Ibid., p. 99.

In Galilean discourse a thing was constituted as a scientific object by mathematical enunciation and by being given to observation. But there was a further condition: scientific objects should in some sense be amenable to manipulation, even if only by analogy. The second letter to Welser includes a technique by which sunspots could be regarded as "manipulable":

> I liken the sunspots to clouds or smokes. Surely if anyone wished to imitate them by means of earthly materials, no better model could be found than to put some drops of incombustible bitumen on a red-hot iron plate. From the black spot thus impressed on the iron, there will arise a black smoke that will disperse in strange and changing shapes. And if anyone were to insist that continual food and nourishment would have to be supplied for the refueling of the immense light that our great lamp, the sun, continually diffuses through the universe, then we have countless experiences harmoniously agreeing in showing us the conversion of burning materials first into something black or dark in color. Thus we see wood, straw, paper, candlewicks, and every burning thing to have its flame planted in and rising from neighboring parts of the material that have first become black. It might even be that if we more accurately observed the bright spots on the sun that I have mentioned, we should find them occurring in the very places where large dark spots had been a short time before.[34]

The strange, changing spots produced on an iron pan were *like* the strange, changing shapes of sunspots, contended Galileo. The conversion of burning materials into something dark was likened to the imagined burning of materials on the surface of the sun, leaving dark spots after combustion. It was even deemed possible that the dark spots appeared exactly in the place of preexisting bright spots, just like the smoke left after the burning of bitumen drops on an iron pan.

Mathematical argumentation enabled Galileo to prove the contiguity of sunspots with the sun. Their appearance and disappearance could thus be interpreted as demonstrating the corruptibility of the heavenly region. Such indirect proof, however, was too abstract for Galileo. It was only through an analogy between the manipulation of earthly matter and the mutability on the sun that the argument of corruption seemed to be validated and the qualitative difference between the heavens and the earth delegitimized:

> In that part of the sky which deserves to be considered the most pure and serene of all – I mean in the very face of the sun – these innumerable multitudes of dense, obscure, and foggy materials are discovered

34. Ibid., p. 140.

to be produced and dissolved continually in brief periods. Here is a parade of productions and destructions that does not end in a moment, but will endure through all future ages, allowing the human mind time to observe at pleasure and to learn those doctrines which will finally prove the true location of the spots.[35]

The dispute on sunspots reveals structural differences between Galileo's discourse and Scheiner's. Galileo, following Copernicus, presupposed the motion of the earth, although he still lacked a fully fledged physical theory that would justify his conviction. To suppose the motion of the earth, however, meant placing an *unobservable* in the center of astronomical discourse. Epicycles and eccentrics – the traditional unobservables in astronomy – had always placed an explanatory role, never that of the thing to be explained. To place an *unobservable* as an explanandum meant it could be legitimated only by *observables*. The whole structure of traditional astronomy was violated by such an act. Instead of focussing on techniques for making observables (the unequal motion of the planets) understood in terms of unobservables (eccentrics and epicycles), it was now necessary to make the unobservable (the motion of the earth) comprehensible in terms of the observable (motion of sunspots). Sunspots seemed to provide Galileo with an outstanding opportunity for justifying the unobservable motion of the earth in terms of the observable motion of the spots, if the analogy between the two could be made manifest to the senses, by changing the conditions of observability and manipulativity. Galileo's main effort in the letters was not dedicated to the geometrical demonstration of contiguity – which was perfectly in tune with the traditions of the "mixed sciences" – but rather to elaborating the analogies between the sun's rotational motion and the earth's, between sunspots and clouds, and between earthly and heavenly matter in order to make the unobservable in some sense observable, and the immutable in some sense manipulable.

The Delineation of Boundaries in Scientific Discourse

The new objects that appeared in the horizons of astronomers did more than merely give rise to various strategies for their appropriation. No less significant was the challenge sunspots presented to the boundaries of astronomy, and the reactions to this challenge. These ranged from affirma-

35. Ibid., p. 119.

tion to transgression and finally, ultimate destruction, and the delineation of new boundaries not to be institutionalized without a violent struggle, i.e., the trial of Galileo.

The *Ratio* of 1599 and Bellarmine's letter to Foscarini of 1615 are two official, or quasi-official documents which testify to precise nuclei on the intellectual map where the Jesuit authorities deliberately took action to preserve boundaries constantly threatened by practice. Both the *Ratio* and Bellarmine's letter endorsed the boundaries of mathematics, natural philosophy, and theology. This endorsement, I have argued in previous chapters,[36] expressed an official commitment to the Thomistic organisation of knowledge, under the pressure of the Dominicans. Thus, the Jesuits' break with the contemplative tendencies of Thomism and their development towards Molinistic theology and practical science was counterbalanced by their avoidance of openly challenging the boundaries dictated by Thomism and guarded by the church's old and established intellectual elite. Clavius's program for a radical reform of the educational system was rejected, and an institutional constraint on astronomical discourse, confining it to the "mixed mathematical sciences" was firmly established. This did not mean that the discussion of physical problems by astronomers was eradicated. The appropriation by mathematicians of certain canonic books traditionally considered part of natural philosophy – e.g., *On the Heavens,* the sixth and seventh book of *Physics,* and the *Meteorology* – signaled the legitimation of such discussion in certain areas. It did mean, however, that the system was sensitive to transgressions of the boundaries between mathematics and physics, and monitored them closely.

The institutional imposition of boundaries constantly threatened by practice, echoes through the texts of many Jesuit mathematicians of the period. Blancanus's *Sphaera mundi* is perhaps the most illuminating, since Blancanus proclaimed himself to be Clavius's follower, and the elaborator of his program. Peter Dear mentioned two instances in this work where Blancanus touched upon physical problems: one concerning the motion of the earth, and the other concerning new stars and whether they signify generation in the heavens. In both cases Blancanus found he had to cross the boundary; in both he excused himself for doing so, and refrained from expressing a judgement on matters which exceeded his authority and fell within the province of the physicist.[37] Dear further mentioned Oratio Grassi discussing the matter of comets and sunspots and then excusing himself abruptly: "But these things are physical rather than mathemati-

36. See my discussion in Ch. 11.
37. Dear, "Jesuit Mathematical Science . . .", pp. 161–162.

cal".[38] Scheiner, in the *Rosa Ursina*, again emphasized the distinction between the astronomer and the physicist, quoting Aristotle and repeating
almost word by word the text of the *Ratio*:

> The astronomer, according to Ptolemy, *Almagest* Book I, chap. I, con-
> siders the quantity of celestial bodies, and indeed Aristotle himself
> declares that [also] in the *Categories*, and teaches in [his] physics that
> quantity, and those things which are connected with it, is considered
> as much by the physicist as by the mathematician, but in different
> ways.[39]

The letters on sunspots and the *Disquisition*, however, written many years
before the *Rosa Ursina*, not only cross the boundaries later confirmed by
Scheiner himself, but include an actual act of transgression.

Scheiner's position may be understood in terms of transgression as it
has been described by Michel Foucault:

> Transgression, then, is not related to the limit as black to white, the
> prohibited to the lawful, the outside to the inside, or as the open area
> of a building to its enclosed spaces. Rather, their relationship takes
> the form of a spiral which no simple infraction can exhaust. Perhaps
> it is like a flash of lightning in the night which, from the beginning
> of time, gives a dense and black intensity to the night it denies, which
> lights the night from the inside, from top to bottom, and yet owes to
> the dark the stark clarity of its manifestation, its harrowing and
> poised singularity; the flash loses itself in this space it marks with
> its sovereignty and becomes silent now that it has given a name to
> obscurity.[40]

The ambiguous nature of transgression probably requires metaphorical
language if it is to be expressed with any accuracy. Thus, transgression is
a momentary illumination of an order of things, immediately swallowed
up by the authoritative hegemony of the very order it defies.

Scheiner's letters on sunspots and his *Disquisition* were an attempt to
describe the motion, location, and substance of sunspots. The investigation of the movement and location of celestial bodies was a legitimate
preoccupation of astronomers, and was achieved through geometrical
representation of observations, calculations, deductions, and analogical
arguments. While using the usual methods of observing mathematicians,
however, Scheiner distinguished between "beliefs" held by traditional as-

38.   Ibid., p. 162.
39.   Ibid., p. 163.
40.   M. Foucault, "A Preface to Transgression" in D. F. Bouchard (ed.), *Lan-
      guage, Counter-Memory, Practice: Selected Essays and Interviews by Michel
      Foucault*, Ithaca, N.Y. 1977, p. 35.

tronomers and newly discovered "facts" – the products of his own efforts, and probably of some other people:

> The fact that these appearances, due to their great power, can be discovered all over the sun . . . is something we are completely convinced of in the most clear light of truth. This fact was hitherto questioned or not known or even perhaps denied altogether by the astronomers . . . this true *fact* was almost more important than any *belief*.[41]

This distinction between facts and beliefs expressed a sense of discontinuity, of a moment of transition in the collective experience of astronomers, and perhaps a recognition of the need to break away from old discursive formations. Faber's testimony that Scheiner agreed with Galileo on the nature of the world system,[42] Kircher's testimony that he adhered to the Ptolemaic system only through constraint and out of obedience,[43] and Descartes' opinion that he had tacitly accepted Copernicanism[44] may account for Scheiner's ambivalence vis-à-vis the tradition, echoing in his distinction between "facts" and "beliefs."

Whereas the greater part of Scheiner's letters was devoted to the movement and location of sunspots, the third part of the third letter was wholly concerned with their substance. Such a discussion clearly lay beyond the boundaries of legitimate astronomical discourse. "What are they [these

41. Scheiner, *Letters on Solar Spots, Opere,* V, p. 25: "Quod eorum ope plurima, hactenus astronomis aut dubitata aut ignorata aut etiam fortassis pernegata, in clarissimam veritatis lucem, per fontem luminis et astrorum ductorem Solem, protrahi posse, plane persuasum habeamus . . . Quia *vero res haec omni fide* prope maior erat . . ." (My italics, R.F.)

42. Stelluti to G. Galileo, *Opere,* XIII, p. 300: "Si trova qui tuttavia il Padre Scheiner Giesuita, che credo stampi le sue osservationi delle macchie solari; e disse alcuni giorni sono al nostro Sig. Fabri che cosa stampava di nuovo V. S.; a che rispose di non saperlo; e lui replicò c'haveva inteso che stampava del flusso e reflusso del mare, e che desiderava di vederlo, e concorre con l'opinione di V.S. circa al sistema mondano".

43. N. Fabri to P. Gassendi, *Opere,* XV, p. 254: "Et toutes foys le bon P. Athanase, que nous avons veu passer icy bien à la haste, ne se peult tenir de nous advoüer, en presence du P. Ferrand, que le P. Malapertius et le P. Clavius mesmes n'improuvoient nullement l'advis de Copernicus, ains ne s'en esloignoient guières, encores qu'on les eusse pressez et obligez d'escrire pour les communes suppositions d'Aristote, que le P. Scheiner mesmes ne suyvoit que par force et par obediance".

44. Descartes to M. Mersennes, *Opere,* XVI, p. 56: "Mais d'ailleurs les observations qui sont dans ce livre [du P. Scheiner] fournissent tant de preuves pour oster au soleil les mouvemens qu'on lui attribuë, que ie ne sçaurois croire que le P. Scheiner mesme en son ame ne croye l'opinion de Copernic".

spots] finally?", he asked, and, unlike Blancanus and Grassi, he proceeded to give an answer, regardless of boundaries.

Scheiner contrived a few strategies to somehow keep himself within the limits of Jesuit astronomical discourse while at the same time allowing for his act of transgression. The effect was truly that of "a flash of lightning in the night". Illuminated by the flash, the historian's gaze is able not only to expose some aspects of the relationships Scheiner was involved in – which is almost all that historians could hope for anyhow – but also the contours of the map which determined those relationships.

First, Scheiner chose to deal with what sunspots were not: they were not clouds, he declared. The legitimacy of this statement was guaranteed a priori, for clouds were earthly substance, and by denying the "cloudiness" of sunspots Scheiner denied their earthly nature. Then he moved to speak in the name of the philosophers, who would surely agree, he said, that sunspots must be "parts of some heaven", or, in other words, stars. Only finally did he arrive at what he thought must be the conclusion of the astronomers:

> That they are bodies existing in themselves, solid and dark, throwing shadows, and for this reason alone they must be stars, no less than the moon and Venus, which appear black in the part turned away from the sun.[45]

Philosophers, he was aware, would deduce the heavenly nature of the spots from the nature of the heaven, of which they were part (he was alluding, here, to the traditional cosmography whereby observed celestial bodies like the planets were carried by material spheres or heavens, filling every spot in space). In contrast, the astronomers' point of departure was phenomena – the observation of dark, shadow-throwing bodies. Such observed "facts" allowed the astronomers to conclude – independently of philosophical speculation on the nature of the heaven – that they were substances, "bodies existing in themselves", and to draw an analogy between spots and others stars. Thus, the astronomers' approach was kept different, and separate from the philosophers'. The astronomer, Scheiner seems to be saying, observes and interprets his observations. Phenomena *indicate* (in the literal sense of the word, namely *point to*) something for the astronomer, although astronomers are not supposed to preoccupy themselves with the essence of the phenomenon observed and with its logical relation to other essences. Scheiner's act of transgression consisted in his audacity in speaking openly about substances which lay beyond

---

45. Scheiner, *Letters on Solar Spots, Opere,* V, p. 30: "Aut sint corpora per se existentia, solida et opaca, et hoc ipso erunt stellae, non minus atque Luna et Venus, quae ex aversa a Sole parte nigrae apparent".

the domain of astronomers, and within that of philosophers. Scheiner, however, was far from defying the boundaries between the two disciplines imposed upon the system. He took care to neutralize the effect of his own transgression by insisting that astronomers talk differently about substances from philosophers, thus leaving the latters' discursive space free from intrusion, and their authority untouched. The question he asked about sunspots: "What are they?" was subverted by him into a question of another kind, namely "What do they signify?" Instead of proceeding from effects (spots) to causes (stars) as a philosopher would do, he interpreted the effect as a sign pointing to, or indicating a phenomenon, which was then woven into a network of other phenomena (the motion of Venus and Mercury, conjunction with the sun, the order of the planets' orbits, etc.) all within the legitimate domain of astronomers. Finally, Scheiner never let his reader forget that real knowledge could only stem from an agreement between philosophers and astronomers, confirming, in fact, the superior position of the philosophers in the disciplinary hierarchy. Until the philosophers' quest for proof was satisfied, no knowledge could be truly legitimized, no "fact" could be recognized as certain:

> It occurs to me that between the sun and Mercury and Venus, in due distance and proportion, there revolve very many wandering stars, of which only those which run into the sun in its motion become known to us. If it so happens – and I have not yet completely abandoned the idea – that we could contemplate only those stars which are close to the sun, only then could this dispute be solved.[46]

The nature of sunspots, then, was not a matter that could be determined yet, either by an individual author, or even by the community of astronomers. It was still left to negotiations between astronomers and philosophers. Scheiner came back to this issue in the *Disquisition*,[47] again referring to it as a case "sub judice", not to be excluded from astronomers' debates, but not to be fully appropriated by them either.

Scheiner's texts reflected institutional constraints even at the moment of transgression. The distinction between mathematics and philosophy in terms of their different methodologies and different themes was maintained even while the methods and objects within each domain were

---

46. Ibid., p. 31: "Subit opinari, a Sole usque ad Mercurium et Venerem, in distantia et proportione debita, versari errones quamplurimos, e quibus nobis soli ii innotescant, qui Solem motu suo incurrant. Si fieri posset, de quo necdum penitus desperavi, ut stellas etiam Soli propinquas contemplaremur, *lis* haec tota decideretur". (My italics, R. F.)

47. Scheiner, *Disquisition, Opere*, V, 65: "De istis vero duobus, corpora haec tenuia esse, at permanentia sive stellas non esse, stronomi certant, et adhuc *sub iudice lis est*". (My italics, R.F.)

undergoing change. Astronomy was a "mixed science" subalternated to mathematics, not wholly autonomous, and unable to objectify its phenomena independently. This could only be done in cooperation with philosophers, charged with the responsibility of determining substances and their natures, through complex negotiations by which truth would finally shine forth, clad in the form of consensus.

The boundaries reproduced in Scheiner's texts provoked Galileo's most severe criticisms of the Jesuit, even though his transgression did not pass entirely unnoticed. In fact, Galileo used Scheiner's assertion of the unpassable limit – the consensus among philosophers and mathematicians – in order to declare the necessity of its destruction.

Galileo first attacked the accepted boundary between astronomy and natural philosophy by remarking upon Scheiner's philosophical interests:

> Yet I seem to see in Apelles a free and not a servile mind. He is quite capable of understanding true doctrines; for, led by the force of so many novelties, he has begun to lend his ear and his assent to *good* and *true* philosophy.[48]

Scheiner was thus praised for his transgression and encouraged to enter into the sphere of the philosophers. Galileo then strove to transform this transgression into a revolutionary act. In this spirit he made his famous distinction between *mathematical astronomers* who utilized fictive inventions like eccentrics, deferents, equants, and epicycles in order to facilitate their calculations, and *philosophical astronomers* who went beyond the attempt at "saving the appearances" and investigated the true nature of the universe. For the first time Galileo defined here the role of the new astronomer in terms of a vocation embodying a set of noble values:

> For such a constitution exists: and it is unique, true, real, and could not possibly be otherwise; and the *greatness* and *nobility* of this problem entitle it to be placed foremost among all questions capable of theoretical solution.[49]

It was only after reading Scheiner's second series of texts that Galileo realized to what extent the Jesuit was confined within the traditional limits of his profession. Galileo's response was to ridicule philosophers ("it is vain to run to such men asking for support"[50]) and delegitimize *traditional* mathematicians dealing with astronomical problems:

> As to the mathematicians, I do not know that any of them have ever discussed the hardness and immutability of the sun, or even that mathematical science is adequate for proving such properties.[51]

48. Galilei, *Letters on Sunspots,* p. 96. (My italics, R.F.)
49. Ibid., p. 97. (My italics, R.F.)
50. Ibid., p. 135.
51. Ibid.

No less vehement was his rejection of the consensus among philosophers and mathematicians:

> Any adversary would resolutely reject . . . the proof Apelles adduces for it, which is that such is the prevailing opinion (according to him) among philosophers and mathematicians.[52]

Galileo's and Scheiner's texts reflect the contradictory interests involved in the boundaries between astronomy as a mixed science subalternated to mathematics, and natural philosophy as a higher discipline aspiring to demonstrated truth. A few letters concerned with solar observations supply additional historical evidence of the Jesuits' consistent attempts to protect the disciplinary structure institutionalized in the *Ratio*. A letter of Welser, describing Clavius's influential role in bringing about his own acceptance of Galileo's discovery of the moons of Jupiter and the phases of Venus, represented an endorsement of these discoveries as a common interest of mathematicians but excluded the possibility of using them as a necessary argument for the centrality of the sun.[53]

The famous letter to Bellarmine, written by the four Jesuit mathematicians of the Roman College, again made a distinction between observation of new objects and new phenomena relating to old objects (such as the mountains of the moon), which they accepted, and discussion of the *nature* or the essence of those objects, which they rejected.[54] Paulo Gualdo, echoing the opinions of many Jesuit mathematicians whom he used to meet regularly, suggested a distinction between observations and the description of them which he deemed a legitimate task for mathematicians, and discussion of the motion of the earth, requiring inferences and causal explanations, for which, he claimed, the time was not yet ripe.[55]

---

52. Ibid., p. 134.
53. Welser to Galileo, *Opere*, XI, p. 52: "Il R. P. Clavio mi scrisse ultimamente, confessando con molto candore che'gli era stato duro et renitente a creder questi miracoli, ma che finalmente, con un buon istromento pervenutogli, si era chiarito talmente a vista d'occhio, che non gli ne restava dubbio alcuno. *Et così dovranno fare poco a poco tutti gli maggiori della professione.* [My italics, R.F.] . . . Se bene non comprendo ancora come se ne inferisca indubitamente la centricità, per così di[re], de'l sole".
54. The mathematicians to Bellarmine, p. 93: "Non si può negare la grande inequalità della luna; ma infin hora noi non habbiamo intorno a questo tanta certezza che lo possiamo affermare indubitamente".
55. P. Gualdo to G. Galileo, *Opere*, XI, p. 100–101: "Che la terra giri, sinhora non ho trovato nè filosofo nè astrologo che si voglia sottoscrivere all'opinione di V.S. . . . A me par che gloria s'habbia acquistata con l'osservanza nella luna, ne i quattro Pianeti, e cose simili, senza pigliar a diffender cosa tanto contraria all'intelligenza e capacità de gli huomini".

The boundaries implied in all these letters, as well as in Scheiner's texts, were imposed by institutional requirements. Galileo's position, on the other hand, was legitimized by his official status as court mathematician and philosopher to Cosimo II. The need to justify the title of philosopher forced Galileo to emphasize the philosophical implications of his work. While Scheiner needed to stress the distinction between mathematics and physics by recognizing the preeminence of the philosophers, Galileo was bound to blur it, explicitly as well as implicitly. He did this explicitly by dividing astronomers into two groups, mathematical astronomers and philosophical astronomers, thus facing Scheiner with an impossible choice; and he did it implicitly by insisting on a discussion of sunspots and their analogy with clouds, or, in other words, with earthly materials.

Galileo's distinction between mathematical and philosophical astronomers left the Jesuit mathematicians with no status within that classification: for mathematical astronomers according to that distinction, only "supposed" the objects of their discourse for purposes of calculations. They had no claims upon reality whatsoever. That, of course, did not apply to Scheiner and the other "Clavisti". Philosophical astronomers, on the other hand, were bound to investigate the structure of the universe. There could be no clearer challenge not only to the boundaries between philosophy and mathematics, but also to the hierarchy between them than this declaration of Galileo, in which he literally dethroned the philosophers from their position, claiming to be their legitimate heir. Although in fact it did no more than dress Clavius's program in the rhetoric of revolution, no Jesuit mathematician could have dreamt of identifying with such an assertion. Thus, it was Galileo who excluded the Jesuits from the cultural field which they both shared, long before they excluded him from the place he aspired to occupy with their cooperation.

Authority and Authorization in Scientific Discourse

The representation of solar phenomena in Scheiner's and Galileo's texts necessarily involved self-representation, testifying to the strategies by which each attempted to establish his own authority.

Galileo's clear differentiation between mathematical astronomy and philosophical astronomy forced him to define his own position: "My opinion", he wrote, "lies midway between that of astronomers . . . and that of philosophers."[56] Thus, he attempted to create a place for the astronomer-philosopher not only by undermining the authority of the Ar-

56. Galilei, *Letters on Sunspots*, p. 97

istotelian philosophers, but also by challenging that of the traditional mathematicians. Establishing his own authority, however, also meant that he had to cope with the canons of proof developed by his interlocutors (the Jesuits) and to modify them in accordance with the needs of his own context.

The story of Galileo's defiance of the Aristotelian philosophers is well known, and hardly needs elaboration here. What does require emphasis, however, is the critical attitude he shared with his Jesuit rival towards the tradition. This common ground was, in fact, exactly the area where the dialogue between Galileo and the Jesuits took place.

Galileo accused the philosophers of preferring Aristotle's erroneous beliefs to conclusions based upon observations and experience:

> It appears to me not entirely philosophical to cling to conclusions once they have been discovered to be manifestly false. . . . People like this, it seems to me, give us reason to suspect that they have not so much plumbed the profundity of the Peripatetic arguments as they have conserved the imperious authority of Aristotle. . . .[57] They will philosophize better who give assent to propositions that depend upon manifest observations, than they who persist in opinions repugnant to the senses.[58]

Instead of looking at the book of nature, Galileo claimed, the philosophers kept looking in the Aristotelian texts: "They wish never to raise their eyes from those pages – as if this great book of the universe had been written to be read by nobody but Aristotle."[59] The evidence derived from nature was rejected by them in the name of opinions drawn from authoritative sources: "It is easier to consult indexes and look up texts than to investigate conclusions and form new and conclusive proofs".[60]

The mathematicians were not spared either. Galileo had not been content to stay within the confines of his profession since he had acquired his first chair as mathematician at the university of Pisa, aided by Clavius's intervention. For, in the newly organized cultural field, mathematicians had an honourable position but strictly limited authority. Indeed, the Jesuit mathematicians themselves had successfully defied the traditional role of the mathematician as confined to "saving the phenomena" by means of fictive mathematical constructions. It was, after all, the mathematicians – Jesuit mathematicians – who confirmed the "reality" of the phenomena discovered by Galileo and the "truth" of his observations. Usually, however, Jesuit mathematicians stopped short of considering the

57. Ibid., pp. 142–143.
58. Ibid., p. 118.
59. Ibid., pp. 126–127.
60. Ibid., p. 142.

implications of the new discoveries. Whether the mountains of the moon proved the imperfect form of the moon's sphere was a question they avoided.[61] Welser's insistence that he did not understand how the centrality of the sun could be deduced from the phases of Venus probably reflected their reluctance to investigate the major question of the system of the world.[62]

Galileo's words in the third letter, questioning the authority of mathematicians to decide upon the nature of sunspots, were not so much a challenge but rather a reflection of the formal status of mathematicians in the context of Jesuit culture:

> As to the mathematicians. I do not know that any of them have ever discussed the hardness and immutability of the sun, or even that mathematical science is adequate for proving such properties.[63]

By speaking of the restricted authority of the mathematicians, Galileo indicated his acquaintance with the constraints imposed on the field by the Jesuits, while hinting at his own much broader claim to mathematical–philosophical knowledge.

Christopher Scheiner could not easily be classified as either a philosopher or a mathematician in Galileo's terms. Not only did he practice observations, using techniques very similar to those described by Galileo, but he also expressed his preference for the "truth" of observations to the traditional "beliefs" of astronomers. Furthermore, he affirmed the "reality" of the newly discovered phenomena, and offered his interpretation of the spots, denying their cloudy nature and treating them as traces of real stars.[64]

Both Galileo and Scheiner transcended the methods of philosophy and the "mixed" mathematical sciences, and crossed their traditional boundaries. The texts of both betray an effort to interpret nature. Each, however, claimed a different status for his interpretation, a claim that was connected to the strategies by which each attempted to establish his authority.

61. See above, Ch. 12.
62. Welser to Galileo, *Opere*, XI, p. 52: "Mons. or Arciprete di Padova mi avisò l'osservatione di V.S. della stella Venere soli quindeci giorni sono: mi parve cosa tanto vaga et curiosa, che nulla più; se bene non comprendo ancora come se ne inferisca indubitamente la centricità, per così di[re], del sole". See also Cigoli's letter to Galileo, where he reports of Grienberger's insistence on arguing hypothetically, *Opere*, XI, p. 319: "Però dice il Sig.r Marchese, il Padre Ganberghiere che non vorrebbe in queste sue oppinioni andasse così a un tratto dichiarandosi, ma *per via di disputa* dicesse lo istesso". (My italics, R.F.)
63. Galilei, *Letters on Sunspots*, p. 135.
64. See above.

Galileo's hermeneutical move was most dramatically expressed in his favourite image of the "book of nature", "the great book of the universe". Reading a book presupposed a system of signs (a language) which stood between a phenomenon and an observer. In Galileo's *Assayer*, published ten years later, this idea was much more explicitly stated:

> Philosophy is written in this grand book, the universe, which stands continually open to our gaze. But the book cannot be understood unless one first learns to comprehend the language and read the letters in which it is composed.[65]

But the idea also permeated Galileo's earlier thinking, though in a less crystallized form. The letter to Grienberger of 1611 contained it in an embryonic form. There Galileo already showed an awareness that there was always some "mediation" between nature and the knowing mind, a mediation performed in and through language:

> How, then, do we know that the moon is mountainous? We know it not simply by the senses, but by joining and linking up discourse with observations and sensate appearance.[66]

Galileo, however, was not, and could not be content with mere interpretation, which he associated with the reading of authoritative texts, but not with that of the book of nature. Rejecting the interpreter's role, he put a great deal of effort into eliminating the role of interpretation from the investigation of nature. The voice of the interpreter had to be disguised, and replaced by the authoritative voice of nature itself, speaking through the text of the mathematical philosopher. Mathematical symbols were conceived not so much as *representing* essential features of phenomena, but rather as a *code* through which nature became accessible to the human mind. The mathematical philosopher had an active but strictly defined role within the system. Galileo's scientist played the role of a code breaker. The mediation between nature and the knowing mind was performed according to rules as precise as those of any mathematical code.

Before analysing the strategies by which Galileo eliminated interpretation from his discourse, it is necessary to examine Scheiner's conception of the astronomer's role.

Scheiner did not agree with merely "saving the phenomena" any more than Galileo did. In the very opening of his first letter, he alluded to the quest for truth and reality underlying his enterprise:

> The fact that these appearances [sunspots] due to their great power, can be discovered all over the sun, the source of light and the leader

---

65. G. Galilei, *The Assayer*, trans. by S. Drake, in *Discoveries and Opinions of Galileo*, pp. 237–238.
66. Galileo to Grienberger, p. 183.

of the stars, is something we are completely convinced of in the most clear light of truth.[67] Throughout the letters and the *Disquisition* Scheiner used a "rhetoric of truth" ("I suffered indeed from the fact that such an excellent occasion for the investigation of the truth . . . would escape me"[68]) expressing his confidence in the ability of the astronomer to establish some hypotheses as true and eliminate others: "With the help of God, I shall explore, in due time and without fail, which of these two hypotheses is true".[69]

As mentioned above, Galileo himself recognized Scheiner's contribution to the truth and reality of astronomical knowledge. Scheiner was praised for establishing the reality of sunspots ("they are real objects and not mere appearances or illusions of the eye or of the lenses of the telescope, as Your Excellency's friend well establishes in his first letter",[70] and for being capable of understanding "true doctrines".[71]

However, Scheiner's discourse of "truth", not unlike Galileo's, was based on the interpretation of phenomena. Indeed, the metaphor of the "book of nature" did not form part of Scheiner's vocabulary. Nonetheless, the need for interpretation stemmed from a new assumption about the investigation of celestial appearances which was common to both. Following a long tradition of astronomical science, both Galileo and Scheiner were well aware that what was visible to the eyes of the observer could not be identified with what really was. Thus, the gap between appearances and reality called for interpretation.

Traditional astronomy avoided interpretation by limiting the role of practitioners to the formulation of hypotheses through which visible appearances could be "saved". The realm of the visible (retrogradation of the planets, for example) did not provide evidence of what really was, but rather called for the formulation of hypotheses (epicycles, eccentrics) whose relation to what really was remained unclear. The rejection by both Scheiner and Galileo of the traditional role of the astronomer stemmed primarily from the belief that although a gap between the visible and the real existed, the visible nevertheless provided evidence for the real. Using the terminology developed by Ian Hacking in *The Emergence of Proba-*

67. Scheiner, *Letters on Solar Spots, Opere,* V, p. 25.
68. Ibid., p. 28: "Dolebam enim mihi eripi tam paratam occasionem *veri inquirendi*". (My italics, R.F.)
69. Ibid., p. 26: "Quorum utrum verum sit, suo tempore utique, Deo iuvante, patefaciam."
70. Galilei, *Letters on Sunspots,* pp. 90–91.
71. Ibid., p. 96.

bility,[72] one may suggest that the old astronomical discourse did not consider appearances as providing "internal evidence" or "the evidence of things" for what really was.[73] According to Hacking, arguing that appearances were "signs" of real entities – "pointing beyond" themselves to something which really existed[74] – was not a legitimate way of explaining nature before approximately the end of the sixteenth century. Hacking said that "pointing beyond" – a nondemonstrative method of reasoning[75] – was finally legitimized around the middle of the seventeenth century. Traces of that process, however, can already be found in the controversy about sunspots. Scheiner, like Galileo, was actually engaged in the interpretation of the heavens – which distinguished him from traditional astronomers – even though he never adopted the vocabulary of "the book of nature".

Following Hacking, I shall use L. Austin's distinction between actually seeing and the collection of evidence as a point of departure for explaining Scheiner's interpretive activity:

> The situation in which I would properly be said to have *evidence* for the statement that some animal is a pig is that, for example, in which the beast itself is not actually on view, but I can see plenty of pig-like marks on the ground outside its retreat. If I find a few buckets of pig food, that's a bit more evidence, and the noises and smell may provide better evidence still. But if the animal then emerges and stands there plainly in view, there is no longer any question of collecting evidence; its coming into view doesn't provide me with more *evidence* that it's a pig, I can now just *see* that it is.[76]

In Scheiner's discourse, as well as in Galileo's, sunspots were treated as providing evidence for something else beyond them, neither directly accessible to the senses nor given to demonstration. Scheiner expressed great enthusiasm about the spots: "Indeed, the door is wide open through which we may freely enter in order to observe the greatness of the sun",[77] he wrote, acknowledging a fact "hitherto questioned or not known or even perhaps denied altogether by the astronomers".[78] Nevertheless, re-

72. I. Hacking, *The Emergence of Probability,* Cambridge, 1975.
73. Ibid., pp. 32–33.
74. Ibid., pp. 34–38.
75. Ibid.
76. Ibid., p. 32, quoted from J. L. Austin, *Sense and Sensibilia,* Oxford 1962, p. 115.
77. Scheiner, *Letters on Solar Spots, Opere,* V, p. 26: "Et vero apertissima est ianua, qua ad Solis quantitatem intuendam liberrime ingrediamur".
78. Ibid., p. 25: "Hactenus astronomis aut dubitata aut ignorata aut etiam fortassis pernegata".

luctant to admit the corruptibility of the heavens he was anxious to show that the spots were not on the sun: "To acknowledge spots on the sun, that very bright body . . . has always been understood by me as inappropriate".[79] Hence he chose to treat the spots as the shadows of stars projected on the sun. The spots, then, were traces of stars, visible marks pointing to celestial bodies which were not yet visible but were certainly real. One enigmatic passage written as a postscript to the third letter seems to confirm my view that Scheiner was aware of doing something different from the astronomers before him: "In all disciplines the road to knowledge remains mysterious and the inventions have to be considered as a minimal part of what needs to be invented".[80] Moreover, he explicitly chose to treat celestial phenomena as signs whose decipherment must lead to truth: "The sun will also give signs; who hears the sun say a falsehood?"[81]

While both Galileo and Scheiner were searching for new ways of bridging the gap between appearances and reality and were beginning to treat celestial phenomena as signs providing "internal evidence", they were also reluctant to ignore the traditional canons of demonstrative truth and replace them with the truth of interpretation. Both attempted to incorporate the "evidence from things" within the commonly accepted form of knowledge called "scientia". However, the status of the "evidence from things" in Galileo's discourse was very different from that it had in Scheiner's. Hence, Galileo accorded a different authority to the astronomer–philosopher than did Scheiner to either the "old" or the "new" mathematician.

In his second letter, Galileo proved the contiguity of the spots with the sun and their rotation with it by pointing out the compatibility between his hypothesis and a series of appearances concerning the changes in the spots' breadths and speeds of motion and the distances between them at the edges of the sun and at its center. The appearances, Galileo concluded, confirmed his hypothesis since all of them could be explained by the motion of points on a rotating sphere. Galileo, however, was aware that he was applying the standard procedures of traditional astronomy and that therefore his hypothesis could only aspire to the status of "saving the phenomena". In this context the hypothesis could be regarded as possible, and even probable, but not absolutely true. Galileo, however, opted for more:

79. Ibid., p. 26: "In Sole, corpore lucidissimo, statuere maculas, easque nigriores . . . mihi inconveniens semper est visum".
80. Ibid., p. 32: "In omnibus disciplinis ingens via restat, et inveniendorum minima pars censeri debent inventa".
81. Ibid.: "Sol quoque signa dabit: Solem quis dicere falsum / Audeat?"

And just as all the phenomena in these observations agree exactly with the spots' being contiguous to the surface of the sun, and with this surface being spherical rather than any other shape, and with their being carried around by the rotation of the sun itself, so the same phenomena are opposed to every other theory that may be proposed to explain them.[82]

Lacking any general principle from which the motion of the spots on the sun and the corruptibility of the heaven could be deduced, the only strategy for claiming the status of scientifically proven knowledge for his conclusions was to eliminate all other possible hypotheses. This was exactly the strategy Galileo decided on. In most things he followed in the footsteps of Scheiner, eliminating some of the possibilities rejected by the Jesuit as well (e.g., could they be in the air? Could they be in the orbit of the moon? etc.) although not always for the same reasons. After a few eliminations of this kind, including the rejection of Scheiner's hypothesis, Galileo concluded: "It would be a waste of time to attack every other conceivable theory".[83]

From the logical point of view, Galileo's effort to gain the status of scientific truth for his hypothesis was doomed to failure, however. Logically speaking, there could always be new hypotheses that would "save the appearances" better than his own. Hence, his assertion that it would be a waste of time to attack all other theories was an ambiguous rhetorical gesture. While all possible hypotheses had supposedly been eliminated and the true one allowed to emerge and be scientifically established, it in fact revealed the inaptitude of this method in the study of nature. It had now become clear that there could be no legitimate transition from the status of a probable hypotheses to the status of a scientifically established truth through an accumulation of empirical evidence. At this stage the expectation that "mathematical astronomy"[84] would become "philosophical astronomy" through more "evidence of things" had to be modified. The rest of the second letter represents a shift in Galileo's strategies. It contains the first enunciation of the principle of inertia, testifying to Galileo's recognition that only by discovering a general principle of motion from which the rotation of the sun would be deduced could the appearances be established as scientific facts.

Now, Galileo's principle of "indifferent" (i.e., inertial) motion, as he himself admitted, was based upon observation and analysis of the motion of heavy bodies on the earth. Also, the example of a ship "having once

82. Galilei, *Letters on Sunspots*, p. 109.
83. Ibid., p. 111.
84. Ibid., p. 97.

received some impetus through the tranquil sea, would move continually around our globe without ever stopping"[85] was invoked in order to legitimate the sun's rotation in terms of an analogy to an earthly body moving on a spherical surface. In the absence of a clear concept of force and gravitation, however, Galileo could not yet establish inertia as the fundamental principle of mechanics. Thus, the validity of his argument about the rotation of the sun became wholly dependent upon the analogy he claimed between the "indifferent" motion of earthly physical bodies and the motion of celestial bodies. This analogy, however, was based on nothing but an act of interpretation, his reading of spots in terms of clouds, and treatment of the relationship between the spots and the sun as being similar to that of the relationship between the clouds and the earth. While the argument from the inertia of earthly bodies was used to support his claim about the rotating sun, the idea of a rotating sun supported the implicit argument about the motion of the earth. Paradoxically, Galileo's attempt to step out of the framework of the "mixed mathematical sciences" towards the formulation of a general principle of motion from which the motion of sunspots could be deduced led him to base his reasoning on an analogy inspired by rhetoric rather than based on "scientific" proof.

The shift from "proof" to "analogy" had far-reaching implications for the authority of the scientist in Galileo's discourse. A proper "demonstration" was in no need of any special authority in order to be accepted as "truth". A "correct analogy", on the other hand, could have no claim to truth apart from the speaker's authority. Thus, Galileo was forced to resort to some radical strategies in order to justify his position as the discoverer of the "true analogy".

From his first letter Galileo represented himself as an individual standing alone against a crowd of persecutors:

> Even the most trivial error is charged to me as a capital fault by the enemies of innovation, making it seem better to remain with the herd in error than to stand alone in reasoning correctly.[86]

At this early stage of his Copernican campaign, this self-representation was quite remote from reality. Galileo had just been cordially received by the mathematicians of the Roman College, who confirmed most of his discoveries.[87] Welser had crowned him as a leader:

> You have led in scaling the walls, and have brought back the awarded crown ... I hope you will be pleased to see that on this

85. Ibid., pp. 113–114.
86. Ibid., p. 90.
87. See above, Ch. 12.

side of the mountains also men are not lacking who travel in your footsteps.[88]

The Linceans were all at his side, expressing the absolute loyalty demanded from them by Prince Cesi: "Every one of us will always write on your behalf" he reassured Galileo in a letter of March 1612.[89] A large number of letters from other sympathizers, the most prominent among them being Cardinal Barberini,[90] Cardinal Francesco di Joyeuse,[91] and the Aristotelian philosopher Lagalla,[92] show that he was not as isolated as he pretended to be.

Even more problematic was Galileo's claim to be right on account of his "correct reasoning". As argued above, Galileo was particularly eager to "prove" his theories, and was not content with the status of "probability" they could secure for themselves in the framework of "the mixed sciences". Only a proof could justify his discourse as philosophical, and not just mathematical, astronomy. His deduction of the motion of sunspots from a general principle of motion – the only way to legitimate his hypothesis scientifically – depended, however, on the implicit analogy between the motion of earthly bodies – to which "indifferent" (inertial) motion was correctly attributed – and celestial bodies, whose laws of motion were as yet completely obscure. The analogy he suggested was a choice arrived at in the context of the new paradigm suggested by Copernicus, but without much knowledge of a physics of the heavens. It did not stem only from correct reasoning.

Galileo's words, however, should not be taken at face value. Rather than reflecting a cultural reality, they expressed a discursive need to legitimize the individual vis-à-vis the consensus. The dichotomy between the righteous individual and the ignorant masses never existed. The opposers were neither ignorant nor wholly against innovation. Rather, it was a fabricated dichotomy, providing the basis for the Galilean myth which was invented in the text of its protagonist. The myth had a function, however. It was meant as an answer to Scheiner, who did not deny the interpretive character of his statements and tended to base the authority of the astronomer on the consensus of the community of astronomers and philosophers. Against that strategy Galileo posited the authority of the individual

88. Galilei, *Letters on Sunspots*, first letter from Mark Welser to Galileo Galilei, p. 89.
89. Cesi to Galileo, *Opere*, XI, p. 283: "Che ciascuno di noi scrivera sempre per lei".
90. Maffeo Barberini to Galileo, ibid., pp. 317–318; 325.
91. Francesco di Joyeuse to Galileo, ibid., p. 373.
92. Giulio Cesare Lagalla to Galileo, ibid., pp. 357–359.

philosopher–astronomer over and above the consensus of the community, and emphasized the superior moral value of his position. The most succinct formulation of the authority of the scientist in Galileo's discourse appeared in the context of his reply to the *Disquisition* (i.e., in the third letter to Welser). Scheiner summarized three arguments in support of Venus's motion around the sun: the consensus among philosophers and mathematicians, the analogy between the expected spots of Venus and other spots (i.e., those seen by Kepler and described by him as the shades of Mercury), and the evidence of things, which he called "experience", and of which the most outstanding example was Galileo's account of the phases of Venus.[93] Here Scheiner explicitly stated what had already been clear from his previous three letters: namely, that the source of authority in his discourse was the consensus of the community of observers, practitioners, mathematical astronomers, and philosophers.[94]

Scheiner's "sociological" insight into the dynamics of scientific communities did not mean that in Jesuit culture there were no canons of proof, including coherent criteria of "internal evidence". Neither did it mean that canonic texts were the only authoritative source of knowledge. Rather, it indicated the high degree of institutionalization which characterized Jesuit science. In that context, scientific activity became collective activity. The collective dimension of his enterprise found its most emphatic expression in the language used in the letters. There, all reports of observations were related in the first person plural: "They (the appearances) provided not only me myself, but also my friends, at first with great astonishment, and thereafter with great delight of the soul",[95] he wrote.

> I myself and one of my friends directed together the optical tube towards the sun. . . . We returned however to this matter last October, and again we perceived those spots. . . . we have consulted the eyes of the most various people.[96]

Also, the conclusions were presented as the result of collective reasoning:

> We have concluded therefore that the imperfection was not in the eyes . . . it was unanimously and correctly concluded by very many

93.   Scheiner, *Disquisition, Opere*, V, p. 46.
94.   See above, pp. 275–276.
95.   Scheiner, *Letters on Solar Spots, Opere*, V, p. 25: "Ea ingentem, non solum mihi, sed et amicis, primum admirationem, deinde etiam animi voluptatem, pepererunt".
96.   Ibid.: "Ego unaque mecum amicus quidam meus tubum opticum . . . in. Solem direximus . . . Redivimus ergo ad hoc negotium mense praeterito Octobri, reperimusque in Sole apparentes maculas . . . Itaque adhibuimus diversissimorum oculos . . ."

people, that these appearances could by no means be situated in the
sun because of a defect of the eye ... and we concluded together
and correctly that in this matter the tube was justly to be cleared of
any blame.[97]

Compared to Scheiner's collective persona, Galileo's individual "I" was
indeed remarkable. True, Galileo too invoked witnesses to confirm his
observations. But even the observations he had performed in front of an
audience were arranged as a scene in which Galileo showed the others
what they should see: "I have observed them for about eighteen months",
he wrote, "having shown them to various friends of mine ...".[98] From
there on it was the all pervading "I" of the writer who sees by himself, "believes", "reasons", "holds", "questions", etc. Even when Galileo
promised to describe a method (discovered "by a pupil of mine"[99]) of
drawing the spots with complete accuracy, he easily slipped from the im-
perative mood into a discursive "I". Hence he began with "Direct the
telescope" and after just three sentences he continued with: "In order to
picture them accurately I first describe ... I find the exact place ... I have
drawn",[100] etc.

In order to establish the authority of an interpreter of natural phenomena, of the scientist who could offer the only true analogy (in this case,
between the physics of the heavens and the physics of celestial bodies)
Galileo resorted to many rhetorical strategies. He observed nature alone
and related his own observations and those of others in the first person.
He created an image of himself as a persecuted individual, and used this
image to suggest his moral superiority. For he claimed that the individual
did not gain any advantage by breaking away from the community.
Rather, setting himself against the "herd" had brought him more vexa-
tions than privileges:

> And I indeed must be more cautious and circumspect than most other
> people in pronouncing upon anything new.
>
> As Your Excellency well knows, certain recent discoveries that de-
> part from common and popular opinions have been noisily denied
> and impugned, obliging me to hide in silence every new idea of mine
> until I have more than proved it.[101]

---

97. Ibid., pp. 25–26: "*Conclusimus* ergo, vitium in oculis non esse ... Oculi
    ergo errore haec in Solem introduci neutiquam posse, *unanimiter a quam-
    plurimus, et recte, est conclusum ... Unde recte pariter conclusimus,* tu-
    bum hac in re omni culpa merito vacare". (My italics, R.F.)
98. Galilei, *Letters on Sunspots*, p. 91.
99. Ibid., p. 115.
100. Ibid.
101. Ibid., p. 90.

Sometimes he acted as an instructor, teaching others the language of nature by telling them how to represent it. At other times he even appropriated nature's own voice to admonish nonauthorized observers that: "Nature, deaf to our entreaties, will not alter or change the course of her effects".[102] No wonder that towards the end of the third letter he arrived at the succinct formulation of the authority of the scientist alluded to above: "For in the sciences the authority of thousands of opinions is not worth as much as one tiny spark of reason in an individual man".[103] This individual, however, was somewhat depersonalized. His reading of natural phenomena was not subjective, neither was it arbitrary. Rather, he was completely tied to nature's own language, which he was obliged to learn before offering his interpretations. And the language, through God's providence, was accessible to anybody ready to acquire it:

> We must recognize divine Providence, in that the means to such knowledge are very easy and may be speedily apprehended. Anyone is capable of procuring drawings made in distant places, and comparing them with those he has made himself on the same days.[104]

The truth Scheiner was speaking about, was the truth of the interpreter of nature. Scheiner did not attempt to disguise or replace interpretation. He contented himself with interpreting "observables" (sunspots) as testifying to the true existence of "unobservables" (tiny stars or some other bodies of heavenly materials) whose precise nature was yet unclear. Thus, in spite of the incorporation of new objects and the application of new methods, he left the structure of his discourse intact. The role of the new astronomer consisted in the true interpretation of the heavens, a stage on the way to "scientific truth". A space was left for the philosophers to apply their own methods and make their own contribution to the debate. "Scientific truth" could only be attained by a consensus among mathematicians and philosophers. The source of authority, in Scheiner's discourse, was the community of interpreters.

In Galileo's discourse the individual, detached from his community but uniquely capable of creating the conditions of enunciation, visibility, and maneuverability in which nature is decoded, was the source of authority. Making unobservables – the motion of the earth, inertial motion – into major objects of explanation, resulted in a radical change in the accepted rules of discourse and the organization of the intellectual map. The inter-

102. Ibid., p. 136.
103. Ibid., p. 134.
104. Ibid., p. 119.

pretation of visibles could not stop short in front of invisibles, whose nature remained unclear. Rather, it became necessary to subordinate the visible to the invisible so that the gap between them was seemingly effaced. In the course of this process, interpretation itself became seemingly superfluous, replaced by the pretence of the human mind to dominate the language of nature. The members of the new community were not interpreters, using their own language to "understand" nature, but rather scientists capable of speaking the language of nature.

Epilogue

The narrative of Galileo and the church, reconstructed in this book, begins by reading the contradictory documents of Galileo's trials as echoing tensions and contradictions in the cultural realities of the Counter-Reformation era. My reading suggests that the different interpretations of the theologians' ban of Copernicanism and of the possible limits of Galileo's campaign expressed two cultural orientations of two rival intellectual elites within the church – the Dominicans and the Jesuits – who attempted to implement the decrees of the Council of Trent and were engaged in a struggle over cultural hegemony.

The traditional elite of the Dominicans remained committed to medieval values throughout the period. In their world the bridge between the transcendental and the mundane was still anchored in the "vita contemplativa". Hence they endorsed a way of life suitable for a small and isolated group, putting emphasis on speculative knowledge as a preparation for the contemplative life. Striving to safeguard the Thomistic synthesis they imposed strict adherence to Thomistic theology and prohibited, under severe penalty, any deviation from Thomistic philosophy. In the second half of the sixteenth century their Thomism became "doctrinaire".

The alternative cultural orientation suggested by the Jesuits posited the "vita activa" at the center of their world view. Education of ever-growing circles of lay people entailed involvement in the world, and was justified in practical terms such as preparation for ecclesiastical and lay careers, and the development of intellectual capacities in general. However, the educational enterprise was also conceived as a kind of apostolic activity helping to achieve salvation. The scope of studies practiced by the Jesuits tended to include less speculative fields – like rhetoric and the "mixed sciences" – and to embrace non-Thomistic opinions. Their Thomism was more "programmatic" than "doctrinaire".

Towards the end of the sixteenth century Jesuit practices in the field of education seemed to endorse new forms of knowledge which threatened to undermine the Thomistic principles of the organisation of culture. Jesuit mathematicians and astronomers under the leadership of Christopher

Clavius claimed the status of *scientia* (i.e., *cognitio certa per causas* ["a certain cognition through causes"]) for their field of studies, regardless of the objections of philosophers, based upon Thomistic arguments, that the abstract objects of mathematics could not indicate causes, and that demonstrations *ex suppositione* did not carry complete conviction. The idea that there could be true knowledge of hypothetical objects infiltrated into the theological discourse, legitimizing the notion of God's *scientia media*, his absolute knowledge of man's future ("hypothetical") acts, not yet predestined by his will.

The Jesuits' alternative theology, which was not unconnected to the new forms of knowledge they practiced, could not fail to raise the deepest objections of traditionalists, especially the Dominicans. Jesuit theological ideas detracted from the omnipotence of God, contended the "pure" Thomists. Obviously they were based on completely wrong philosophical notions. The violation of the traditional canons of knowledge could not but stem from the subversion of Thomism, due to the Jesuits' involvement in the variegated scientific culture of the time. The reaction of the Dominicans was not limited to censure of Jesuit theology. The old "guardians of the faith" were determined to uproot evil from the ground and launched their attack against the educational programme of the Jesuits by submitting the early versions of the *Ratio studiorum* to the Inquisition.

In response the Jesuits resorted to the legitimation they could find in Thomism. The last version of the *Ratio* (1599) is a surprisingly conservative text. In contrast to Clavius's attempt to modify the Thomistic principles of the organisation of knowledge and change the hierarchy of the disciplines, the *Ratio* reasserted a clear demarcation between physics and mathematics, metaphysics and theology, and used these demarcations to control the intellectuals of the society from within, and to gain legitimation from without.

The publication of the *Sidereus nuncius* (1609) signaled the emergence of a new scientific discourse which applied new strategies in the constitution of its objects, and attempted to redraw the boundaries of the cultural field and to establish its authority by new means. My preliminary investigations of the Dominicans' cultural orientation indicate that their "doctrinarian" Thomism, paradoxically imbued with sceptical undertones, could have led to the theologians' rejection of Copernicanism in 1616, and to the pope's arguments against its defence later on, up to the trial of 1633. The Jesuits' reaction, however, was much more complex, and endangered their delicate position within the church establishment.

Recent work on Jesuit scientific practices has drawn attention to the emergence of an amorphous, relatively undefined discourse of the Jesuit mathematicians, forming part of the tradition of the schools and yet exceeding its boundaries. This discourse, precisely because of its fluidity and uncertain status within Jesuit culture itself, was open to dialogue with speakers from other intellectual and institutional environments. The dialogue was constrained, however, by the need to formally maintain the traditional boundaries between fields of knowledge imposed by the *Ratio studiorum*. My analysis of the dispute on sunspots between Christopher Scheiner and Galileo points out how closely related their respective discourses were, and yet how much they grew apart, as a result of the different constraints to which they were subject.

Scheiner, following in the footsteps of Galileo, embraced the new observational possibilities offered by the telescope, and attempted to incorporate the newly observed phenomena into his discourse. He claimed sunspots to be real material objects indicating true motions in heaven. The practices of establishing the existence of sunspots were likewise similar for both Galileo and the Jesuit. They both insisted on the continuous and repetitive nature of their observations, recruited witnesses to confirm their statements, devoted much space to descriptions of their mode of observation, and were concerned to specify the techniques of representing their observations. Being anchored in the "mixed mathematical sciences", they both used mathematical proofs to substantiate their claims. The differences between Scheiner's discourse and Galileo's cannot therefore be easily reduced to general terms such as the mathematical or experimental nature of their arguments. Scheiner's discourse was mathematical and experimental, and aspired for truth, refusing to remain within the boundaries of hypotheses not carrying full conviction. Galileo's discourse was not only mathematical and experimental, but also claimed the status of necessary demonstrations for its conclusions, aspiring for legitimation in terms of the canons of knowledge accepted by the schools.

Galileo's discourse, however, deliberately defied the boundaries of the "mixed mathematical sciences" which Scheiner only transgressed and which were truly reaffirmed by the very act of transgression. Both Galileo and Scheiner aspired to establish the relevant properties of sunspots in mathematical language in order to constitute them as objects of science. In this they were both building upon the work of Clavius, assuming that mathematical arguments were capable of establishing true arguments about the real world. Galileo, however, went further in his attempt to change the status of astronomical "invisibles", and to imagine the possible implications of manipulation of heavenly materials. The strength of his arguments was then used to erase the boundaries between mathematics and physics, and to establish the authority of a new kind of astronomer–philosopher. Scheiner confined his strategies to the "mixed mathematical sciences", although he shifted from the hypothetical mode

of argumentation to an interpretive one. By suggesting a kind of "herme-
neutics of the heavens", he managed to differentiate his discourse from
that of the philosophers without renouncing his claims to truth. The final
authority of judgement, he then suggested, was the consensus of astronomers and philosophers.

Scheiner's discourse, no less than Galileo's, suggested new patterns of
constituting phenomena as scientific objects and alternative patterns of
authorization. Those, however, were different from Galileo's, and although they transgressed the traditional boundaries between astronomy
and philosophy, they by no means destroyed them. Thus, he was able to
offer new options without challenging the accepted organisation of culture, which secured the peculiar status of the Jesuits within the Catholic
establishment.

Scheiner's reaction to the publication of the *Dialogue Concerning the
Two World Systems* has been one of the episodes told and retold by historians of the Galileo affair. Scheiner was among the first to see the new
book upon its publication. He came out furious, after quickly perusing
its pages. His fury has always been interpreted in terms of his quarrel with
Galileo over the priority of the discovery of sunspots. In the light of my
analysis there was much more to his anger. In the last few years two scholars have shown that among the three arguments for the Copernican system presented by Galileo on the third day, the one based on the motion
of sunspots was the most conclusive, though the most difficult to clarify.
Galileo based his argument on Scheiner's discovery of the tilting axis of
the sun, published only three years before the *Dialogue*. Moreover, in the
general framework of his book, Galileo claimed no more than a probable
status for his arguments for Copernicanism. This claim could be interpreted as part of a calculated strategy to use the concept of the probable
as a mechanism for the legitimation of new ideas – an emulation of the
mechanism of inclusion frequently applied by the Jesuits as an efficient
educational practice. In these circumstances the Jesuits could no longer
ignore the dangers they were facing while being so implicated with Galileo's arguments. Thus they were forced to retreat from backing Galileo,
and cooperated with the traditionalists in his condemnation.

Index